Hydrogen Materials Science and Chemistry of Metal Hydrides

NATO Science Series

A Series presenting the results of scientific meetings supported under the NATO Science Programme.

The Series is published by IOS Press, Amsterdam, and Kluwer Academic Publishers in conjunction with the NATO Scientific Affairs Division

Sub-Series

I. Life and Behavioural Sciences	IOS Press
II. Mathematics, Physics and Chemistry	Kluwer Academic Publishers
III. Computer and Systems Science	IOS Press
IV. Earth and Environmental Sciences	Kluwer Academic Publishers
V. Science and Technology Policy	IOS Press

The NATO Science Series continues the series of books published formerly as the NATO ASI Series.

The NATO Science Programme offers support for collaboration in civil science between scientists of countries of the Euro-Atlantic Partnership Council. The types of scientific meeting generally supported are "Advanced Study Institutes" and "Advanced Research Workshops", although other types of meeting are supported from time to time. The NATO Science Series collects together the results of these meetings. The meetings are co-organized bij scientists from NATO countries and scientists from NATO's Partner countries – countries of the CIS and Central and Eastern Europe.

Advanced Study Institutes are high-level tutorial courses offering in-depth study of latest advances in a field.
Advanced Research Workshops are expert meetings aimed at critical assessment of a field, and identification of directions for future action.

As a consequence of the restructuring of the NATO Science Programme in 1999, the NATO Science Series has been re-organised and there are currently Five Sub-series as noted above. Please consult the following web sites for information on previous volumes published in the Series, as well as details of earlier Sub-series.

http://www.nato.int/science
http://www.wkap.nl
http://www.iospress.nl
http://www.wtv-books.de/nato-pco.htm

Series II: Mathematics, Physics and Chemistry – Vol. 82

Hydrogen Materials Science and Chemistry of Metal Hydrides

edited by

T. Nejat Veziroglu

International Association for Hydrogen Energy,
University of Miami,
Miami, Florida, U.S.A.

Svetlana Yu. Zaginaichenko

Institute of Hydrogen and Solar Energy,
Kiev, Ukraine

Dmitry V. Schur

Institute of Problems of Materials Science of NAS,
Kiev, Ukraine

Victor I. Trefilov

Institute of Problems of Materials Science of NAS,
Kiev, Ukraine

Kluwer Academic Publishers

Dordrecht / Boston / London

Published in cooperation with NATO Scientific Affairs Division

Proceedings of the NATO Advanced Research Workshop on
Hydrogen Materials Science and Chemistry of Metal Hydrides
Alushta, Crimea, Ukraine
16–22 September 2001

A C.I.P. Catalogue record for this book is available from the Library of Congress.

ISBN 1-4020-0868-6

Published by Kluwer Academic Publishers,
P.O. Box 17, 3300 AA Dordrecht, The Netherlands.

Sold and distributed in North, Central and South America
by Kluwer Academic Publishers,
101 Philip Drive, Norwell, MA 02061, U.S.A.

In all other countries, sold and distributed
by Kluwer Academic Publishers,
P.O. Box 322, 3300 AH Dordrecht, The Netherlands.

Printed on acid-free paper

Printed in the Netherlands.

Contents

Preface

The 2001 International Conference «Hydrogen Materials Science and Chemistry of Metal Hydrides» (ICHMS'2001) was held in the picturesque town Alushta (Crimea, Ukraine) on the bank of Black Sea in September 16-22, 2001. In the tradition of the earlier ICHMS conferences, the 7th ICHMS'2001 provided an international forum for the presentation and discussion of the latest research on transition to hydrogen-based energy systems, technologies for hydrogen production, storage, utilization, materials, energy and environmental problems. The aim of ICHMS'2001 was to provide an overview of the latest information on research and development in the different topics cited above. The representatives from industry, public laboratories, universities and governmental agencies could meet, discuss and present the most recent advances in hydrogen concepts, processes and systems, to evaluate current progress in these areas of investigations and to identify promising research directions for the future.

The ICHMS'2001 was the first conference in this series, where a related new important topic of considerable current interest on fullerene-related materials as hydrogen storage was included into the conference program. The hydrogen sorbing properties of newly discovered carbon nanostructural materials inspire hydrogen scientists with optimism. Thus, the ICHMS'2001 conference was unique in bringing together hydrogen and carbon materials researchers and engineers from developed countries of Europe and America, new independent states of FSU and other countries for discussions in advanced materials development and applications.

The ICHMS'2001 format consisted of invited lectures, oral and poster contributions and also the conference representatives took part in the exhibition of new materials and equipment.

This book with ICHMS'2001 Proceedings brings together the research papers that were presented. We hope that they will serve as both a useful reference and resource material for all the participants and for those whose interest in the subject matter may develop after the event.

Finally, this conference was generously supported by the Scientific and Environmental Affairs Division of NATO as an Advanced Research Conference within the Physical and Engineering Science and Technology Area of the NATO Science Programme. Their contribution is gratefully acknowledged and the Organizing and all ARW participants want to overflow with effusive thanks to NATO Committee for the financial support of our 7th ICHMS'2001 Conference and to Mr. Jean Fournet, Assistant Secretary General, Chairman of NATO Science Committee, and Mr. Fausto Pedrazzini, Programme Director, NATO Scientific Affairs Division, for the displayed mutual understanding and the comprehension of significance of problems under discussions at the ICHMS'2001 conference.

T. Nejat Veziroglu
Svetlana Yu. Zaginaichenko
Dmitry V. Schur

INTERNATIONAL ADVISORY AND ORGANIZING COMMITTEE OF ICHMS'2001

Chairperson		
Prof. V.I.Trefilov	(in memory)	Ukraine
Honour Chairperson		
Prof. T.N.Veziroglu	President of International Association for Hydrogen Energy, Coral Gables	USA
Co-Chairperson		
Prof. A.P.Shpak	Vice-President of NAS of Ukraine	Ukraine
Prof. O.M.Nefedov	Vice-President of RAS	Russia
Members		
Prof. Yu.A.Osipiyan	Russian Academy of Science	Russia
Prof. V.V.Skorokhod	Institute for Problems of Materials Science of NAS, Kiev	Ukraine
Prof. S.A.Firstov	Institute for Problems of Materials Science of NAS, Kiev	Ukraine
Prof. V.I.Shapovalov	Materials & Electrochemical Research (MER) Corporation	USA
Prof. Z.A.Matysina	Dnepropetrovsk State University	Ukraine
Prof. L.Owsiacky	Science and	Canada
Prof. A. Du Charme	Technology	USA
Dr. B.A. Atamanenko	Center in Ukraine	Ukraine
Dr. A.L.Shilov	Inst. of General & Inorganic Chemistry, Moscow	Russia
Prof. U.M.Mirsaidov	Academy of Sciences of Tajikistan	Tajikistan
Prof. M.Groll	University of Stuttgart	Germany
Prof. I.R. Harris	University of Birmingham	UK
Prof. M.A. Fetcenko	Ovonic Battery Company, Troy	USA
Prof. B.Baranowski	Inst.of Physical Chemistry of PAS, Warsaw	Poland
Prof. F.A. Lewis	Queen's University, Belfast	UK
Prof. Dr. J.Schoonman	Delft University of Technology, Delft	Netherlands
Dr. H.J. Bauer	Ludwig-Maximilians-Universität, Munich	Germany
Prof. M.D.Hampton	University of Central Florida, Orlando	USA
Prof.Dr.A.Mekhrabov	Middle East Technical University, Ankara	Turkey
Prof. B.Ibrahimogly	Middle East Technical University, Ankara	Turkey
Prof. D. Hui	University of New Orleans	USA
Prof. L.O.Valoen	Norwegian University of Science and Technology, Trondheim	Norway
Prof. L.A.Avaca	Inst. de Quimica de Sao Carlos, Universidade de Sao Paulo	Brasil
Prof. V.V.Lunin	Moscow State University	Russia
Prof. M.Genovese	European Commission, INCO COPERNICUS, Brussels	Belgium

Prof. V.A.Yartys	Physico-Mechanical Institute UAS, Lvov	Ukraine
Prof. D.Tománek	Michigan State University, East Lansing	USA
Prof. D.K.Slattery	Florida Solar Energy Center	USA
Prof. V.Sh.Shehtman	Institute of Solid State Physics of RAS, Chernogolovka	Russia
Prof.S.K.Dolukhanyan	Institute of Chemical Physics, Yerevan	Armenia
Prof. O.A.Ivashkevich	Belarusian State University, Minsk	Belarus
Prof. E.M.Shpilevsky	Belarusian State University, Minsk	Belarus
Prof. P.Catania	International Energy Foundation	Canada
Prof. S.K.Gordeev	Central Research Institute of Materials, St. Petersburg	Russia
Prof. I.E.Gabis	St. Petersburg State University, Institute of Physics	Russia
Prof. I.Sh.Normatov	Institute of Chemistry of TAS, Dushanbe	Tajikistan
Prof. O.Savadogo	Ecole Polytechnique de Montreal	Canada
Prof. D.Srivastava	NASA Ames Research Center, Moffett Field	USA
Prof. B.Rao	Virginia Commonwealth University, Richmond	USA
Prof. M.Z.-C. Hu	Oak Ridge National Laboratory, Chemical Technology Division	USA
Dr. V.V.Kartuzov	Institute for Problems of Materials Science of NAS, Kiev	Ukraine
Dr. B.P.Tarasov	Institute of Problems of Chemical Physics of RAS, Chernogolovka	Russia
Prof. M.Paolucci	Terra Universita degli Studi di Roma	Italy
Prof. J.C.Bolcich	Centro Atomico Bariloche	Argentina
Dr. T.A.Iljinykh	Cabinet of Ministry of Ukraine	Ukraine
Dr. A.P.Mukhachev	Plant «Zirconium», Dneprodzerjinsk	Ukraine
Dr. A.P.Pomytkin	National Technical University of Ukraine "Kiev Politechnic Institute"	Ukraine
Prof. V.N.Verbetsky	Moscow State University	Russia
Prof. C.Sholl	University of New England, Armidale	Australia
Prof. A.Switendick	ACS Associates, Albuquerque	USA
Dr. O.N. Efimov	Institute of Problems of Chemical Physics of RAS, Chernogolovka	Russia
Dr. Yu.M. Shulga	Institute of Problems of Chemical Physics of RAS, Chernogolovka	Russia
Prof.B.G.Timoshevskiy	Ukrainian State Maritime University, Nikolaev	Ukraine
Dr. T.Nietsch	Rolls-Royce pcl, Industrial Power Group	UK
Prof. J.-Y.Lee	Advanced Institute of Science and Technology, Taejon	Korea
Dr. Yu.F.Shmal'ko	Institute of Mechanical Engineering Problems of NASU, Kharkov	Ukraine
Prof. A.F.Prisevok	Belorussian State Polytechnical Academy, Minsk	Belarus

PROGRAM COMMITTEE

Dr. D.V.Schur	- chairperson	(Ukraine)
Dr. S.Yu.Zaginaichenko	- vice-chairperson	(Ukraine)
Dr. B.P.Tarasov	- vice-chairperson	(Russia)
Prof. V.V.Kveder	- vice-chairperson	(Russia)

Mr. V.K.Pishuk
Mr. A.Yu.Vlasenko
Ms. K.A.Lysenko
Mr. V.A.Bogolepov
Mr. V.M.Adejev
Mrs. L.I.Kopylova
Mrs. T.I.Shaposhnikova
Mr. A.D.Zolotarenko
Mr. A.P.Simanovsky
Dr. M.R.Tkach
Mrs. O.G.Radchenko
Dr. V.B.Chernogorenko
Dr. Yu.M.Lytvynenko
Dr. V.D.Dobrovolsky
Dr. K.B.Isayev
Mr. V.I.Tkachuk
Dr. I.G.Slys
Dr.V.B.Voitovich
Mr.V.I.Bilik
Dr. R.S.Biryukova

international association for hydrogen energy
POST OFFICE BOX 248266 • CORAL GABLES • FLORIDA 33124 • USA

WELCOME TO THE PARTICIPANTS OF ICHMS'2001

The twenty-first century will be known as the century during which the Hydrogen Energy System replaces the present fossil fuels system. Hydrogen is going to be the permanent answer to the twin global problems: (1) the rapid depletion of fossil fuels, and (2) the environmental problems caused by their utilization, such as the greenhouse effect, climate change, acid rains, ozone layer depletion, pollution and oil spills.

Hydrogen is already making inroads into the fossil fuel realm in every direction. Especially over the past two years, there has been an increase in activities. Siemens-Westinghouse announced that they will have a 1 MW H_2 power plant available for sale soon, having a 70% efficiency. There are hydrogen fueled bus demonstration projects in several cities of the world. A H_2 fueled Mercedes bus will be on the market in two years' time. All of the major car companies have announced that they will offer H_2 fueled cars to the public by the year 2004. Hydrogen hydride electric batteries are already available for lap-top computers and electric cars. The Airbus Company is developing a H_2 fueled air transport. The United States and Japan are working on H_2 fueled hypersonic passenger planes. The Shell Oil company has established a Hydrogen Division. No doubt the other petroleum companies will follow suit.

International conferences, such as the ICHMS'2001, will help speed up this transformation. At the conference, recent research findings on hydrogen materials science and metal hydrides chemistry will be presented and discussed. The chemistry of metal hydrides and hydrogen materials science will play an important role in hastening the conversion to the Hydrogen Energy System. The research endeavors of the scientists and engineers participating in this conference will make significant contributions to facilitate this milestone conversion.

I take this opportunity to congratulate the organizers of this important series of International Conferences on Hydrogen Materials Science and Chemistry of Metal Hydrides, and wish all of the participants a very productive conference and pleasant days in the beautiful Crimea.

T. Nejat Veziroglu
Honorary Chairman, ICHMS'2001
President, International Association for Hydrogen Energy

NATIONAL UKRAINIAN ACADEMY OF SCIENCE

Ukraine, 252601, Kiev, Vladimirskay str. 54
Tel: 380-44-226-2347, Fax: 380-44-228-5522

Dear colleagues!
Delegates of ICHMS'2001!

I, as President of National Academy of Science of Ukraine, with great pride and cordiality welcome You, delegates of the conference. I am glad that our Academy was able to keep this conference in heavy years for Ukraine and did not lose its positions and reputation as one of the organizations occupying the leading place in the world in this field of knowledge.

Hydrogen materials science supposes investigations and applications of the materials that may interact with hydrogen, and consequently accumulate and storage it, and also investigations and applications of the materials stable in the hydrogen medium and do not form any chemical compounds with it. The latters are used to produce devices and equipment operating in the hydrogen medium and with hydrogen. Our Academy gives much attention to the investigation and development of materials of both types. Such large institutes of Academy of Sciences of Ukraine as Institute for Problems of Material Science, Institute of Metal Physics, Institute for Problems of Engineering, Physico-Mechanical Institute, Institute of Welding and others make a valuable contribution to solving the problems of hydrogen materials science.

Hydrogen sorbing properties of newly discovered carbon nanostructural materials inspire hydrogen scientists with certain optimism. If high hydrogen capacity of those materials is confirmed, humanity will take an important advance in many fields of science and technology. The use of materials of high hydrogen sorbing ability will allow the change of the relationship between cars in the streets in favour of electric cars resulting in improvement of the human environment of big cities.

Dear participants of the conference I would like to congratulate you on the important event in the scientific life and assure you that Ukraine is always glad to see you and will support such forums in the future.

President of National Academy of
Science of Ukraine **B.E. Paton**

STCU УНТЦ

SCIENCE AND TECHNOLOGY CENTER IN UKRAINE
УКРАЇНСЬКИЙ НАУКОВО-ТЕХНОЛОГІЧНИЙ ЦЕНТР

LABORATORNYJ PROVULOK, 3, KYIV, 252133, UKRAINE, TEL./FAX: +380 (44) 227-8150/ 227-8156
ЛАБОРАТОРНИЙ ПРОВУЛОК, 3, КИЇВ, 252133, УКРАЇНА, ТЕЛ./ФАКС: +380 (44) 227-8150/ 227-8156

Dear ladies and gentlemen!

I am glad to greet You on behalf of the Science and Technology Center in Ukraine (STCU) and it is unusually pleasant in view of the fact that You all make a contribution by your's work to the construction of new hydrogen energy systems that are the basis of the energetics of the future.

A new energy paradigm must be developed in response to growing worldwide demand for energy and commitment to environmental concerns. In the development of all new energy options, hydrogen necessarily will play an important role because of its ability to supplement any energy stream and to be applied to any load. Given the significance of energy in the environmental problems of our world, it is urgently necessary that the leaders in civic and industrial societies have a more thorough understanding and appreciation of the existing states of energy systems and their related technologies. Such knowledge is all the more pertinent knowing that these energy disciplines are subject to rapid change as the emphasis shifts to solving the global environmental problems facing us.

Environmental and energy problems occupy the more and more important place in our life and consequently in the life of every State. These problems challenge the countries having highly developed industry and large cities-megapolices with great numbers of cars.

The foresight of leaders of the vast majority of large automotive companies permitted to bring the level of harmful materials ejection almost to zero. Today many firms have experimental models of electric cars and refine their constructions. However, these models are still expensive and have a number of drawbacks.

The ICHMS'2001 Conference gathered scientists who are active in the many aspects of hydrogen in metals, alloys, carbon nanomaterials from fundamental to applications. Hydrogen sorbing properties of newly discovered carbon nanostructural materials inspire hydrogen scientists with certain optimism.

Reports of very high, reversible adsorption of molecular hydrogen in pure nanotubes, alkali-doped graphite, and pure and alkali-doped graphite nanofibers (GNFs) have aroused tremendous interest in the research community, stimulating much experimental work and many theoretical calculations worldwide.

An economical, safe, hydrogen-storage medium is a critically needed component of a hydrogen-fueled transportation system. Hydrogen storage in a carbon-based material offers further advantages associated with its low mass density. Furthermore, fuel cell technology involving the conversion of hydrogen into protons, or hydrogen and oxygen into electric current, is being vigorously researched for both transportation and small power-plant applications.

Within the limits of the industrial production the scientists and investigators efforts go into the development of non-power intensive and ecologically clean productions.

In the developed countries social organizations play an important role in the solution of these problems.

The Science and Technology Center in Ukraine concentrates most attention on scientific investigations in this field and render support in every possible way to the promising ideas starting from their development and ending with promotion stage.

Considering the fact that the most advanced technologies and experienced specialists worked at the war-industrial complex designing weapons of mass destruction, one can hope that many energy and ecology problems may be quickly solved using this potential.

Hydrogen is still awaiting wide-scale utilization. Only in aerospace does it have a firmly established role. New energies always need time. And that is precisely why we must take the lead, confidently, courageously and without hesitation. Our environment and climate need hydrogen, and ICHMS'2001 will help to build a bridge from Today to Tomorrow by designing the energy link!

In the time since the STCU was established we have already provided $ 50 million to finance more than 300 projects, supporting the conversion of more than 6200 nuclear scientists and scientists who worked at the military - industrial complex.

We provide financial support to hold a number of the conferences devoted to consideration of environmental and energy problems.

The present conference is one of the first to be supported by us since 1995 and I think we will keep it in the center of our attention.

Let me wish you fruitful work and success in discussing vital important problems for all humanity.

STCU
Deputy Executive Director **Arthur Du Charme**

VII INTERNATIONAL CONFERENCE
"Hydrogen Materials Science & Chemistry of Metal Hydrides"

Alushta, Crimea, UKRAINE, September 16 – 22, 2001

Dear colleagues !

Organizing Committee of the VIIth International conference "Hydrogen materials science and chemistry of metals hydrides" is glad to welcome you in Alushta!

We have not seen for two years. It seems that only yesterday we parted in Katsiveli and today meet in Alushta.

We begin the new millennium not only in the new place, but with the renewed subjects.

Besides the traditional subjects, the new class of materials based on fullerenocarbons is in the special place in the conference program. According to some experimental data the applications of this materials may promise us the splendid prospects. However optimism in this problem should be supported by experiments and well-reproduced results.

The Development of fuel cells was stimulated due to their applications in electric cars and electric power stations. These subjects are also reflected in the conference program.

We hope that the conference will allow you not only to work fruitfully and gain much new interesting information, but give you opportunity of resting well and taking away the charge of cheerfulness and indelible impressions on Crimean pyramids.

Organizing Committee will exert every effort that you should have comfortable days and feel cosily in our forum.

Organizing Committee of ICHMS'2001

A.P. Shpak, **S.Yu. Zaginaichenko,**

D.V. Schur, **B.P. Tarasov**

RESEARCH OF FULLERITES HYDROGEN CAPACITY

D.V. SCHUR, B.P.TARASOV, Yu.M. SHUL'GA,
S.Yu. ZAGINAICHENKO, Z.A. MATYSINA
Institute of Hydrogen and Solar Energy, P.O. Box 150, Kiev, 03150 Ukraine
Institute of Problems of Chemical Physics of RAS, Chernogolovka, 142432 Russia
Institute for Problems of Materials Science of Ukrainian Academy of Sciences, 3, Krzhizhanovsky str., Kiev, 03142 Ukraine
Dnepropetrovsk State University, 49000 Ukraine

1. Introduction

Discovery of fullerenes [1-7] and features of their formation, structure and physical and chemical properties are of great interest for scientific groups. Fullerenes have made the investigators pay attention to their unusual hydrogen sorbing properties [8-29].

The high hydrogen capacity supposed for these materials has also stimulated the hydrogen scientists pay attention to the synthesis of the new carbon form [28, 30-35].

The fourth allotropic modification of carbon (fullerenes) is unique molecules showing peculiar capabilities of reacting with other substances [36-39]. Under certain conditions fullerenes can accept and donate hydrogen atoms [40-44]. The research into the processes of their reversible reaction with hydrogen is of interest in connection with the development of systems for the hydrogen accumulation and storage [45-48]. The high potential in practical applications of fullerenes will promote the progress in the studies. This will inevitably result in developing the materials of new types and generations.

2. Experimental studies into the hydrogen sorption properties of fullerites

The most wide-spread (due to its technological availability) and therefore the most stable modification of fullerenes is a C_{60} molecule. For this reason it has been chosen by scientists-experimenters as a model variant to conduct the studies into physical and chemical properties and other operating characteristics of compounds from the given class.

Being considered in detail, physico-chemistry of each fullerene modification shows intrinsic serious differences. However, when we look at this class of materials in general, we can find much in common for all the fullerene molecules: features of their formation, solubility in hydrocarbons, their electrical neutrality, donor-acceptor reactions, their exo- and endocabability etc.

Theoretically, hydrogen capacity prospects for new materials seems rather optimistic. When one hydrogen atom is joined to each carbon atom (what is quite probable), there appears possibility for producing the sorbing matrix based on these materials. It allows the storage up to 7.7 wt. % H_2.

N. Veziroglu et al. (eds.), Hydrogen Materials Science and Chemistry of Metal Hydrides, 1–18.

This value meets and exceeds all the requirements (both USA energy department and other international organizations) to the materials from this class.

Therefore, the search of the way for conducting the reversible and complete reaction

$$C_x + \frac{x}{2} H_2 \leftrightarrow C_x H_x \qquad (1)$$

and considering features of structures and properties would allow the application of the materials as systems for hydrogen accumulation and storage in many fields of engineering and technologies. As known, the abcense of materials from this class restrains the wide application of hydrogen as fuel and the energy carrier.

3. Methods for hydrogenation of fullerene molecules

Before considering the features of the hydrofullerene formation, one should note that hydrogen forming these fullerenes exists in the lattice in different energy states and geometric positions. At least, two hydrogen states are known in the hydrofullerene: lattice and fullerenated hydrogen.

a) Lattice hydrogen is hydrogen existing in the state of the solid solution and distributed in the interstices. Its content corresponds to the $C_{60}H_x$ compound where x - prime number (<6) depending on the crystal structure.

Geometry and thermodynamics of cubic lattices allow the introduction of some hydrogen atoms into the interstices under certain conditions (in the fcc lattice - up to 2 hydrogen atoms, in the bcc lattice - up to 6 hydrogen atoms per a lattice point or a fullerene if hydrogen atoms are located in the tetrapositions). Therefore, considering the solid solution of hydrogen (i.e. lattice hydrogen) in the C_{60} fullerite with the fcc lattice, maximum hydrogen content per a fullerene molecule equals to 2 atoms what corresponds to $C_{60}H_2$, and for the bcc lattice this corresponds to $C_{60}H_6$, i.e. 6 atoms per a fullerene molecule. Such structures are stable below 293 K [72-75].

b) Fullerenated hydrogen is hydrogen which atoms are chemically combined with carbon atoms. It is the product formed on the direct hydrogenation of each fullerene molecule present in the lattice points. Such a bond is formed in the fcc lattice of $C_{60}H_x$ hydrofullerene at x>2, and in the bcc lattice it is formed at x>6.

As found experimentally, fullerenated hydrogen manifests itself by several stable hydrofullerene modifications or hydride phases. At present, three of them are observed experimentally. These are $C_{60}H_6$, $C_{60}H_{18}$ and $C_{60}H_{36}$. All the intermediate phases are less stable, although they possess intrinsic thermodynamics, bond energy with carbon atoms within the fullerene molecule due to their position in the carbon lattice, and energy of interatomic interaction when a hydrogen atom is in the interstice.

Thus, hydride formation occurring in the fullerites consists of two stages: 1. Saturation of the fullerite lattice with mobile hydrogen; 2. Hydrogenation of fullerene molecules.

Being formed at the first stage, the compound is unstable under normal conditions and at temperatures above 293 K, when the hydrogen pressure is released, hydrogen is desorbed from the hydrofullerene lattice. However, hydrogenation of the fullerene molecules proceeds through saturation of the fullerite lattice with hydrogen.

The first attempts to prepare the stable hydrofullerene with the $C_{60}H_x$ composition through the attachment of a hydrogen atom to each carbon atom (x≥60) were unsuccessful.

The direct synthesis of hydrofullerenes from the gase phase was complicated with a number of technological problems. The process proceeded at high hydrogen pressures (50-85 MPa) and temperatures (573-623 K) and only one third the carbon atoms available took part in the reaction.

There were other attempts to prepare hydrofullerenes. Tarasov et al. have reported the results in detail [29]. The recent research into hydrofullerenes has been reported [49-53].

To stress the scale of the work done we should enumerate the general directions in the research into the methods for the hydrofullerene synthesis. First of all the following methods should be listed (Table 18.1):

1) reaction of gaseous hydrogen with palladium fulleride, $C_{60}Pd_{4.9}$;
2) catalytic hydrogenation of fullerenes in toluene in the presence of Ru/C;
3) radical hydrogenation with gaseous hydrogen at 723 K and 6.9 MPa with promoter, C_2H_5I;
4) fullerene reduction with lithium in ammonia in the presence of t-BuOH;
5) reduction in toluene through hydroborating or hydrozirconating;
6) hydrogen transfer on the fullerene from the dihydroanthracene melt at 623 K;
7) fullerene hydrogenation in the toluene solution at the moment of hydrogen evolution during the reaction of zinc with concentrated hydrochloric acid;
8) electrochemical hydrogenation in the 30% KOH solution.

TABLE 1. Known methods for hydrofullerene synthesis [29]

Method	Chemical reaction	Conditions: pressure P, temperature T
Direct non-catalytic hydrogenation (C. Jin et al., 1994)	$C_{60} + H_2 \Rightarrow C_{60}H_{2-18}$	P_{H_2} = 50-85 MPa, T = 573-623 K
Reaction of gaseous hydrogen with $C_{60}Pd_{4.9}$ (A.S. Lobach, B.P. Tarasov et al., 1996)	$C_{60}Pd_{4.9} + H_2 \Rightarrow C_{60}H_{2-26}$	P_{H_2} = 2.0 MPa, T = 473-623 K
Catalytic hydrogenation in toluene solution in the presence of Ru/C (K. Shigematsu et al., 1993)	$C_{60} + H_2 \Rightarrow C_{60}H_{36-48}$	P_{H_2} = 2-12 MPa, T = 383-553 K
Radical hydrogenation with promoter, C_2H_5I (M.I. Attala et al., 1993)	$C_{60} + H_2 \Rightarrow C_{60}H_{\sim 36}$	P_{H_2} = 6.9 MPa, T = 723 K
Reduction with lithium in ammonia in the presence of t-BuOH (R.E. Haufler et al., 1990)	$C_{60} + H_2 \Rightarrow C_{60}H_{18-36}$	T = 78 K
Reduction in toluene solution through hydroborating (C.C. Henderson et al., 1993) or hydrozirconating (S. Ballenweg et al., 1993)	$C_{60} + H_2 \Rightarrow C_{60}H_{2-4}$	T = 278 K
Hydrogen transfer on the fullerene from the dihydroanthracene (C. Ruchardt et al., 1993)	$C_{60} + H_2 \Rightarrow C_{60}H_{18-36}$	T = 623 K
Fullerene hydrogenation in the Zn-concentr.HCl-toluene system (A.D. Darwish et al., 1995)	$C_{60} + H_2 \Rightarrow C_{60}H_{18-36}$	T = 293 K
Electrochemical hydrogenation in the 30% KOH solution (R. Nosu et al., 1996)	$C_{60} + xH_2 + xe \Leftrightarrow C_{60}H_{\sim x} + xOH$	Under normal conditions

All the methods have advantages and disadvantages. However, because of difficulties in organizing the reversible hydrogen sorption by above methods, many of them will not be devoted proper attention.

Since the application of hydrogen as the energy carrier and fuel suggests its utilization mainly in the gaseous state, the gas-solid processes (as the most acceptable alternative for the applied utilization) will be paid a special attention.

4. Hydrogenation of mechanical mixtures

Reaction of fullerite with gaseous hydrogen (deuterium), as more acceptable alternative for the applied use, requires a special attention and discussion. As noted previously [15-18], chemical transformations occurring in the mechanical mixtures of fullerite and hydride-forming metal and alloy powders (Pd, V, $LaNi_5$, $LaNi_{4.65}Mn_{0.35}$, $CeCo_3$ and others) under the action of hydrogen (deuterium) have been studied in a wide pressure and temperature range.

Hydrogenation at 1.0-5.0 MPa and 573-673 K results in mixtures of metal hydrides, MH_y, and hydrofullerenes, $C_{60}H_x$. $C_{60}H_x$ is formed in two stages. First, the solid solution of hydrogen in the fullerite lattice is formed, then the fullerene molecules each are hydrogenated.

$$\{C_{60} + M\} + H_2 \xrightarrow{1-5\text{MPa},\ 570-670\text{ K}} \{C_{60}H_x + MH_y\}. \tag{2}$$

Further heating the reaction products, $\{C_{60}H_x+MH_y\}$, in the inert atmosphere up to 800 K results in their step dehydrogenation to form fullerite-metal compositions

$$\{C_{60}H_x + MH_y\} \xrightarrow{<600\text{ K}} \{C_{60}H_x + M\} + 0.5y\ H_2 \xrightarrow{\sim 800\text{ K}}$$

$$\longrightarrow \{C_{60} + M\} + 0.5(x+y)\ H_2. \tag{3}$$

Further heating over 950 K leads to carbidizing some elements of the mixture.

5. Reaction of hydrogen with metallofullerides

Depending on the temperature, reaction of hydrogen (deuterium) with metallofullerides, $C_{60}Pt$ and $C_{60}Pd_{4.9}$, results either in metallofulleride hydrides, $<C_{60}MH_x>$, or in hydrofullerene mixtures, $C_{60}H_x$ (x=2-26) with Pt and/or PdH_y, respectively

$$C_{60}Pt + H_2 \xrightarrow{1\text{-}3\text{ MPa, }400\text{-}550\text{ K}} <C_{60}PtH_x>, \tag{4}$$

$$C_{60}Pt + H_2 \xrightarrow{1\text{-}3\text{ MPa, }600\text{-}700\text{ K}} C_{60}H_x + Pt, \tag{5}$$

$$C_{60}Pd_{4.9} + H_2 \xrightarrow{1\text{-}3\text{ MPa, }400\text{-}550\text{ K}} <C_{60}Pd_{4.9}H_x>, \tag{6}$$

$$C_{60}Pd_{4.9} + H_2 \xrightarrow{1\text{-}3\text{ MPa, }600\text{-}700\text{ K}} C_{60}H_x + PdH_y \tag{7}$$

Upon heating to 800 K, almost all hydrogen is evolved from the reaction products, and the resulting mixture of fullerite and metal can be reversibly hydrogenated at 400-700 K to form mixtures of hydrofullerene and metal hydrides (Pt or Pd) [10,23,24].

6. Hydrogenation of fullerites with atomized hydrogen

Hydrogen desorbed from the hydrides is in the active partly atomized state. Activity of hydrogen atoms consists in their high reaction capability. Lifetime of hydrogen atoms reaches 3 c. Their recombination on the fullerite surface gives rise to the energy evolution, the hydrogen molecule is desorbed from the surface. Active centers on the surface are subjected to the local heating what favors other non-activated hydrogen atoms to penetrate into the subsurface layers of the crystal to fill the vacant interstices and to react with carbon atoms under the action of the positive pressure of lattice gas.

The fullerite hydrogenation with hydrogen desorbed from the hydrides results in hydrofullerenes that contain from 10 to 30 hydrogen atoms per a fullerene molecule, C_{60} [19,20]. Cycling at 300-673 K gives rise to the increase in the hydrogen capacity of hydrofullerenes.

Cycling also favors a deeper process and a more homogeneous distribution of hydrogen over the fullerite volume. The IR spectra of these samples show intense absorption bands in the region of valence C-H vibrations (2800-3000 cm^{-1}) and in the region of C-D vibrations (2050-2200 cm^{-1}) in the case of deuterofullerenes. The relative intensities of the absorption bands for the starting C_{60} decrease abruptly (Fig. 1).

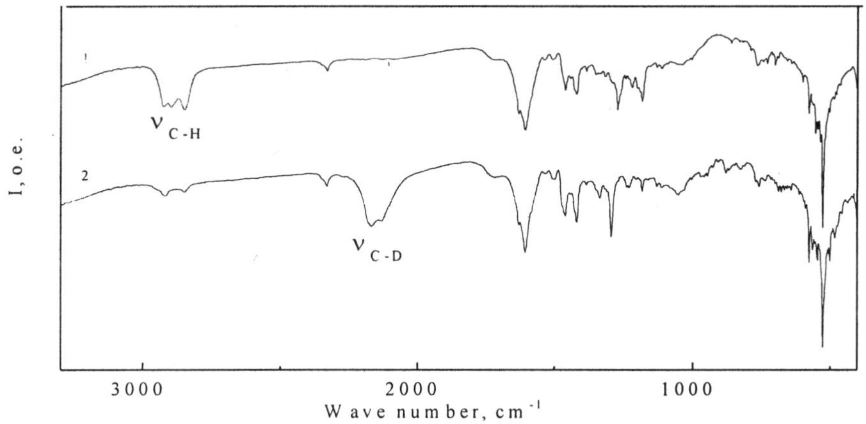

Fig. 1. IR spectra of hydro- (1) and deuterofullerenes (2)

X-Ray diffraction analysis of the reduction product shows that hydro(deutero)fullerene ($C_{60}H_x$ or $C_{60}D_x$) remains the single-phase system with the starting fullerite structure. The a_o parameter of the face-centered cubic (fcc) lattice increases monotonically with the amount of hydrogen or deuterium in the hydrofullerene (Fig. 2).

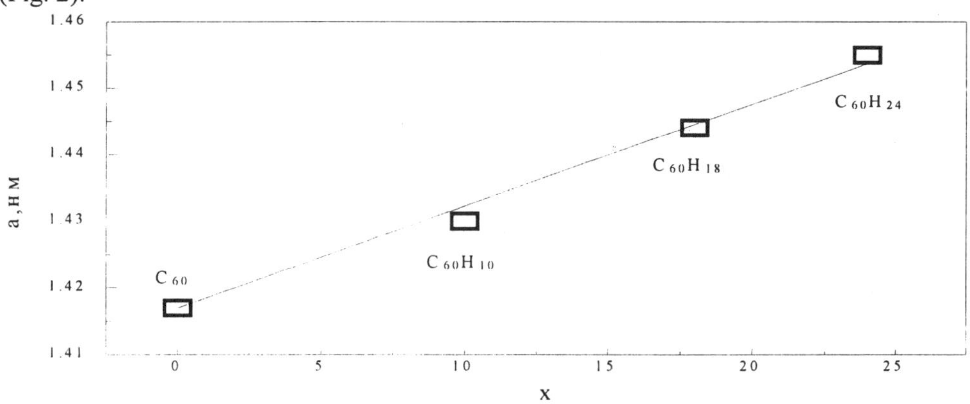

Fig. 2. Dependence of the fcc lattice parameter, a_o, of hydrofullerite on the number of hydrogen atoms

Heating the fullerene hydrides results in the intense gas evolution. The IR spectrum of the solid product from the thermal hydrofullerene decomposition shows the bands corresponding to C_{60} and a series of bands which positions are close to those of photo- or thermobarically polymerized C_{60} with the cyclobutane rings between the fullerene molecules [14]. This indicates thermal decomposition of hydrofullerenes accompanied by polymerization of fullerene clusters (Fig. 3). Using hydrogen from the hydrides, the samples dehydrogenated at 800 K can be hydrogenated again to form hydrofullerenes. However, repeated cycling increases amount of polymerized fullerene [25].

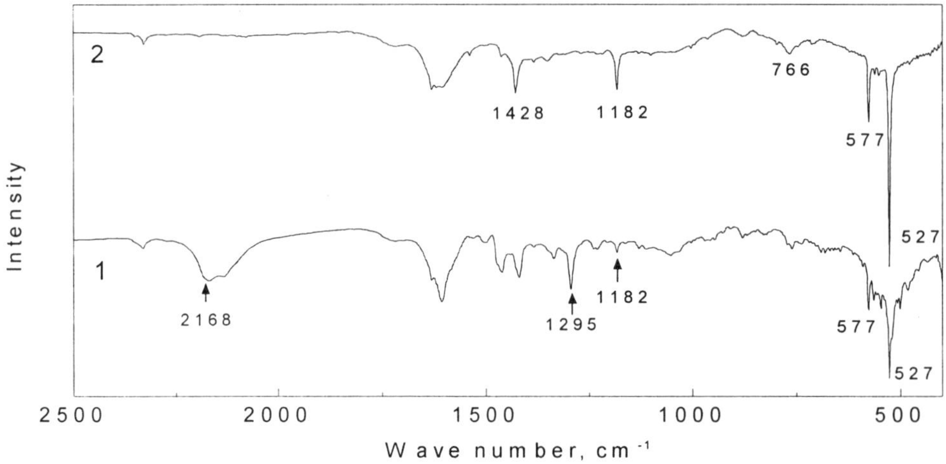

Fig. 3. IR spectra of $C_{60}D_{24}$ (1) and the decomposition product (2)

The addition of NH_3, HI, C_2H_5I (5-10 wt. %) to the hydrogen evolved from the metallohydride accelerates considerably the fullerite hydrogenation but also increases the degree of polimerization of fullerene molecules in the lattice of the dehydrogenated product [60-61].

Hydrogenation of fullerites as an attractive method for hydrogen storage and synthesis of individual hydrofullerenes are being studied intensively. Deuterides with $C_{60}D_8$, $C_{60}D_{18}$ and $C_{60}D_{36}$ compositions (fcc structure, a_o = 1.423, 1.448 and 1.500 nm, respectively) have been produced at 373-773 K and p_{D2} = 15 MPa [62]. Controlled doping of fullerites with palladium decreases (down to 373 K) the temperature of hydrogen sorption relative to that for the undoped C_{60} sample at the same gas pressure (H_2 or D_2) 100 MPa [63]. Neutron diffraction has shown that after hydrogen sorption the samples retains the fcc lattice. Neutron diffraction of palladium-doped fullerene has shown no characteristic peaks of metallic Pd either for starting samples or after their treatment with hydrogen. $C_{60}H_x$ hydrides with the low hydrogen content (x=1÷6) can be prepared by the treatment of fullerene-containing carbon black with hydrogen plasma [26]. This is attributed to the low partial pressure of hydrogen during synthesis what hinders overcoming the activation barrier for hydrogenation of the fullerene molecule. However, this pressure is sufficient to form the solid solution. Dynamic equilibrium occurring at x = 1÷6 is due to balancing between the flow of atomic hydrogen formed in plasma and the flow of molecular hydrogen desorbed from the volume. This is attributed to the fact that energy of atomic hydrogen is sufficient to penetrate into the crystallite and to fill interstices. However, the absence of protecting oxide, oxycarbide and other films on the crystallite surface does not allow the creation of the excess pressure of lattice hydrogen that stimulates the formation of fullerenated hydrogen. Hydrogen atoms recombine at the grain boundaries and imperfections in the crystal. Molecules are freely desorbed from the volume to form the flow counter to atomic hydrogen.

7. Liquid-phase hydrogenation

An interesting feature of physical and chemical properties of fullerenes is their solubility in hydrocarbons.

Being in the weak solution, the fullerene molecule reacts with different reagents to form fullerene hydrides up to $C_{60}H_{60}$.

Being in the solution, molecules of fullerene hydride tend to form clusters as temperature increases.The size of clusters depends on the concentration and the temperature of the solution. The change of the solution temperature results in the solution supersaturation and the formation of hydrofullerens, i.e. the solid phase.

This mechanism may be used to prepare hydrofullerenes from the liquid solutions in hydrocarbons through the formation of fullerene hydrides by the catalytic method.

Therefore, the problem on liquid-phase hydrogenation of fullerenes needs to be paid due attention although the mechanism for the creation of reversible hydrogen sorption by this method is still unknown.

Progress has been made in the liquid-phase hydrogenation of fullerenes using heterogeneous catalysts: the cost for catalysts has been cut, selectivity and degree of hydrogenation of dissolved fullerene molecules have been enhanced. For example, interesting data on C_{60} hydrogenation on the Ni/Al_2O_3 catalyst in toluene at 423-523 K and 2.5-7.5 MPa have been obtained [27]. Under these milder conditions compared with those for the Ru/C catalyst [8], C_{60} fullerene is hydrogenated selectively to form $C_{60}H_{36}$. The decrease in the reaction temperature results in reducing selectivity by $C_{60}H_{36}$. More severe conditions may contribute the enhancement in hydrogenation of the fullerene molecule at the presence of Ni/Al_2O_3 [27].

Liquid-phase hydrogenation of C_{60} with hydrogen (at atmospheric pressure in the presence of 4% Pd/SiO_2, 5% Ru/Al_2O_3 or 5% Rh/Al_2O_3) or with hydrogen donors (cyclohexane, decalin etc.) at 323-343 and 348-433 K results in $C_{60}H_x$ hydrofullerenes [11]. In some experiments Pd/SiO_2 was used as the catalyst for hydrogenation of fullerene by hydrogen transfer from the donor (cyclohexane, cyclohexene) to the acceptor (C_{60}). Basing on data obtained by chemical (titration with hydrogen peroxide) and spectrum (IR and NMR) analyses, authors [11] have concluded that the products from the C_{60} reduction correspond to $C_{60}H_{42\text{-}46}$. Unfortunately, the results on the mass-spectrum determination of fullerene hydrides have not been reported; they might provide more exact information on their compositions.

In searching the systems that would meet the requirements for the practically applicable hydrogen accumulators, authors [64] have examined the possibility for using the soluble catalysts and salt melts as the media for the catalytical hydrogenation of unsaturated C=C bonds. Alkoxy compounds of metals from group IV have been used as catalysts for fullerene hydrogenation and hydrofullerene dehydrogenation. Dehydrogenation of $(C_{60}/C_{70})H_{36}$ proceeding at 573 K for 6 h has given rise to $(C_{60}/C_{70})H_{22}$ with the loss of 1.74 wt. % H_2, while hydrogenation of $(C_{60}/C_{70})H_{22}$ (423 K,

12 h, 3.6 MPa) has resulted in $(C_{60}/C_{70})H_{\sim45}$. This study has shown that the use of certain soluble catalysts allows the considerable decrease in the activation energy of hydrogenation and the process temperature down to the record low values. Another approach proposed (doping the fullerene with alkali metal) also leads to the considerable decrease in the activation barrier of the fullerene hydrogenation. Dehydrogenation in this system occurs at considerably lower temperatures than those for the non-catalytic dehydrogenation of solid hydrofullerites. The use of operating temperatures below 400 K is important for the competitivity of the hydrogen storage system. The approaches developed in [64] demonstrate the possibility for achieving the temperature range acceptable in practice.

Formation of $C_{60}H_2$ upon electrochemical hydrogenation of C_{60} in 30 % aqueous KOH was also reported. Hydrogenation of C_{60} and dehydrogenation of $C_{60}H_2$ on electrolysis are reversible [12]:

$$C_{60} + xH_2O + xe^- \rightleftarrows C_{60}H_x + xOH^-. \quad (8)$$

In this case the question is the first stage of hydrogenation analogous to that of transition metals and alloys. It consists in the formation of the hydrogen sublattice in the parent lattice of fullerite (or saturation of interstices in the fullerite lattice with lattice hydrogen).

8. Other features of the reaction of fullerites with hydrogen (deuterium)

The work [65-68] is dedicated to application of mechanochemistry to activate reactions of fullerites with graphite.

In the majority of experiments on the fullerene reduction the mixtures of fullerene hydrides are produced.

Hydrofullerites with the same empirical formula, for example $C_{60}H_{36}$, but synthesized by different methods are drastically distinguished by their solubilities, light and air resistances, abilities to withstand the prolonged storage and other properties [13,69].

Mass-spectrum analysis at the electron impact and field desorption has shown that $C_{60}H_{18}$ is the first stable product from thermal decomposition of $C_{60}H_{36}$ prepared under conditions of non-catalytic hydrogen trasfer from the donor to the acceptor. Intense peaks of $C_{60}H_{18}$ and C_{60} appearing in the mass spectrum of the $C_{60}H_{36}$ sample after its prolonged storage indicate that $C_{60}H_{36}$ undergoes decomposition. Formation of $C_{60}H_{36}$ as the only product from C_{60} decomposition proceeding under non-catalytic hydrogen transfer conditions is confirmed by the data obtained using matrix-assisted laser desorption ionization mass spectrometry (MALDI-MS).

As noted above, in inert or hydrogen atmosphere hydrofullerites decompose to form hydrogen and fullerite [19,70]. The solid product from thermal decomposition of $C_{60}D_{24}$ deuteride at 823 K mainly consists of fullerite with the expanded lattice. The a_0 constants of the fcc lattice are 1.452 and 1.417 for the decomposition product and C_{60}, respectively. Presumably, the increase in the fullerite crystal lattice constant resulting from the reversible hydrogenation can be useful in the synthesis of metal fullerides with superconducting properties [70].

The position of the $C1_s$ peak maximum in the X-ray photoelectron spectrum of $C_{60}D_{24}$ deuteride is shifted towards the lower bond energies in comparison with those for

C_{60} what allows the conclusion about a small positive charge present on the deuterium atoms. The peak becomes narrower as deuteride undergoes thermal decomposition, and its 'center of gravity' shifts towards the higher bond energies [70] (Fig. 4).

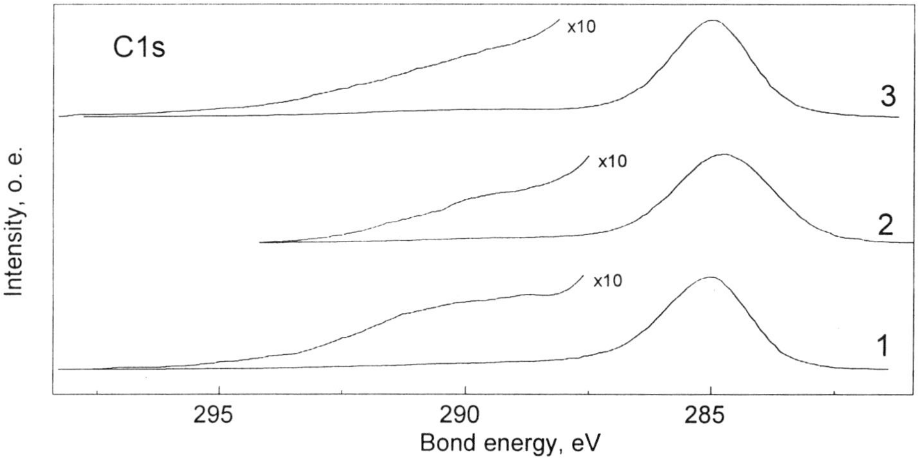

Fig. 4. X-ray photoelectron spectra of C_{60} (1), $C_{60}D_{24}$ (2) and the decomposition product (3)

Particular interest in the scientific and applied aspects is felt in the possibility for hydrogen localization inside the fullerene cluster as method for increasing the hydrogen capacity of the product prepared. It has been found that $C_{60}H_{52}$ and $C_{60}H_{48}$ fluoresce in the long-wave region. Possibility for penetrating the H_2 molecules into the closed hydrofullerene cavity has been considered to interpret this phenomenon [71]. Vibration spectra of the hydrofullerene prepared under ultrahigh hydrogen pressures, P_{H2} = 0.6 or 3.0 GPa, have been studied by inelastic neutron scattering (INS) spectroscopy [72]. The spectra of hydrofullerenes have shown that $C_{60}H_x$ synthesis (P_{H2} = 0.6 or 3.0 GPa, 620 K, 24 h) followed by cooling the sample to 77 K under a high hydrogen pressure results in the formation of the solid solution of hydrogen in the $C_{60}H_x$ lattice ($x \approx 24$ and 32-46 for P_{H2} = 0.6 or 3.0 Gpa, respectively) [72,73]. The main features of the INS spectrum of $C_{60}H_{32}$ prepared at P_{H2} = 3.0 GPa are similar to those for $C_{60}H_{\sim 23}$ prepared at P_{H2} = 0.6 Gpa [74,75]. Presumably, a considerable fraction of hydrogen (~3 H_2 molecules per C_{60} molecule) is located in the interstices in the $C_{60}H_x$ lattice and leaves the sample with heat evolution upon heating to 293 K. INS method allows the detection of dissolved molecular hydrogen by the rotational transition peaks in the H_2 molecule (~15 and 30 meV) in the difference INS spectra of the 'quenched' (H_2-containing) and heated (H_2-free) products from fullerite hydrogenation. The peak present near 13 meV in the INS spectrum suggests the possibility for polimeric binding the $C_{60}H_{32}$ hydrofullerite at high H_2 pressure (3.0 GPa). The low-energy part of the INS spectrum (< 9 meV) for $C_{60}H_{32}$ is believed to indicate the presence of endohedral hydrogen atoms unbound with carbon [72,73].

Hydrofullerenes prepared at P_{H2} = 0.6 and 3.0 Gpa, T = 520-620 K and then quenched at T<120 K exhibited ferromagnetic properties up to T≈300 K. Their magnetization was as high as M=1.2 emu g^{-1} at H=10 kÖe [76]. According to the data on chemical and mass-spectrum analyses, the atom concentration of elements from the iron group in hydrofullerenes is insignificant. Magnetization that disappears upon heating the samples above 340 K or upon prolonged storage at room temperature is a property inherent for hydrofullerenes prepared by different methods [76,77]. Probably, their magnetization is caused by the presence of unstable transient radical-type electronic states. The hydrogen penetration into the cage, polymerization of the fullerene cages occurring upon dehydrogenation and other high-energy transformations of the C-C and C-H bonds under the sterically complicated conditions in the fullerene lattice could be a reason for generating quasistable carbon atoms with the unsaturated bond.

Thermodynamic properties of the $C_{60}H_{36}$ hydrofullerene prepared by non-catalytic hydrogen transfer have been studied in the temperature range from 4.8 to 340 K [78,79]. The following values of thermal capacity, enthalpy, entropy and Gibbs free energy were found for the hydrofullerene prepared at T=298.15 K and p=101.325 kPa:

C_p^o (298.15) = 690 J $K^{-1}mol^{-1}$;
ΔH^o(298.15) = 84.94 kJ mol^{-1};
S^o(298.15) = 506 J $K^{-1}mol^{-1}$;
ΔG^o(298.15) = -66.17 kJ mol^{-1}.

Quantun-chemical calculations of C-H bond energies in the fullerene hydrides performed by the semi-empirical AM1 method give the values of 306.27, 298.32, 295.8 and 293.72 kJ mol^{-1} for $C_{60}H_{12}$ (T_h), $C_{60}H_{18}$ (C_3) and $C_{60}H_{36}$ (T and T_h), respectively; symmetry of molecules is given in parentheses[80]. Heating of $C_{60}H_{36}$ under isothermal conditions (594 and 610 K, 20 h) results in dehydrogenation of the sample and evolution of gaseous H_2. In this case the solid residue has the'$C_{60}H_6$' composition: $C_{60}H_{36} \rightarrow C_{60}H_{18} \rightarrow$ '$C_{60}H_6$'. Dehydrogenation occurs stepwise [81] and is accompanied by sublimation of $C_{60}H_{36}$ and $C_{60}H_{18}$. The estimated saturated vapour pressures of $C_{60}H_{36}$ and $C_{60}H_{18}$ at 594 K are $9.8 \cdot 10^{-9}$ and $5.4 \cdot 10^{-9}$ MPa, respectively.

Deuteride, $C_{60}D_{24}$, prepared by deuteration of fullerite with gaseous D_2 (according to [19]) is crystalline powder with the fcc lattice (a_o = 1.455 nm) and the crystallite size of 51-56 nm [25]. Synthesis of $C_{60}D_{36}$ with the fcc lattice (a_o = 1.455 nm) has been reported [25]. The $C_{60}H_{36}$ sample prepared by hydrogenation under high pressure [82] or in the presence of iodine has the body-centered cubic (bcc) lattice [84].

Theoretical and experimental studies into the structures of the most stable T, T_h, S_6 and two D_{3d} isomers of the $C_{60}H_{36}$ hydrofullerene have been carried out. Comparison of the experimental ^{1}H and ^{13}C NMR spectra of the solid $C_{60}H_{36}$ with the calculated ones allows the assumption about the T symmetrical structure for $C_{60}H_{36}$ prepared by non-catalytic hydrogen transfer. According to the X-ray emission and photoelectron spectra, quantum-chemical calculations and electrochemical behavior, the most probable structure for the $C_{60}H_{36}$ hydrofullerene is that with the T symmetry and with four isolated benzenoide rings located on the surface of the $C_{60}H_{36}$ hydrofullerene in the vertices of the tetrahedron (Fig. 5) [90,91].

Comparison of the CuK_α spectra for C_{60}, $C_{60}H_{36}$ and C_6H_6 prompts the conclusion that π-electron systems of hydrofullerenes and benzene are similar [90,92].

Studies into the electrochemical behavior of $C_{60}H_{36}$ have shown that the $C_{60}H_{36}$ molecule should be 'hard' with low reactivity in redox reactions. Electrochemical activity is diminished in the series $C_{60} > C_{60}H_2 > C_{60}H_{36} \approx C_6H_6$ and much the same for the two latter compounds what can be also evidence in favor of the T-symmetrical structure for the $C_{60}H_{36}$ molecule [91].

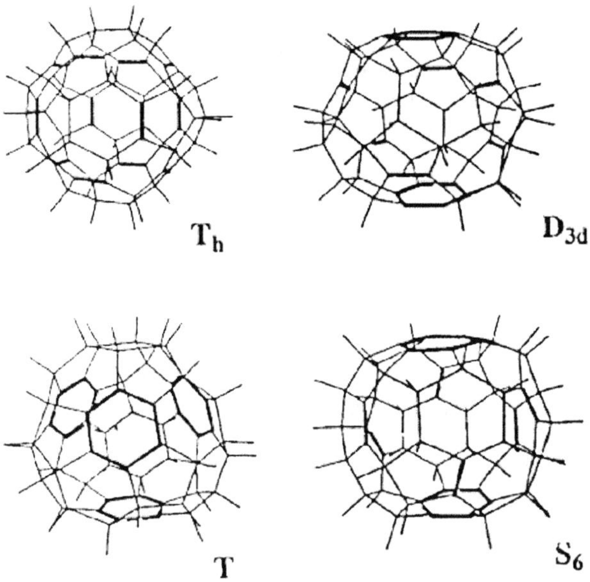

Fig. 5. The most stable $C_{60}H_{36}$ isomers

The presence of a large number of reaction centers in the C_n fullerene molecules results in a large number of possible isomers of hydro- and fluorofullerenes. Based on the combination of the MO LCAO method and molecular mechanics, it has been predicted [93] the structures of the most stable C_nX_k isomers with high hydrogen and fluorine contents (X=H, F; n = 76,78,84) and compared the theoretical data with the available experimental results.The deuterium derivative of the $C_{60}D_{36}$ fullerene has been prepared by method [83] that consists in C_{60} hydrogenation with deuterium at 623 K and D_2 pressure 3.4 MPa in the presence of iodine as promoter. According to the neutron diffraction structural data [94], the T-symmetrical structure proves to be the most stable for $C_{60}D_{36}$. As noted above, the analogous structure is also preferable for $C_{60}H_{36}$ [83].

According to [95], hydride with the net composition, $C_{60}H_{18.7}$, is stable up to 703 K in Ar atmosphere. Further increase in the temperature results in hydrogen evolution accompanied by decomposition of the part of fullerene molecules and formation of graphite platelets and methane. It should be noted that the data on the values of the lattice constants and behavior of thermal hydrofullerene decomposition obtained by different authors are often distinguished. It has been clearly found that properties of hydrofullerenes considerably depend on the synthesis conditions and the fullerite prehistory.

The use of catalytic amounts of ruthenium or platinum does not decrease the temperature of hydrofullerene dehydrogenation. Thermogravimetric (TG) and differential thermogravimetric (DTG) studies of $C_{60}H_{36}$ in the presence of oxygen were carried out [27]. $C_{60}H_{36}$ proved to be partially oxidized at 550 K to form $C_{60}H_{36}O_x$ (x~4.8). $C_{60}H_{36}$ combustion occurs at 745 K. Analogous values for C_{60} equal to 705 and 888 K, respectively [27].

Complexes of polyaromatic compounds and fullerene with platinum immobilized on silica exist in the temperature range 323-373 K and can be repeatedly hydrogenated and dehydrogenated [96] what is of interest from the viewpoint of developing the practical systems for hydrogen storage. They must show good reproducibility of hydrogenation-dehydrogenation cycles in the temperature range as low as possible.

9. Conclusions

In conclusion it should be stressed that although some methods for fullerite hydrogenation have yet to give the maximum hydrogen content for some reasons, nevertheless, they are competitive at this stage of studies.

As show data from Table 2, even C_{60} fullerite hydrogenated by 50% can easily compete with many metallic and intermetallic compounds which bright representatives are intermetallides of AB_2 and AB_5 types extensively used at present.

TABLE 2. Metallic and organic hydrides as hydrogen accumulators [18]

Reactions	Volume of reversible hydrogen, (at.H/cm^3) 10^{-22}	Weight content of reversible H_2, %	Electro-chemical capacity, MAh/g
$Pd + 1/2H_2 \leftrightarrow PdH_{\sim 0.8}$	5.30	0.75	200
$V + H_2 \leftrightarrow VH_2$	10.50	3.75	990
$Ti + H_2 \leftrightarrow TiH_2$	9.05	4.00	1050
$Mg + H_2 \leftrightarrow MgH_2$	6.60	7.60	2000
$CeCo_3+2H_2 \leftrightarrow CeCo_3H_4$	4.10	1.25	330
$LaNi_5+3H_2 \leftrightarrow LaNi_5H_6$	4.30	1.40	360
$TiFe + H_2 \leftrightarrow TiFeH_2$	5.90	1.90	500
$Mg_2Ni+2H_2 \leftrightarrow Mg_2NiH_4$	8.0	3.60	950
$C_6H_6 + 3H_2 \leftrightarrow C_6H_{12}$	3.35	7.15	1880
$C_7H_8 + 3H_2 \leftrightarrow C_7H_{14}$	2.85	6.15	1620
$C_{10}H_8 + 5H_2 \leftrightarrow C_{10}H_{18}$	4.50	7.25	1900
$C_{60} + 30H_2 \leftrightarrow C_{60}H_{60}$	8.25	7.70	2000
$C_{70} + 35H_2 \leftrightarrow C_{70}H_{70}$	8.25	7.70	2000
$C_{60} + 18H_2 \leftrightarrow C_{60}H_{36}$	5.40	4.50	1250

The problem on the reaction of fullerenes with hydrogen is as the infinite field of science as the fullerene science itself.

What has been done by now in this field of knowledge is not the first step but the fleeting glance at this problem.

At present, scientists have made the breath-taking success in the research of hydrogen capacity of carbon nanotubes. However, this is a separate serious problem that is not covered in the present work.

10. References

1. Kroto, H.W., Heath, J.R., O'Brien, S.C. and Smalley, R.E. (1985) C_{60}: Buckminsterfullerene, *Nature* **318**, 162.
2. Smalley, R.E. (1988) Otkryvaya fullereny, *Usp. fiz. nauk* **168**(3) 323-330.
3. Kerl, R.F. (1988) Istoki otkrytiya fullerenov: eksperiment i gipoteza, *Usp. fiz. nauk* **168**(3), 331-342.

4. Kroto, G. (1988) Symmetry, outer space, stars and C_{60}, *Usp. fiz. nauk* **168**(3), 343-344.
5. Osawa, E. (1970) Kogaku (Kyoto) **25**, 854 (in Jap.).
6. Osawa, E. (1993) Philos. Trans. R. Soc. London Ser. A – Phys. Sci. Eng., 3431.
7. Bochvar, D.A. and Gal'pern, E.G. (1973) O gipoteticheskikh sistemakh: karbododekaedre, s-ikosaedre i karbo-s-ikosaedre, Dokl.AN SSSR **209**(3), 610-612.
8. Shigematsu, K., Abe, K., Mitani, M. and Tanaka, K. (1992) Catalytic hydrogenation of fullerene C_{60}, *Chem. Express* 7(12), 37-40.
9. Hirsch, A.H. (1994) The chemistry of the fullerenes, *Georg. Thieme Verlag. Stuttgart*, 203 p.
10. Lobach, A.S., Tarasov, B.P., Shul'ga, Yu.M., Perov, A.A. and Stepanov, A.N. (1996) The D_2 reaction with palladium fulleride $C_{60}Pd_{4,9}$, *Izv. RAN., Ser. khim.*, 483-484.
11. Drelinkiewicz, A., Byszewski, P. and Bielanski, A. (1996) Catalytic hydrogenation of C_{60} fullerene, *React. Kinet. Catal. Lett.* **59**(1), 19-27.
12. Nozu, R. and Matsumoto, O. (1996) Hydrogenation of C_{60} by electrolysis of KOH-H_2O solution, *J. Electrochem. Soc.* **143**(6), 1919-1923.
13. Goldshleger, N.F. and Moravsky, A.P. (1997) Gidridy fullerenov, *Uspekhi khimii* **66**(4), C.353-375.
14. Rao, A.M., Eklund, P.C., Venkateswaran, U.D., Tucker, J., Dunkan, M.A., Bendele, G.M., Stephens, P.W., Hodeau, J.-L., Marques, L., Nunez-Regueiro, M., Bashkin, I.O., Ponyatovsky, E.G. and Moravsky, A.P. (1997), *Applied Physics A64*, 231.
15. Tarasov, B.P., Fokin, V.N., Moravsky, A.P. and Shul'ga, Yu.M. (1997) Hydrogenation of fullerite in the presence of intermetallic compounds, *Izv. RAN., Ser. khim.* **4**, 679-683.
16. Tarasov, B.P., Fokin, V.N., Moravsky, A.P., Shul'ga, Yu.M. and Yartys', V.A. (1997) Hydrogenation of fullerenes C-60 and C-70 in the presence of hydride-forming metals and intermetallic compounds, *J. Alloys and Comp.* **25**, 253-254.
17. Tarasov, B.P. (1998) Mechanism of hydrogenation of fullerite-metallic compositions, *Zhurn obshchei khimii* **68**, 1245-1248.
18. Tarasov, B.P., Fokin, V.N., Moravsky, A.P., Shul'ga, Yu.M., Yartys', V.A. and Schur, D.V. (1998) Promotion of fullerene hydride synthesis by intermetallic compounds, *Proc. of 12th World Hydrogen Energy Conference*, Buenos Aires, Argentina **2**, 1221-1230.
19. Tarasov, B.P., Fokin, V.N., Fokina, E.E., Rumynskaya, Z.A., Volkova, A.P., Moravsky, A.P. and Shul'ga, Yu.M. (1998) Sintez gidridov fullerena vzaimodeistviem fullerita s vodorodom, vydelyayuschimsya iz gidridov intermetallicheskikh soedineniy, *Zhurn obshchei khimii* **68**(10), 1585-1589.
20. Tarasov, B.P., Fokin, V.N., Moravsky, A.P. and Shul'ga, Yu.M. (1998) Sintez i svoistva kristallicheskikh gidridov fullerenov, *Izv. RAN., Ser. khim.* **10**, 2093-2096.
21. Dresselhaus, M.S., Williams, K.A. and Eklund, P.C. (1999) Hydrogen absorption in carbon materials, *MRS Bulletin* **24**(11), 45-50.
22. Hirsch A. (1999), *Topic in Current Chemistry* **199**, 1.
23. Goldshleger, N.F., Tarasov, B.P., Shul'ga, Yu.M., Perov, A.A., Roschupkina, O.S. and Moravsky, A.P. (1999) Vzaimodeistvie fullerida platiny $C_{60}Pt$ s deiteriem, *Izv. RAN, Ser. khim* (5), 999-1002.
24. Goldshleger, N.F., Tarasov, B.P., Shul'ga, Yu.M., Roschupkina, O.S., Perov, A.A.

and Moravsky, A.P. (1999) In Fullerenes, Recent Advances in the Chemistry and Physics of Fullerenes and Related Materials, – Vol.7 (Eds. Kadish K.M., Kamat P.V. and Guldi D.), Pennington, NJ: The Electrochemical Society, 647.

25. Shul'ga, Yu.M., Tarasov, B.P., Fokin, V.N., Shul'ga, N.Yu. and Vasilets, V.N. (1999) Deiterid fullerena $C_{60}D_{24}$: issledovanie spektral'nymi metodami, *Fizika tverdogo tela* **41**(8), 1520-1526.
26. Ugovenok, T.S., Udalov, Yu.P. and Gavrilenko, I.B. (1999) Vzaimodeistvie fullerenovoy sazhi s vodorodnoy neravnovesnoy plazmoy, *Zhurn. prikl. khimii* **72**(10), 1728-1730.
27. Osaki, T., Tanaka, T. and Tai, Y. (1999) Phys. Chem. **1**, 2361.
28. Kornilov, M.Yu., Ljubchuk, T.V., Isaev, S.D., Panasyuk, H., Chernogorenko, V.B. and Schur, D.V. (2000) Perhydrofullerenes as accumulators of hydrogen, *Molecular Materials* **13**, 285-290.
29. Tarasov, B.P., Goldshleger, N.F. and Moravsky, A.P. (2001) Vodorodosoderzhaschie soedineniya uglerodnykh nanostruktur, Usp. khimii **70**(2), 149-166.
30. Shulga, Yu.M., Tarasov, B.P., Koshyk, D., Drahan, V. and Zaginaichenko, S.Yu. (2000) Hydride method of C_{60} fullerite production with "enlarged" fcc lattice, *Abstr. Int. Symp. "Metal-Hydrogen Systems"*, Australia, Queensland, Noosa, 196.
31. Anikina, N.S., Schur, D.V., Simanovsky, A.P., Dubovoy, A.G. and Ivanchenko N.V. (2001) *Problem on fullerene production by electric arc method*, Proc. VII Intern. Conf. «Hydrogen Materials Science and Chemistry of Metal Hydrides», Alushta, 590-593.
32. Lavriv, L.V., Bogdanenko, S.N., Palanchuk, I.V., Bilyk, V.I. and Schur, D.V. (2001) *Features of fullerene synthesis by arc method*, Proc. VII Intern. Conf. «Hydrogen Materials Science and Chemistry of Metal Hydrides», Alushta, 594-595.
33. Lavriv, L.V., Anikina, N.S., Simanovsky, A.P. and Schur (2001) *Features of fullerene extraction with toluene*, Proc. VII Intern. Conf. «Hydrogen Materials Science and Chemistry of Metal Hydrides», Alushta, 596-597.
34. Trefilov, V.I., Schur, D.V., Tarasov, B.P., Shulga, Yu.M., Chernogorenko, A.V., Pishuk, V.K. and Zaginaichenko, S.Yu. (2001) Fullerenes is future materials basis, Kiev: Izd. ADEF – Ukraine, 148 p.
35. Trefilov, V.I., Schur, D.V., Zaginaichenko, S.Yu., Dubovoy, A.G., Pishuk, V.K., Schur, D.V., Rogozinskiy A.A. Patent on invention «Fullerene-making process from solutions in toluene», Ukr. Institute Industrial Property, Application N2001085692 of 11.04.2002.
36. Wudl F. (1992) *Acc. Chem. Res.* **25**, 157.
37. Taylor, R., Walton, D.R.M. (1993) The chemistry of fullerenes, *Nature (London)* **363**, 685.
38. Sokolov, V.N. and Stankevich, I.V. (1993) Fullereny – novye allotropnye formy ugleroda, Usp. khimii **62**, 455-472.
39. Sokolov, V.I. (1999) Khimiya fullerenov, *Izv. AN. Ser. khim.* N 7, 1211.
40. Fagan, P.J., Krusic, P.J., Evans, D.H., Lerke, S.A. and Johnston, E.J. (1992) Synthesis, chemistry and properties of a monarlkylayed buckminsterfulleren derivative, t-BuC_{60} anion, *Am. Chem. Soc.* **114**, 9697-9699.
41. Attalla, M.I., Vassallo, A.M., Tattam, B.N. and Hanna, J.V. (1993) Preparation of hydrofullerenes by hydrogen radical induced hydrogenation, *J. Phys. Chem.* **97**,

6329-6331.
42. Henderson, C.C. and Cahill, P.A. (1993) Synthesis of the simplest C_{60} hydrocarbon derivative, *Science* **259**, 1885-1887.
43. Gerst, M., Beckhaus, H.-D., Ruchardt, C., Campbell, E.E.B. and Tellgmann, R. (1993) *Tetrahedron Lett.* **34**, 7729.
44. Tarasov, B.P., Fokin, V.N., Moravsky, A.P. and Shul'ga, Yu.M. (1996) O vzaimodeistvii fullerenov s deiteriem v prisutstvii palladiya, Izv. RAN. Ser. khimii, 1867-1868.
45. Baum, R.M. (1993) *Chem. Eng. News* **22**, 8.
46. Hirsch, A. (1994) *Chemie in unserer Zeit* **28**, 79.
47. Avent, A.G., Darwish, A.D., Heimbach, D.K., Kroto, H.W., Meidine, M.F., Parsons, J.P., Remars, C., Roers, R., Ohashi, O., Taylor, R. and Walton, D.R.M. (1994) *J. Chem. Soc. Perkin Trans.* **2**, 15.
48. Lobach, A.S., Perov, A.A., Rebrov, A.I., Roschupkina, O.S., Tkacheva, V.A. and Stepanov, A.N. (1997) Production and investigation of C_{60} and C_{70} fullerenes hydrides, *Izv. RAN. Ser. khimii*, 671-678.
49. Tarasov, B.P., Fokin, V.N., Moravsky, A.P., Shul'ga, Yu.M., Yartys', V.A. and Schur, D.V. (1998) *Promotion of fullerene hydride synthesis by intermetallic compounds*, Hydrogen Energy Progress XII Proceedings of the 12th Word Hydrogen Energy Conference, Buenos Aires, Argentina, 21-26 June 1998, **2**, 1221-1230.
50. Tarasov, B.P., Fokin, V.N., Moravsky, A.P., Shul'ga, Yu.M., Yartys', V.A. and Schur, D.V. (1998) Synthesis, structure and properties of crystalline fullerene hydrides, 4th Int. Conference on Nanostructured Materials, Stockholm, Sweden, 15-19 June, 1998, Abstracts, 1-92.
51. Tarasov, B.P., Fokin, V.N., Moravsky, A.P., Shul'ga, Yu.M., Schur, D.V., Yartys', V.A. (1998) Hydrogen sorption-desorption in fullerene-metals(intermetallics) systems, XII Int. Symposium on metal hydrogen systems: fundamentals and applications, Hangzhou, China. October 4-9, 1998: Abstract Booklet, A4:14-P.
52. Tarasov, B.P., Fokin, V.N., Moravsky, A.P., Shul'ga, Yu.M., Yartys', V.A., Schur, D.V. (1998) Study of fullerene-metals-hydrogen and fullerene-hydrogen systems, Journees d'Automne 1998, Paris, 27-29 Octobre 1998. Programme et considenses des Communications of Autumn Meeting SF2M, 119.
53. Tarasov, B.P., Fokin, V.N., Shulga, Yu.M., Schur, D.V., Yartys, V.A., Koshyk, D. and Kudarevko, V. (2000) Hydrogen storage in carbon nanostructures, Abstr. Int. Symp. “Metal-Hydrogen Systems”, Australia, Queensland, Noosa, 195.
54. Tarasov, B.P., Shulga, Yu.M., Fokin, V.N., Fokina, E., Moravsky, A.P., Schur, D.V., Turok, R., Zaginaichenko, S.Yu., Yartys V.A. and Kucheryavenko, N. Investigation of fullerite – deuterium systems. Ibid., 143.
55. Kornilov, M.Yu., Isaev, S.D., Chernogorenko, V.B. and Schur, D.V. (2000) Perhydrofullerenes as accumulators for hydrogen, *Molecular Materials* **13**, 285-290.
56. Tarasov, B.P., Shul'ga, Yu. M., Fokin, V.N., Vasilets, V.N., Shul'ga, N.Yu., Schur, D.V. and Yartys, V.A. (2001) Deuterofullerene $C_{60}D_{24}$ studied by XRD, IR and XPS, *Journ. Alloys and Compounds* **314**, 296-300.
57. Shul'ga, Yu.M., Tarasov, B.P., Fokin, V.N., Schur, D.V., Volkov, G.A., Rubtsov, V.I., Krasochka, G.A., Chapysheva, N.V., Shevchenko, V.V. (2001) *Preparation*

and properties of deuterofullerenes: new data, Proc. VII Intern. Conf. «Hydrogen Materials Science and Chemistry of Metal Hydrides», Alushta, 682-685.

58. Tarasov, B.P., Goldshleger, N.F., Schur, D.V. (2001) Application of carbon nanotubes for hydrogen storage, Ibid, 678-681.
59. Shul'ga, Yu.M., Martynenko, V.M., Tarasov, B.P., Fokin, V.N., Rubtsov, V.I., Shul'ga, N.Yu., Krasochka, G.A., Chapysheva, N.V., Shevchenko, V.V. and Schur, D.V. (2002) O termicheskom razlozhenii deiterofullerita $C_{60}D_{19}$, *Fiz. tv. tela* **44**(3), 522-523.
60. Tarasov, B.P. (2000) Hydrogen accumulation in carbon nanostructures, *In «Fullerenes and fullerenelike structures»*, Minsk, 113-120.
61. Tarasov, B.P. (2000) *Int. Sci. Journ. Alternative Energy and Ecology* **1**, 168.
62. Bezmelnitsyn, V.N., Glazkov, V.P., Zhukov, V.P., Somenkov, V.A. and Shilstein, S.Sh. (1999) *Neutron diffraction study of structure of the deuterated fullerenes*, 4th Biennial International Workshop in Russia "Fullerene and Atomic Clusters" (St.Petersburg.) Abstract Book, 141.
63. Glazkov, V.P., Zhukov, V.P. and Somenkov, V.A. (1999) *Interaction of palladium doped fulleren with hydrogen*, 4th Biennial International Wokrshop in Russia "Fullerene and Atomic Clusters" (St. Petersburg). Abstract Book, 142.
64. Loutfy, R.O., Veksler, E.M. and Moravsky, A.P. (1999) *Hydrogen storage on fullerenes: catalytic approach and analysis*, 4th Biennial International Wokrshop in Russia "Fullerene and Atomic Clusters" (St. Petersburg), Abstract Book, 226.
65. Murata, Y., Kato, N., Fujiwara, K. and Komatsu, K. (1996) *J. Org. Chem.* **4**, 3483.
66. Komatsu, K., Murata, Y., Wang, G.-W., Tanaka, T., Kato, N. and Fujiwara, K. (1999) *Fullerene Sci. Technol.* **7**, 609.
67. Komatsu, K., Fujiwara, K., Murata, Y. and Braun, T. (1999) Aqueous solubilization of crystalline fullerenes by supramolecular complexation with γ-cyclodextrin and sulfocalix [8] arene under mechanical high-speed vibration milling, *J. Chem. Soc., Perkin Trans.* **20**, 2963.
68. Orimo, S., Majer, G., Fukunaga, T., Züttel, A., Schlapbach, L. and Fujii, H. (1999) Hydrogen in mechanically prepared nanostructured graphite, *Applied Phys. Lett.* **75**(20), 3093-3095.
69. Bensasson, R.V., Hill, T.J., Land, E.J., Leach, S., McGarvey, D.J., Truscott, T.G., Ebenhoch, J., Gerst, M. and Rüchardt, C. (1997) Spectroscopy and photophysics of $C_{60}H_{18}$ and $C_{60}H_{36}$, *Chem. Phys.* **215**, 111-123.
70. Shul'ga, Yu.M. and Tarasov, B.P. (1998) The C_{60} fullerite with extended FCC lattice, *Pis'ma v Zhurn.exp.teor.fiz.* **68**(3) 239-242.
71. Darwish, A.D.M., Taylor, R. and Loutfy, R. (2000) *In Proceeding of 197th Meeting of Electrochemical Society. Toronto*. Abstract. N 693.
72. Bashkin, I.O., Antonov, V.E., Kolesnikov, A.I., Ponyatovsky, E.G., Mayers, J., Parker, S.F., Tomkinson, J., Moravsky, A.P. and Shul'ga, Yu.M. (2000) Hydrogen in the vibrational spectra of high-pressure hydrofullerite, *Molecular Materials.* **13**(1-4), 251-256.
73. Bashkin, I.O., Antonov, V.E., Boltalina, O.V., Gol'd, I.V., Gurov, A.F., Kolesnikov, A.I., Krinichnaya, E.P., Moravsky, A.P., Khasanov, S.S., Shul'ga, Yu.M., Ponyatovsky, E.G. (2000) *Int. Sci. Journ. for Alternative Energy and Ecology* **1**, 134.
74. Kolesnikov, A.I., Antonov, V.E., Bashkin, I.O., Grosse, G., Moravsky, A.P.,

Muzychka ,A.Yu., Ponyatovsky, E.G. and Wagner, F.E. (1997) Neutron spectroscopy of C_{60} fullerene hydrogenated, *Phys J. Condenced Matter.* **9**, 2831.
75. Kolesnikov, A.I., Antonov, V.E., Bashkin, I.O., Li, J.C., Moravsky, A.P., Ponyatovsky, E.G. and Tomkinson, J. (1999) *Physica B.* **263-264**, 436.
76. Bashkin, I.O., Antonov, V.E., Ossipyan, Yu.A., Ponyatovsky, E.G., Moravsky, A.P., Shulga, Yu.M. and Morozov, Yu.G. (2000) Magnetic properties of high-pressure hydrogenated and polymeric fullerites, *Molecular Materials.* **13**(1-4), 263-268.
77. Shul'ga, Yu.M., Lobach, A.S., Morozov, Yu.G., Spektor, V.N. and Ovchinnikov, A.A. (1998) X-ray photoelectronic spectrums and magnetic properties of $C_{60}H_{36}$ and $C_{70}H_{36}$ fullerenes hydrides, *Zhurn.fiz.khim.* **72**(1), 115-118.
78. Lebedev, B.V., Bykova, T.A., Lobach, A.S. (1999) Thermodynamic properties of hydrofullerene $C_{60}H_{36}$ in the range of 4,8-340 K, 4th Biennial International Wokrshop in Russia "Fullerene and Atomic Clusters". St. Petersburg, Abstract Book, 258.
79. Lebedev, B.V., Bykova, T.A. and Lobach, A.S. (1999) Thermodynamic properties of hydrofullerene $C_{60}H_{36}$ in the range of 0-340 K, *Doklady AN* **368**, 629-631.
80. Rathna, A. and Chandrasekhar, J. (1993) *Chem. Phys. Lett.* **206**, 217.
81. Dorozhko, P.A., Lobach, A.S., Popov, A.A., Senyavin, V.M., Yushenko, V.V. and Korobov, M.V. (1999) Sublimation and de-hydrogenation of fullerene hydride $C_{60}H_{36}$, 4th Biennial International Wokrshop in Russia "Fullerene and Atomic Clusters". St. Petersburg, Abstract Book, 257.
82. Jin, C., Hettich, R., Compton, R., Joyce, D., Blencoe, J. and Burch, T. (1994) Direct solid-phase hydrogenation fullerenes, *J. Phys. Chem.* **98**, 4215.
83. Attalla, M.I., Vassallo, A.M., Tattam, B.N. and Hanna, J.V. (1993) Preparation of hydrofullerenes by hydrogen radical induced hydrogenation, *J. Phys. Chem.* **97**(24), 6329-6333.
84. Hall, L.E., McKenzie, D.R., Attalla, M.I., Vassallo, A.M., Davis, R.L., Dunlop, J.B. and Cockayne, D.J.H. (1993) The structure of $C_{60}H_{36}$, *J. Phys. Chem.* **97**, 5741-5744.
85. Kepert, D.L. and Crare, B.W. (1996) Fulleren hydrides based on skew pentagonal pyramidal arrangements of hydrogen atoms, *Theochem.* **363**(2), 179-190.
86. Book, L.D. and Scuseria, G.E. (1994) Isomers of $C_{60}H_{36}$ and $C_{70}H_{36}$, *J. Phys. Chem.* **98**, 4283.
87. Dunlap, B.I., Brenner, D.W. and Schriver, G.W. (1994) Symmetric isomers of hydrofullerene $C_{60}H_{36}$ *J. Phys. Chem.* **98**, 1756.
88. Bühl, M., Thiel, W. and Schneider U. (1995) Magnetic properties of $C_{60}H_{36}$ isomers, *J. Am. Chem. Soc.* **117**(16), 4623-4627.
89. Lobach, A.S., Shul'ga, Yu.M., Rebrov, A.I., Perov, A.A. and Morozov, Yu.G. (1997) Nuclear magnetic resonance, X-ray photoelectron and electron energy loss spectroscopies and magnetization studies of fullerene hydride $C_{60}H_{36}$, In Fullerenes. Recent Advances in the Chemistry and Physics of Fullerenes and Related Materials Vol.4 (Eds. K.M. Kadish and R.S. Ruoff), The Electrochemical Society. Pennington, NJ, 246.
90. Okotrub, A.V., Bulusheva, L.G., Asanov, I.P., Lobach, A.S. and Shulga, Yu.M. (1999) X-ray spectroscopic and quantum-chemical characterization of hydrofullerene $C_{60}H_{36}$, *J. Phys. Chem. A.* **103**(6), 716-720.

91. Lobach, A.S. and Strelets, V.V. (1999) Electrochemical behaviour and parameters of hydrofullerene $C_{60}H_{36}$, *Izv. RAN. Ser. khimii* N 11, 2083-2087.
92. Okotrub, A.V., Bulusheva, L.G., Asanov, I.P., Lobach, A.S. and Shul'ga, Yu.M. (2000) Study of electronic structure of $C_{60}H_{36}$ hydrofullerene isomers series, *Zhurn. neorg. khimii* **45**(3), 523-532.
93. Breslavskaya, N.N. and D'yachkov, P.N. (2000) Isomerism of C_nX_k (X = H, F; n=76, 78, 84) fullerenes hydrides and fluorides with great content of hydrogen and fluorine, *Izv. RAN. Ser. khimii* **45**, 499.
94. Hall, L.E., McKenzie, D.R., Davis, R.L., Attalla, M.I. and Vassallo, A.M. (1998) *Acta Cryst.* **B.54**, 345.
95. Brosha, E.L., Davey, J., Garzon, F.H. and Gottesfeld, S. (1999) *J. Mater. Res.* **14**, 2138.
96. Davydov, V.Ya., Sheppard, N. and Osava, E. (2000) *Int. Sci. Journ. for Alternative Energy and Ecology* **1**, 183.

PHYSICAL-CHEMICAL ANALYSIS IN SYNTHESIS OF TETRAHYDROBORATES AND TETRAHYDROALUMINATES OF METALS

U.MIRSAIDOV
President of the Academy of Sciences of the Republic of Tajikistan,
Director of the V.I.Nikitin Institute of Chemistry of Academy of Sciences of the Republic of Tajikistan
33 Rudaki Ave., 734025, Dushanbe, Tajikistan,
E-Mail: academy@science.tajik.net

The processes of synthesis and conversion of Tetrahydroborates and Tetrahydroaluminates, studied in this work, have been investigated from positions of theoretical rules of physicochemical analysis according to experimental data on study of phase equilibria, rules of donor-acceptor theory and ideas of Luis about acid-based interaction.

1. Tetrahydroborates

Up to present time we have data about Tetrahydroborates of the majority of metals of Periodic System. However, many of them still are poorly accessible because of imperfect methods of their synthesis. The overwhelming number of compounds have been synthesized in solvated states. And in the case of heavy transitive metals, as a rule, only Tetrahydroborates with mixed ligands have been synthesized.
The methods of synthesis with participation of Diboran served as a key to invention of many kinds of Tetrahydroborates, but these methods do not provide the synthesis of pure substances [1,2]. The most acceptable among known methods of synthesis of Tetrahydroborates of transitive metals is the reaction of halogenides of transitive metals with Tetrahydroborates of Alkaline metals in solutions or in solid phase.
A considerable number of Tetrahydroborates of Metals have been synthesized as solvates in the media of oxygen-containing and nitrogen-containing solvents of donor types(THF, Ethers of Polyglycols, Pyridine, Aromatic and Aliphatic Amines). Exchange interaction of MBH_4 with M^nCl_n in organic media is considerably connected with nature of solvent and interaction of components in solution. That is why mainly our efforts were made to investigation of phase equilibrium in the systems with participation of Tetrahydroborates of Metals.
Firstly for exchange reaction

$$M^nCl_n + nLiBH_4 \xrightarrow{THF} Mn(BH_4)_n + nLiCl \qquad (1)$$

N. Veziroglu et al. (eds.), Hydrogen Materials Science and Chemistry of Metal Hydrides, 19–28.

we have used Sodium Tetrahydroborate that allowed us to synthesise products of reaction with admixture of Lithium Chloride. The replacement of LiBH4 with Sodium Tetrahydroborate at first seemed us unacceptable because of practical insolubility of $NaBH_4$ in THF(Tetrahydrofuran). But we have determined that when there is abundance of Sodium Tetrahydroborate and if we increase time of reagents' contact we are able to finish the exchange reaction with almost full conversion of MCl_3 to $M(BH_4)_3$ [3.4]. We managed to intensify the interaction using special pound devices like planetary mill.
Use of Sodium Tetrahydroborate in exchange reaction (1) ensure the synthesis of pure product because contrary to Lithium Chloride, NaCl is almost unsolvable in THF. In addition to this, unlike $LiBH_4$, $NaBH_4$ has a weak tendency to complex-formation with $M(BH_4)_3$ [5,7].
Using NaBH4 we managed to synthesis for the first time free of Chlorine THF-solvates of Tetrahydroborates of the whole row of Rare-Earth Metals:

$$MCl_3 + 3NaBH_4 \longrightarrow M(BH_4)_3 + 3NaCl (M=\text{Rare-Earth Metal}) \qquad (2)$$

The problem of synthesis of free of solvent Tetrahydroborates of majority transitive metals is very difficult and yet is too complex to be solved. By the method of thermo-vacuum drying in major cases we only managed to have as a result a partial dissolution. For example, bistetrahydrofuranate of Scandium Tetrahydroborate, being converted into monosolvate
$So(BH_4)_3 \cdot$ THF during drying it does not lose farther THF until its thermo-destruction. The data we have show that Tristerahydrofuranates of Tetrahydroborates of Yttrium Group's Lanthanoids are able to lose THF only up to moment they do not reach the following composition
$Ln(BH_4)_3 \cdot 0.8THF$. We received the analogous result during salting-out of Tetrahydroborates from solutions of donor solvent while heating it.
The key to synthesis of pure non-solvated Tetrahydroborates could be the complete carrying out of exchange reaction (2) in non-solvating media or in solvents with weak donor ability.
We have determined that the four members of Lanthanoids row(La, Ce, Pr, Nd) have low solubility (up to 0.5) in benzol and toluol. And even such low solubility allowed carrying out the exchange reaction (2) in the media of these solvents intensively mixing it. In this way for the first time have been synthesized non-solvated Lanthanoids' Tetrahydroborates of the beginning of the row of Lanthanoids [8]. THF-solvates of these Tetrahydroborates are able to desolvate during salting-out. In this way have been synthesized non-solvated Tetrahydroborates of La, Ce, Pr, Nd.
The diagrams very clearly illustrate the abilities of THF-solvates of Tetrahydroborates of La, Ce, Pr, Nd to step by step giving of THF during heating. The diagrams also illustrate a considerable thermal stability of non-solvated compounds[9,10].
Solubilities of Tetrahydroborates are directly correlated to character of bounds in them. The most covalent Tetrahydroborates of Zirconium, Hafnium dissolve in non-polar solvents. Less covalent Tetrahydroborates, for example, Tetrahydroborates of Thorium, Gadolinium are solvable in solvated media. Some Tetrahydroborates of Metals with mixed ligands have ionic characters and well dissolve in liquid ammonia [9].

The necessity of finding of optimal conditions for synthesis of Tetrahydroborates of Metals, for possibility of complex-formation, and for changing of solubility character and solvation of components of synthesizing components prompt us to investigate systematically the phase equilibria in the systems of $M(BH_4)_3$-solvent, $M(BH_4)_3$-$M'(BH_4)_m$-solvent(M=Rare-Earth Metal, M'= Alkaline or Alkaline-Earth Metal) in the large interval of temperatures [5,6,11-13]. We selected THF as general solvent for all Tetrahydroborates of Rare-Earth Metals.

According to constructed by us the solubility polytherms of $M(BH_4)_3$-THF the highest chemical activity to THF among many kinds of solvates is expressed in Scandium Tetrahydroborate that forms okta-, geksa-, and tetrasolvates[11,14]. We determined that Tetrahydroboratesc of Rare-Earth Metals of Cerium group and Yttrium group have differences in abilities in solvatation. Compounds of Cerium group at normal temperature form tertasolvates; and Tetrahydroborates of Metals of usually form trissolvates. During increasing of temperature the solubility of Tetrahydroborates of Cerium group increases more intensively than the one of Tetrahydroborates of Yttrium group. Tetrahydroborates of light lanthanoides in the interval of 293-323 K increase their solubility three times; and heavy Lanthanoids in the same interval increase their solubility only two times. This difference we can use during separation of Tetrahydroborates of Lanthanoids[9].

In the threefold systems of Tetrahydroborate of Rare-Erth Metal – Tetrahydroborate of Alkaline metal – THF we have determined chemical interaction with formation of binary Tetrahydroborate complexes 1:1 – $M'[M(BH_4)_4]$ (M=Rare-Earth metal, M'= Li, Na) [2, 13, 15-17]. of Tetrahydroborates of Rare-Earth metals of Cerium and Yttrium Groups is became to be quite different. And in each group (La, Y) the ability to complex-formation depends from selection of external-sphere cation of Alkaline Metal. Binary complexes with Lithium Tetrahydroborate $LI[LN(BH_4)_4]$ have been synthesized and characterized for Tetrahydroborates of Rare-Earth Metals of the whole row of Yttrium group. These compounds are able for equilibrium crystallization from saturated solutions as THF-solvates; they have characteristic areas in IR- and NMR-spectra and they become thermodestructed at 373 K. Tetrahydroborates of Rare-Earth Metals and Sodium Tetrahydroborate form unstable binary complexes that exist only in solution [18]. Complex Tetrahydroborates of Rare-Earth Metals and Lithium Tetrahydroborate are concentrated sources of BH_4 and may be interesting as selective reductors of functional groups and effective agents of catalytic conversions in thin organic synthesis. Tetarhydroborates of Rare-Earth Metals of Cerium Group are not able to form stable enough binary Tetrahydroborates neither with Sodium Tetrahydroborate nor with Lithium Tetrahydroborate[3].

Analysis of received solubility diagrams helped us to determine the existence and synthesized unknown until that time complex Tetrahydroborates of Rare-Earth Metals and Alakline Metals. This analysis was very informative for selection of reagents for synthesis of Tetrahydroborates of Rare-earth Metals. For example, comparison of isotherm of the systems of $LN(BH_4)_3$ – MBH_4 – THF with participation of Tetrahydroborates of Lithium, Sodium and Potassium points to expediency of selection as a reagent (for synthesis of Lanthanoid Tetrahydroborate by exchange of $LnCl_3$ – MBH_4 – THF) precisely Sodium Tetrahydroborate. We cannot use for this purpose Lithium Tetrahydroborate because it forms the stable complexes Li $[Ln(BH_4)_4]$; so much is impossible to use Potassium tetrahydroborate which does not dissolve in the presence of Ln(BH4)3. This approach allowed us to synthesis pure Tetrahydroborates of Lanthanoids.

Observation of two types of signals in ^{11}B • NMR spectra of THF-solutions of $Lu(BH_4)_3$ and $LiBH_4$ at decrease of temperature prove the existence for the complexes of $M[Ln(BH_4)_4]$ (M=LI, Na). This observation testifies that the solution becomes equilibrium

$$M(BH_4)3 + M'BH_4 \rightleftarrows M'[M(BH_4)_4] \quad (M''=LI, Na) \qquad (3)$$

In the system of $Lu(BH_4)_3$ – $LiBH_4$ – THF the existence of molecules of $LiBH_4$, $Li[Lu(BH_4)_4]$, $Lu(BH_4)_3$ have been determined. Chemical shift of observed quintet is on more weak area than calculated on additivity the average number for ratio 1:1 of Tetrahydroborates of Lithium and Lutetium in solution [19,20]. In the case of Sodium and Rare-Earth Metals of Yttrium group (and compounds of Lithium, Sodium, and Rare-Earth Metals of Cerium group) the given equilibrium is considerably displaced to left and that is why it takes place only partial formation of binary Tetrahydroborates in solution. Because of this fact during synthesis of Tetrahydroborates of Lanthanoids by the method of replacement of their halogenides with Sodium Tetrahydroborate the product, exuded from initial THF-solutions, contains admixtures of Sodium Tetrahydroborate.

Tetrahydroborates of Titan, Hafnium, Circonium, of the whole row of Rare-Earth Metals and actinoides are able to form complexes with Tetrahydroborate of Tetraalkylammonia. We have synthesized and characterized binary Tetrahydroborates of Rare-Earth Metals and Tetrabutylammonia. Tetrahydroborates of Circonium and Hafnium have the ability to add only one mole of R_4NBH_4 [21]. At the same time Tetrahydroborates of Rare-Earth metals and actinoides add two moles Bu_4NBH_4 [22]. Obviously, the high tendencies to complex-formation in Tetrahydroborates of Rare-Earth Metals and actinoides (in contrary to d-elements), are based on the abilities of atoms of f-elements to expand their coordinational spheres at the expense of remote f-electrons.

It is interesting to trace changes of BH_4 group under influence of acceptors of various capacities. BH_4 group has radius of

2.3 Å. Proton, being the strongest acceptor, displaces diboran from Tetrahydroborates:

$$2NaBH_4 + H_2SO_4 \rightarrow B_2H_6 + 2H_2 + Na_2SO_4 \qquad (4)$$

The first act is a deformation of $BH_4 \longrightarrow H\text{-}\partial ......BH_3$, which is followed by geterocyclic decay and dimerization of Borine in Diboran:

$$2H^{-\partial} BH_3 \longrightarrow 2H- + 2BH_3 \longrightarrow 2H^- + B_2H_6 \qquad (5)$$

Triftoride of Boron – Luis acid reacts analogically to proton:

$$3NaBH_4 + 4BF_3 \longrightarrow 2B_2H_6 + 3NaBF_4 \qquad (6)$$

Conversions (4) and (6) are very profitable energetically: the energy of crystallic bars of salts forming. In both cases the entropy of the system increases (especially by equation (4)).
Though trichloride of Aluminium is an electrophfile reagent its acceptor capacity is lower than that of BF_3 and at first under its influence the displacement of Diboran does not take place:

$$3NaBH_4 + AlCl_3 \longrightarrow 3NaCl + Al(BH_4)_3 \qquad (7)$$

But Tetrahydroborate of Aluminium is unstable and if we keep or increase the temperature it will decompose with formation of complex mixture of Boron Hydrides and Aluminium. The first stage, probably, is:

$$Al(BH_4)_3 \longrightarrow AlH_3 + 3BH_3 \qquad (8)$$

Halogenides of Rare-Earth Metals. Being weak acceptors, do not repeat properties of AlCl3. They do not promote geterolitic decomposition of BH^-_4 group (5) and their interaction with $NaBH_4$ is limited with exchange reaction

$$LnCl_3 + 3NaBH_4 \longrightarrow Ln(BH_4)_3 + 3NaCl \qquad (9) \qquad (Ln\text{=Rare-Earth Metals})$$

The increase of charge in the row Rare-Earth Metals La → Lu increases their acceptor abilities. But this is expressed in the ability of Tetrahydroborates of Rare-Earth Metals of Yttrium group form complexes with Lithium Tetrahydroborates and in considerable less stability of THF-solvates Tetrahydroborates of Rare-Earth Metals of Cerium group (have been synthesized only non-solvated Tetrahydroborates of La, Ce, Pr, Nd).

For BH^-_4 group are known mono- H-BH_3, di- 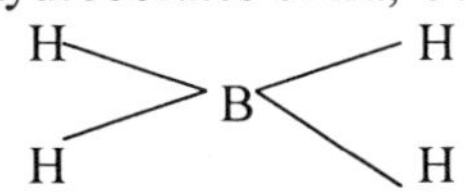 and

tridentant 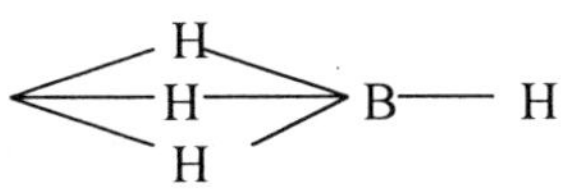 coordinations with atoms of

Metal that forms the complex. The direct structural on monocrystals gives a conclusion about geometry of bound of BH_4 group. Such investigations have been carried out only for some Tetrahydroborates of transfer metals.

2. Synthesis of Hydroaluminates and their Derivatives

Hydroaluminates, the group of complex Tetrahydroaluminates, have common formulas $M^n[(AlH_4)_n]$, Polyhydroaluminates $M^n[(AlH_4)_n]_n{\cdot}AlH_3$ and Hexohydroaluminates $M^{3n}[(AlH_6)n$. Tetrahydroaluminates $M^n[(AlH_4)_n]$ is one of the most important group of Hydrogen compounds.
In the beginning of our research Tetrahydroaluminates of Alkaline metals have been closely investigated. But even for metals of IIA Group the data about synthesis of

Hydroaluminates, conversions and properties of Hydroaluminates has remained incomplete.

The method of direct synthesis of Tetrahydroaluminates in melts

$$M(liquid)+Al(solid)+2H_2(gas) \longrightarrow MAlH_4(liquid) \quad (10) \quad (M=Na, K, Rb, Cs)$$

$$Na(liquid) + Al(solid) + 2H_2(gas) + MCl(solid) \xrightarrow[180\ atm]{523\ K} MalH_4(solid) + NaCl(solid) \quad (11)$$

is the most effective method (high speed of reaction and high output of reagents) in comparison with methods of synthesis in organic media. Methods of synthesis in organic media are multi-stages, have low speeds and have potentially danger of solvent to decompose with explosion. But constructional difficulties of creating the apparatus for direct synthesis make it necessary to look for perspective methods of synthesis of Tetrahydroaluminates. These perspective methods should exclude use of high parameters.

Key reaction of synthesis of Tetrahydroaluminates

$$4M^nH+nAlCl \longrightarrow M^n[(AlH_4)_n]+3M^nCl \quad (12)$$

has been carried out in the media of solvating solvent, that dissolves synthesized Tetrahydroaluminate. By this way $LiAlH_4$(in the media of Et_2O and $NaAlH_4$) and $Ca(AlH_4)_2$ in THF have been synthesized. Using this method we could not synthesized Tetrahydroaluminates that do not dissolve in this solvent (e.g. $NaAlH_4$ in Et_2O). Actually, reaction (12) consists of two stages:

$$3M^nH_n+nAlCl_3 \longrightarrow nAlH_3+3M^nCl_n \quad (13)$$

$$M^nH_n+nAlH_3 \longrightarrow M^n[(AlH_4)_n] \quad (14)$$

Conversion with formation of Tetrahydroaluminate (13) is a limitative conversion. When we dose $AlCl_3$ on the crystal's surface of hydride of metal, i.e. when there is a plenty of it the synthesized AlH_3 reacts by reaction (14). In classical case of use of LiH in the media of Et_2O the products of reaction are:

1. Well-soluble $LiAlH_4$ (in solutions)
2. Practically insoluble Lithium Chloride (in deposit)

The reaction was intensificated by the way of energetic mixing and adding of prepared $LiAlH_4$.

Tetrahydroaluminates of Aklkaline-Earth metals as well as their Chlorides are insoluble in Et_2O. These results with "double"-brake during their synthesis and reactions (13) and (14), practically, do not take place.

We determined that in the case of Hydrides of Alkaline Metals the key to investigation of reactions (13) and (14) was the synthesis of complex forms of Tetrahydroaluminates, soluble in ether. We determined the existence and synthesized soluble in ether complexes of Tetrahydroaluminates of Alkaline-Earth Metals with Tetrahydroaluminate and Tetrahydroborate of Lithium of the following composition:

$M(AlH_4)_2{\cdot}2LiEH_4$ (M=Ca, Sr, Ba; E=B, Al)

according to the equations:

$$MCl_2+2LiAlH_4+2LiBH_4 \longrightarrow M(AlH_4)_2 \cdot 2LiBH_4+2LiCl \quad (15)$$
$$MH_2+2AlH_3+2LiEH_4 \longrightarrow M(AlH_4)_2 \cdot 2LiEH_4 \quad (16)$$

From saturated solutions have been crystallized complexes $M(AlH_4)_2:LiEH_4=1:2$. We can see partial dissociation with splitting of Ligand during dilution:

$$M(AlH_4)_2 \cdot 2LiEH_4 \rightleftarrows M(AlH_4)_2 \cdot (2-n)LiEH_4 + nLiEH_4 \quad (17)$$

In this way we synthesized products of the following compositions:

$Ca(AlH_4)_2 \cdot 0.5\ LiBH_4 \cdot 0.3\ Et_2O$
$Sr(AlH_4)_2 \cdot 0.3\ LiBH_4 \cdot 0.3\ Et_2O)$
The distinctive feature of IR-spectra of synthesized complexes is presence of intensive strip of absorption in the area of 1960-1970 sm-1. We can classify this strip of absorption to Al-H-B bridge fluctuations. In spectra display also Al-H and B-H-valent fluctuations in typical areas for them. Also there are strips pointing on octahedron surrounding of Aluminium atom. This gives a basis to imagine their probable structure:

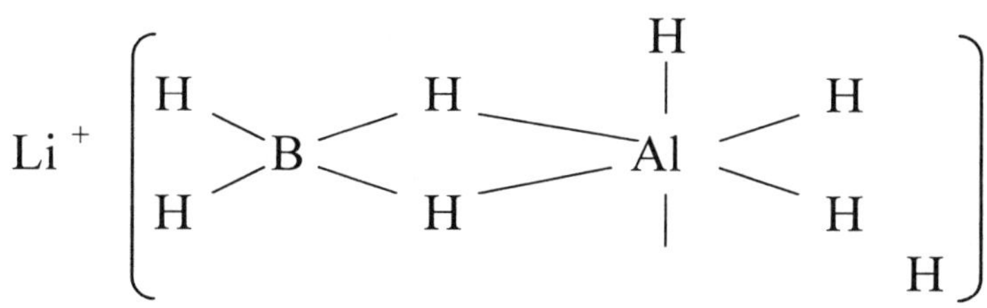

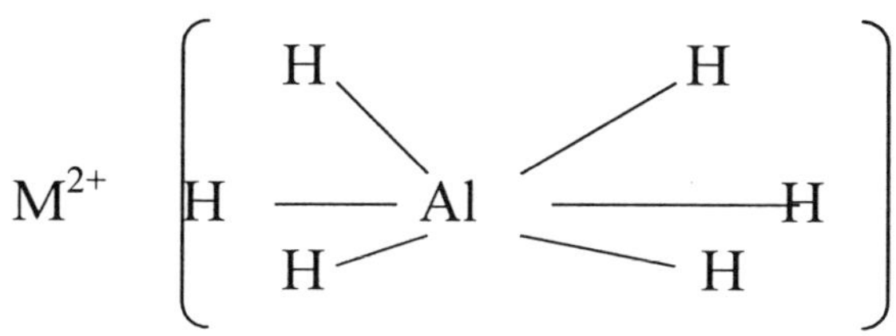

Difractograms of the complexes are different from difractograms of their components: $LiBH_4$ or $LiAlH_4$, $M(AlH_4)_2$. On sum of physicochemical properties the synthesized compounds, probably, are close to binary salts. They have a low solubility in Et_2O:
0.7 of mass for $Ca(AlH_4)_2 \cdot 2LiBH_4$,
1,2 of mass for $Sr(AlH_4)_2 \cdot 2LiBH_4$,
1,7 of mass for $Ba(AlH_4)_2 \cdot 2LiBH_4$

The latter compound and complexes $Ba(AlH_4)_2 \cdot AlH_3$ (see below) are first known soluble forms of Barium Tertahydroaluminate; for Barium Tetrahydroaluminate have not been yet found solvents [23].
During investigation of phase equilibria in the system of $Ca(AlH_4)_2 - LiBH_4 - Et_2O$ at 298 K were determined areas of crystallization not only of complex of composition $Ca(AlH_4)_2 \cdot 2LiBH_4 \cdot 4Et_2O$ but also of complex $Ca(AlH_4)_2 \cdot 8LiBH_4 \cdot 10Et_2O$. The existence of such compounds indicates on considerable acceptability of Potassium ion having charge $^{2+}$ and ability of Aluminium atom to expand coordinational sphere up to octahedral one.
Determination of existence of labile complexes of Tetrahydroaluminates of Alkaline Metals with Tetrahydroaluminate and Tetrahydroborate of Lithium prompted us, on the one hand, to investigate complex-formation of Binary Hydrides and Tetrahydroaluminates with AlH_3 (AlH_3 in comparison with $LiAlH_4$ and $LiBH_4$ is coordinationally less saturated); on the other hand the task was set to investigate interaction of Binary Hydride and Alumohydride of metal with Aluminium Chloride, particularly, using Sodium Hydride, Potassium Hydride and other Binary Hydrides that had been considered not being able to react in Et_2O.
Interaction of MH and $MalH_4$ with $AlCl_3$ has been investigated till nowadays for Lithium Hydride in direction of synthesis of Lithium Tetrahydroaluminate and Aluminium Hydride on classical for today Schlezinger equation. Really, in this process there is a complex variety of reactions,; total combinations of these reactions we can see in chemical diagram of V.T.Micheeva.
We generalized and expanded the Physicochemical Diagram of interaction in mutual system of MH-Et_2O (M, Al // Cl, H). You can see this diagram in figure. The diagram was drawn according to the methodic of physicochemical analysis and model the direction of interaction as well as stekhiometry of some reactions.
The conditions of most of reactions shown on physicochemical diagram have not been determined yet. Conversions of Lithium Hydride have been investigated more detailed than conversions of other Hydrides.
The interaction of Binary Hydride of Metal with Binary Hydride of Metal-complexformer is a classical way of synthesis of complex hydrides;

$$nMH + mEH_3 \longrightarrow mMH \cdot mEH_3 \text{ (M-metal equivalent, E=B. Al, Ga)} \quad (18)$$

Hydrides of elements of III Group have the ability to form such hydride anion complexes. Actually, the interaction of Diboran with MH became the key to invention of Tetrahydroborates of most of metals. But for Aluminium Hydride the number of known reactions with formation of Tetrahydroaluminates is limited.

We investigated the interactions of metal and Aluminium Binary Hydrides according to physicochemical diagram (see chart). For the first time has been determined that solvate of Aluminium Hydride is a very active reagent for conversion of Binary Hydrides of Alkaline and Alkaline-Earth Metals to Tetrahydroaluminates in the media of Et_2O according to reaction (18). For the first time in the media of Et_2O have been synthesized Hydroaluminates of Na, Ca, Sr, Ba; Hexohydroaluminate of Sodium Na_3AlH_6, not known before Polyhydroaluminates of Alkaline-Earth Metals that have the general formula $M(AlH_4)_2 \cdot nAlH_3$[23, 24]

Having investigated phase equilibria in the system of $Ca(AlH_4)_2$ – AlH_3 – Et_2O we determined increase of solubility of insoluble by itself $Ca(AlH_4)_2$ in the presence of AlH_3 in the media of Et_2O; have been determined the boards of equilibrium crystallization of Ployhydroaluminates with gepta- and decahydroaluminate anions $Ca(Al_2H_7)_2 \cdot nEt_2O$, $Ca(Al_3H_{10})_2 \cdot mEt_2O$ [25].

In the whole we can conclude that complex view of task of synthesis of complex hydrides of Aluminium from positions of physicochemical analysis and investigation of phase equilibria allowed to determine and realize directions of interactions in the system Binary Hydride- Halogenide of Aluminium , to understand more properly the mechanisms of reactions , to elaborate effective methods of their synthesis.

3. References

1. Kadlec, V., Kadlecova, H., and Mosek J. (1976) *Chemistry listy*, **70**, 673-698.
2. Semenenko, K.N., Kravchenko, O., and.Polyakova V. (1973), *Russian Journal: Succeses of Chemistry*, **42**, No:1, 3-26.
3. Mirsaidov, U., Kurbonbekov, A., Rottenberg, T., and Juraev, H. (1978) *Proceedings of the Academy of Sciences of the USSR, Inorganic Materials*, **14**, No: 9, 1722-1724.
4. Mirsaidov, U.,.Rahimova, A., Dymova, T. (1979) *Proceedings of the Academy of Sciences of the USSR, Inorganic Materials,* **15**, No:9, 1585-1588.
5. Mirsaidov, U., Kurbonbekov, A., (1980) *Russian Journal of Inorganic Chemistry,* **25**, No:5, 1386-1389.
6. Mirsaidov, U., Kurbonbekov, A., Hikmatov M. (1981) *Russian Journal of Inorganic Chemistry,* **26**, No:10, 2875-2876.
7. Mirsaidov, U., Rahimova, A., Aliev, H. (1980) *Reports of the Academy of Sciences of the Tajik SSR,***23**, No:3, 150-153.
8. Mirsaidov, U., Kurbonbekov, A., Hikmatov M. (1982) *Russian Journal of Inorganic Chemistry,* **27**, No:2, 2436-2439.
9. Mirsaidov, U., Dimova, T. (1984) *Borohydrides of Transitive Metals*, Dushanbe Publishing house “Donish”,307 pages.
10. Mirsaidov, U., (1983) *Borohydrides of Rare-Earth Metals*, Dushanbe, 20 p., Saved in VINITI, 26.05.83, No:2863-83.
11. Mirsaidov, U., Gatina, R.,.Rottenberg, T. (1980) *Russian Journal of Inorganic Chemistry,* **25**, No:6, 1620-1623.
12. Mirsaidov, U.,.Rottenberg, T., Samiev, Ya. (1978) *Russian Journal of Inorganic Chemistry,***24**, No:7, 1995-1996.

13. Mirsaidov, U., Rahimova, A., Dymova T. (1977) *The Reports of the Academy of Sciences of USSR,* **236**, No:1, 120-123.
14. Mirsaidov, U., Kurbonbekov, A., Juraev, H., Rahimova, A. (1976) *The Reports of the Academy of Sciences of the Tajik SSR,* **19**, No:2, 32-36.
15. Mirsaidov, U., Rahimova, A., Dymova T. (1978) *Russian Journal of Inorganic Chemistry,* **23**, No:12, 3326-3329.
16. Mirsaidov, U., Rahimova, A. (1979) *Proceedings of the Academy of Sciences of the Tajik SSR, Department of Physical. Mathematical, Geological, and Chemical Sciences,* No:2, 34-37.
17. Mirsaidov, U., Rahimova, A. (1978) *Russian Journal of Inorganic Chemistry*, **.23**, No:7, 1931-1934.
18. Turova N.(1984) *Russian Journal of Coordinational Chemistry*, **10**, No:9, 1165-1174.
19. Mirsaidov, U., Rahimova, A., Kurbonbekov, A., Boiko G. (1982) *Proccedings of the First Soviet-Indian Symposium on Actual Problems of Magnesium Resonance Spectroscopy, Inorganic materials, Dushanbe,* 113-114.
20. Mirsaidov, U., Rahimova, A., Kurbonbekov, A., Boiko G. *Ibid.*, No:7, 115-116.
21. Hoestra, H., Katz, J. (1949) *Journal of American Chemical Society*, **71**, 2488-2492.
22. Mirsaidov, U., Titov, L., Gavrilova L., and others (1982) *Russian Journal of Coordinational Chemistry*, **27**, No:8, 1953-1958.
23. Mirsaidov, U., Pulotov, M., Aliev, Kh., Alikhonova, T. (1985) *Proceedings of the Academy of Sciences of the Tajik SSR,* **28**, No:8, 21-22.
24. Mirsaidov, U., Gatina, R., Khudoydodov B. and others (1986) *Proceedings of the Academy of Sciences of the SSSR,* **289**, No:2, 241-243.
25. Mirsaidov, U., Gatina, R., Khudoydodov, B. (1981) *Tetrahydroaluminates of IIA Group, Dushanbe*, 15, saved in VINITI 05.09.86, No: 65

A NOVEL NEGATIVE ELECTRODE MATERIAL FOR MH-Ni BATTERY AT LOW TEMPERATURE

Minshou ZHAO□ and Changying SUN
Key Laboratory of Rare Earth Chemistry and Physics Changchun Institute of Applied Chemistry, 159 Ren Min Street 130022 CHINA

A novel non-AB_5-type alloy [$(RENiAlMnCoA)_x$; RE = La or La-rich mischmetal; $2 \leq x < 5$; A = a trace element = has been synthesized by arc melting of constituent elements, and characterized by XRD and electrochemical properties first time. The results showed the alloy has multi-phase structures, and has discharge capacity more 10 times than that of commercial AB_5-type alloy as negative electrode material for MH-Ni battery at low temperature, even at -40^0C. The effect of discharge current and temperature on the electrochemical properties has been studied. Some hypotheses have been preliminarily proposed for synthesis of negative electrode material used at low temperature.

1. Introduction

MH-Ni battery with AB_5-type alloy as negative electrode material has been commercially produced and applied in many field as power source, for example, portable phone, cellar phone, video camera, personal computer of note book type, electric bicycle, motor bike, pure electric vehicle, hybrid electric vehicle and so on. However, the battery is required to be used at the span of temperature from 40^0C to -18^0C. The discharge capacity of the battery would dramatically decrease below -20^0C.

N. Veziroglu et al. (eds.), Hydrogen Materials Science and Chemistry of Metal Hydrides, 29–34.

The performance of the battery to be improved at low temperature is necessary at North–East and North–West in China, North America, Europe and Russia, which include more than the third of land area over the world. The temperature at the places mentioned above in winter would decrease down to - 40^0C.

The electrochemical characteristics of hydrogen storage alloy was investigated at low temperatures [1-4], and a few researchers dealt with the performance of MH-Ni battery at various low temperatures[5-6], but the electrochemical properties of hydrogen storage alloys as negative electrode materials and the performance of MH-Ni battery haven't been dealt yet at - 40^0C.

In this paper the electrochemical properties of the novel negative electrode material have been performed at room temperature and - 40^0C, respectively.

2. Experimental detail

The sample alloy was synthesized by arc melting of constituent elements of RE (RE = La, mischmetal, rich-La mischmetal), Ni, Al, Mn, Co and trace other elements on a water-cooled copper crucible under argon atmosphere. The Ar gas was purified by sieve and silicon gel. The sample alloy was turned over and remelted three times at least in order to improve the alloy homogeneity. The alloy ingot was mechanically pulverized to be 300 meshes for XRD analysis, and The pulverized powder of 200-250 meshes was used for the preparation of metal hydride electrode. The metal hydride electrodes were prepared by pressing the mixtures of alloy powder with nickel powder in the weight ratio of 1:5 to form electrode pellets of diameter of 10mm at the compacting pressure of 200 MPa. The electrochemical tests were carried out in an open cell using 6 M KOH electrolyte solution and sintered $Ni(OH)_2$/NiOOH plat is used as counter electrode with sufficiently large discharge capacity at room temperature (20^0C) and low temperature. The cut-off voltages for discharge were 1.0V at room temperature and 0.9V at -40^0C, respectively. The charge/discharge test was conducted on DC-5 Battery Testing instrument controlled by a computer.

3. Results and discussion

3.1 XRD ANALYSIS

The X-ray diffraction pattern of the sample alloy indicated that the sample alloy has multiphase structure. The second phase in the multiphase structure may be beneficial to enhance the discharge capacity, as shown in non-stoichiometric AB_5-type alloy[7].

3.2 DISCHARGE CAPACITY VARIED WITH CYCLE NUMBER

The discharge capacity of the sample alloy electrode *vs*. cycles was shown in Fig.1 at room temperature. With cycle numbers, the discharge capacity of the sample alloy electrode was almost not changed for 45 cycles and attained about 300mAh/g.

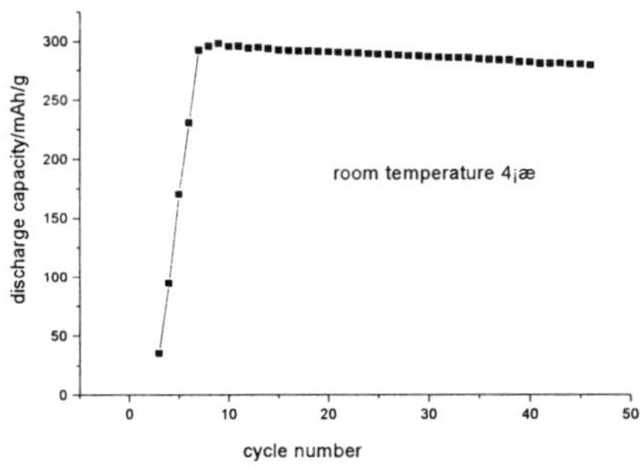

Fig.1 Discharge capacity of sample alloy electrode as function of cycle number at room temperature

3.3 THE EFFECT OF TEMPERATURE ON DISCHARGE CAPACITY OF THE METAL HYDRIDE ELECTRODE

The discharge capacities of the sample alloy and commercial AB_5-type alloy at low temperatures were listed in table 1 to compare each other. It was found that the discharge capacities of the sample alloy would decrease with decreasing temperature and extending cycle numbers, but very slow and keep large enough capacity,. as shown in table 1. However, the discharge capacities of the commercial AB_5-type alloy very sharply decrease with decreasing temperature.

TABLE 1. discharge capacity at –26^{0}C and-33^{0}C, mAh/g ($I_c = I_d = 65mA/g$)

Cycle number	3	6	10	25	45
Sample at -26^{0}C	264	263	261	254	244
Sample at -33^{0}C	246	240	219	220	193
AB_5-type alloy at -33^{0}C	149	74	45	32	31

3.4 THE EFFECT OF DISCHARGE CURRENT ON DISCHARGE CAPACITY OF THE METAL HYDRIDE ELECTRODE

At -40^{0}C the discharge capacity as function of discharge current was listed in table 2. As we known, the discharge capacity of any metal hydride electrode would decrease with increasing discharge current, the sample alloy so behaved. However, the discharge capacity of sample alloy electrode decreased slowly and kept considerable large value to be used, even at discharge current of 145mA/g. On contrast, the discharge capacity of commercial AB_5-type alloy electrode decreased dramatically and can't been used at all.

TABLE2. Discharge capacity as function of I_d at -40^{0}C, mAh/g

Discharge current/ mA/g	65	100	145
Alloy sample	214	193	158
Commercial AB_5-type alloy	14.4	16.2	8.8
Discharge capacity ratio	14.4	16.2	17.9

3.5 DISCHARGE CHARACTERISTICS OF METAL HYDRIDE ELECTRODE

Typical galvanostatic discharge curves of the metal hydride electrodes were shown in

Fig.2. The middle voltage of the sample alloy electrode would been determined from curve 1 to be 1.089V, but that of commercial AB_5-type alloy electrode could be determined from curve 2.

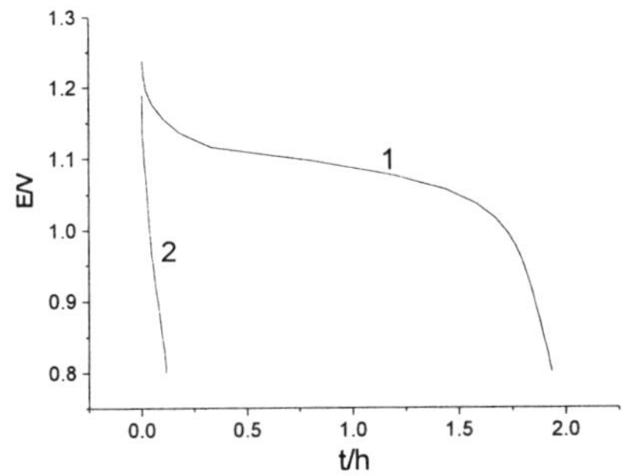

Fig. 2. Discharge characteristics of metal hydride electrodes at -40^0C
Curve 1 sample alloy; Curve 2 AB_5-type alloy

4. Hypotheses For Synthesis Of Negative Electrode Alloy Used At Low Temperature

In general the position of a plateau in P – C –T plot for hydrogen storage alloy would move down with decreasing temperature, and desorbing hydrogen is an endothermic reaction. So desorbing hydrogen and the diffusion of hydrogen in the alloy will be very difficult at deeply low temperature circumstance. So the structure and composition of the metal hydride alloy used in MH-Ni battery at low temperature should be essentially considered. Based on our experimental results, the following hypotheses have been proposed preliminarily.

(1) Desorbing hydrogen is a limiting step during discharge process.

(2) Compositions and structure of the alloy should be chosen to be beneficial for hydrogen diffusion in the alloy.

(3) Effect of the alloy corrosion in alkali solution on electrochemical characteristics of negative electrode is a factor, but not a very key factor yet.

Of course, the hypotheses should be corrected, explained and demonstrated by more many experiment results in future.

Acknowledgments

This work was supported by the National Natural Science Foundation of China (No.20171042) and National Defend Foundation of Chinese Academy of Sciences.

5. References

1. Zhao, M. S and Sun, C. Y. (2000) Electrochemical properties of rare earth AB_3-type alloy as negative electrode material in MH-Ni battery. Hydrogen Energy Progress XIII Proceedings of the 13th World Hydrogen Energy Conference; Beijing, China, June12-15; P. 1106- 1110.

2. Sakai, T. and Ishikawa, H. (1990) Metal hydride anodes for nickel-hydrogen secondary battery. J. Electrochem. Soc.; 137(3):795- 799.

3. Zhao, M. S and Sun, C. Y. (2001) Negative electrode material in MH-Ni battery at low temperature. Proceedings of the 3rd China National Hydrogen Energy Conference, 13-15 April; Hangzhou China; P. 220-223.

4. Yie, H. and Zhang, H. (2001) Study of the low temperature electrochemical performance of hydrogen storage alloy. ibd.; P. 127- 134.

6. Guo H. and Pan C. (2000) Discharge performance of MH-Ni battery at low temperature. Chinese Journal of Power Sources;24(4):189- 191

7. Furukawa, N. and Nogami, M.(1995) Development of high capacity hydrogen absorbing alloys for high performance nickel-metal hydride batteries. 46th ISE Annual Meeting, Xiamen, China, , K-8-2

THERMODYNAMICS OF HYDROGEN ORDERING IN THE RARE-EARTH SUPERSTOICHIOMETRIC DIHYDRIDES

I.G. RATISHVILI [(1)], N.Z. NAMORADZE [(2)]
[(1)] *E.Andronikashvili Institute of Physics, Georgian Academy of Sciences 6 Tamarashvili str;, 380077 Tbilisi, Republic of Georgia*
[(2)] *Institute of Cybernetics, Georgian Academy of Sciences 5 Sandro Euli str;, 380086 Tbilisi, Republic of Georgia*

RH_{2+c} ordering interstitial alloys (R - rare-earth atoms, H - hydrogen or deuterium atoms) characterized by a pair of long-range-order (LRO) parameters are considered. Two-parameter superstructures are known to be realized, particularly, in CeH_{2+c}, TbH_{2+c} and LaH_{2+c} compounds. Theoretical treatments of equilibrium ordering processes in these systems have shown that the pair of LRO-parameters provide a number of peculiar properties unexpected in the systems with one LRO-parameter. These anomalous ordering phenomena are reviewed briefly below.

1. Introduction

The system under consideration. Dihydrides of rare-earth elements RH_2 are known to have a FCC metal lattice the tetrahedral interstitial sites of which are occupied by hydrogen atoms. In superstoichiometric dihydrides RH_{2+c} it is assumed that the FCC lattice of N rare-earth atoms (R-atoms) contains : $N_T = 2N$ hydrogens located in 2N tetrahedral interstitial sites (they are called H_T-atoms), and $N_O = cN$ ($0< c < 1$) H-atoms located in N octahedral interstitial sites (they are called H_O-atoms). $c = N_O/N$ is called the concentration of octa-hydrogen atoms [1].

In a number of RH_{2+c} compounds the subsystem of H_O-atoms at $T = T_{tr}$ undergoes a disorder-order transition and the totality of H_O-atoms and their vacancies can be treated as a two-component ordering *substitutional alloy* with a simple FCC lattice consisting of cN H_O-atoms and $(1-c)N$ vacancies. It was established neutronographycally [2-4] that in LaH_{2+c}, CeH_{2+c} and TbH_{2+c} the superstructures formed in the subsystem of H_O-atoms are characterized by *two* long-range-order (LRO) parameters, η_1 and η_2. In the subsequent theoretical treatments [5-7] it was shown that these superstructures are described by a similar distribution function,

$$n(x,y,z) = c + \eta_1\gamma \exp[i\, 2\pi x] + 2\eta_2\gamma \cos[\pi (x+2y)], \qquad (1)$$

where n(x,y,z) characterizes the probability that the sites with coordinates x,y,z are occupied by H_O-atoms and $\gamma = 0.25$ is a normalizing factor. On the set of FCC lattice sites function (1) takes three different values - n_1, n_2 and n_3, defined as

$$n_1 = c + \eta_1\gamma + 2\eta_2\gamma, \quad n_2 = c + \eta_1\gamma - 2\eta_2\gamma, \quad n_3 = c - \eta_1\gamma. \qquad (2)$$

N. Veziroglu et al. (eds.), Hydrogen Materials Science and Chemistry of Metal Hydrides, 35–44.

The properties of the ordering subsystems are determined usually by the corresponding free-energy functions. Within the frames of the mean field approximation of the Static Concentration Waves model [8] the free-energy function of an ordering system with the distribution function (1) is written as [5] :

$$F(\eta_1, \eta_2; T) = N k_B \{(1/2) [V_0 c^2 + V_1 (\gamma\eta_1)^2 + 2 V_2 (\gamma\eta_2)^2] + \\ + T \Sigma_{1\le i\le 3} \nu_i [n_i \ln n_i + (1-n_i) \ln(1-n_i)]\}, \qquad (3)$$

where V_0, V_1 and V_2 are Fourier-components of H–H interaction potential, taken with wave vectors $\mathbf{k}_0 = 0$, $\mathbf{k}_1 = (2\pi/a)$ (100) and $\mathbf{k}_2 = (2\pi/a)$ (1/2 1 0), respectively, and given in temperature units (see [5]). $\nu_1 = 0.25$, $\nu_2 = 0.25$ and $\nu_3 = 0.5$ are parts of the FCC lattice sites where function (1) equals to n_1, n_2 and n_3, respectively. (Here and below occupation numbers n_1, n_2 and n_3 are assumed to be given by expressions (2).)

The object of our consideration is the ordering subsystem of H_O-atoms in RH_{2+c} compounds, described by the distribution function (1) and free-energy function (3).

The problems under consideration. Two-parameter ordering in RH_{2+c} compounds induces many questions concerning the development of equilibrium redistribution of H_O-atoms. Particularly, what peculiar phenomena characterize the ordering processes in the case of two-parameter superstructures ? how are they revealed in the experiment ? what are the main parameters of RH_{2+c} compounds, which determine the sequence of equilibrium ordered states (the trajectory of the ordering process in the two-dimensional (η_1,η_2)-space) ?

These and similar problems can be discussed basing on the analysis of the sequences of equilibrium ordered states formed in the temperature range below the disorder-order transition point, i.e. on investigation of thermodynamics of the ordering process.

2. The method of investigation

We calculate the equilibrium ordered states in RH_{2+c} compounds. Equilibrium states of the ordering subsystem at temperatures T are defined as those $(\eta_1(T),\eta_2(T))$-states which minimize the free-energy function $F(\eta_1, \eta_2; T)$. More precisely, equilibrium values of order parameters $\eta_1(T)$ and $\eta_2(T)$ are determined as the (η_1,η_2)-coordinates of the free-energy absolute minimum at the given temperature T. Thus, we have to determine the free-energy extrema and to find among them the deepest minimum.

Definition region of the free-energy function in .the (η_1,η_2)-plane. Occupation numbers n_1, n_2 and n_3 must satisfy the inequalities

$$0 \le n_1, n_2, n_3 \le 1 \qquad (4)$$

These conditions and relations (2) restrict the region P in the (η_1,η_2)-plane, where the free-energy function is determined and where the possible equilibrium values of variables η_1 and η_2 can be located. If we additionally impose the conditions $\eta_1 > 0$ and $\eta_2 > 0$, then the part of the region P under consideration comes to that presented in Figure 1. Here the conditions $n_2 = 0$, $n_1 = 1$ and $n_3 = 0$ are presented by the boundary lines *cb*, *bd* and *da*, respectively. Intersections of these lines are corner points *b* and *d* .

Determination of the free energy extrema within the region P. There exists a standard way (see e.g. [5-7]). It is based upon the numerical solution of equations

$$\partial F(\eta_1, \eta_2) / \partial \eta_1 = 0, \qquad \partial F(\eta_1, \eta_2) / \partial \eta_2 = 0, \tag{5}$$

and on the sign-analysis of the determinant $\Delta(\eta_1,\eta_2)$,

$$\Delta(\eta_1,\eta_2) = (\partial^2 F(\eta_1,\eta_2)/\partial\eta_1^2)\,(\partial^2 F(\eta_1,\eta_2)/\partial\eta_2^2) - (\partial^2 F(\eta_1,\eta_2)/\partial\eta_1\,\partial\eta_2)^2 . \tag{6}$$

In the case of the free-energy function (3) equations (9) transform into

$$\ln[(n_1\, n_2\, (1-n_3)^2) / ((1-n_1)(1-n_2)\, n_3^2)] = -\eta_1\, V_1 / T , \tag{7}$$

$$\ln[(n_1(1-n_2)) / ((1-n_1)\, n_2)] = -\eta_2\, V_2 / T . \tag{8}$$

It must be stressed that function (3), equations (8) and their numerical solutions are valid in the whole temperature range $0 \le T \le T_{tr}$. Parameters $\eta_1(T)$ and $\eta_2(T)$ calculated in this way for compounds $TbH_{2.18}$ and $LaH_{2.25}$ agree with the corresponding experimental values in the full temperature range (see [5, 7]).

Main energy parameter p. Our treatments [5-7] reveal the importance of the parameter

$$p = V_2 / V_1 . \tag{9}$$

Different RH_{2+c} hydrides are characterized by different values of p-parameter. It was established that $p(TbH_{2+c}) = 1$ [5], $p(CeH_{2+c}) \cong 1.25$ [6] and $p(LaH_{2+c}) = 0.7743$ [7]. The influence of p-parameter on the ordering process is illustrated by the sequences of equilibrium ordered states in (η_1,η_2)-plane, calculated for CeH_{2+c}, TbH_{2+c} and LaH_{2+c} with the same concentration $c = 0.35$ (see Figure 1). In all cases the disordered state is represented by the zero-point (point $\{\eta_1 = 0, \eta_2 = 0\}$) and the low-temperature limit ordered state - by one of the corner points, *b* or *d* points.

3. Main results of thermodynamic consideration of the ordering RH_{2+c} compounds

3.1. STEPS OF ORDERING PROCESSES

In binary compounds with *one order parameter* η the ordering process is represented by the sequence of ordered states characterized by η values within the range $0 \le \eta \le \eta^{max}$. Different systems can differ only by the type of $\eta(T)$ dependence, while for all of them the equilibrium $\eta(T)$-values are located along the same trajectory containing the points of η-axes, between $\eta = 0$ (the disordered state) and $\eta = \eta^{max}$ (the low-temperature state of maximum order). But if the binary system is characterized by *two order parameters*, η_1 and η_2, then the pairs of equilibrium order parameters $\{\eta_1(T), \eta_2(T)\}$ can form, in general, different sequences, different trajectories in the (η_1,η_2)-plane, terminated in different corner points of the region P. A typical example of different scenario of the process development is presented in Figure 1.

We see that the ordering in CeH_{2+c} ($p = 1.25$) is described by the curve 0 - *b*. The ordering in LaH_{2+c} ($p = 0.7743$) is given by a trajectory consisting of two parts divided by a kink at the point $(\eta_1^0, 0)$. This is the point of an additional, order-order transition. We say that in CeH_{2+c} we have a *one-step* process and in the second case (in LaH_{2+c}) - a *two-step* process. The third trajectory, 0 - *m* - *d*, represents a two-step process as well.

The order-order transformation can be associated with a kink of the continuous trajectory (as for $p = 0.7743$), or with a finite jump between the ordered states ($p = 1$),

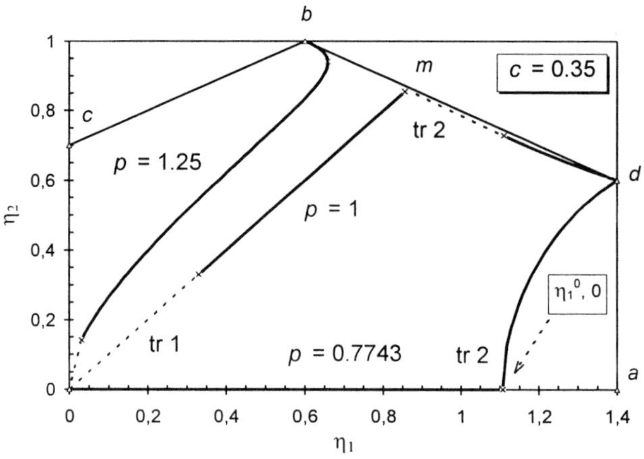

Figure 1. Sequences of equilibrium ordered states in the case of ordering systems RH_{2+c} ($c \doteq 0.35$) characterized by different values of the energy parameter p. (η_1^0, 0) - the point dividing the neighboring trajectory steps.

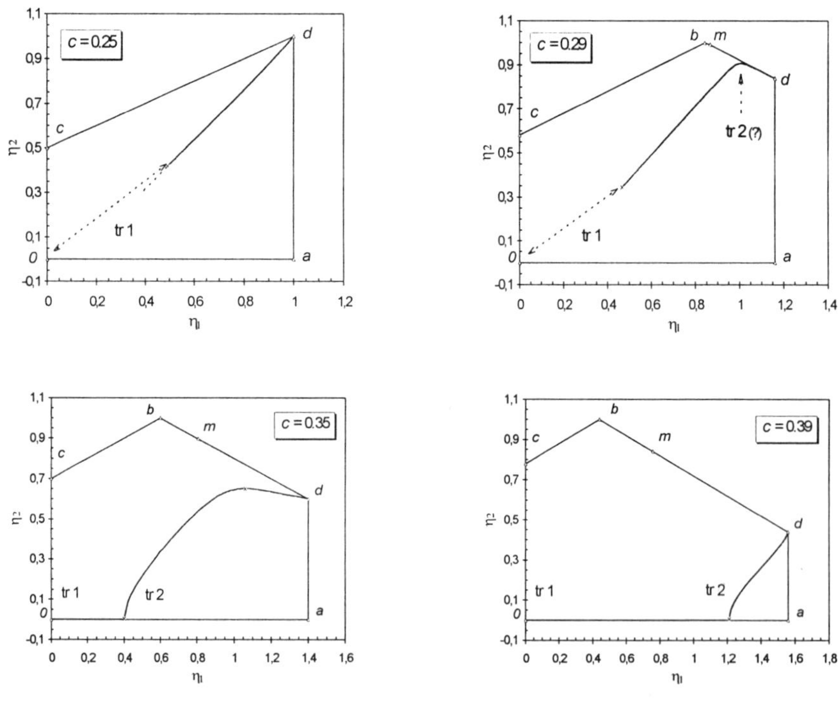

Figure 2. Variation of the ordering process caused by hydrogen concentration changes in a hypothetical RH_{2+c} compound with an energy parameter $p = 0.9$.

representing continuous and discontinuous order-order phase transitions, respectively. Similarly, the disorder-order transition is continuous for $p = 0.7743$ and discontinuous in the cases of $p = 1.25$ and $p = 1$.

Calculations performed in [6, 7] within the range $0.1 < c < 0.5$ had shown as well that the number of ordering steps in the given RH_{2+c} systems depends on the concentration c. Namely, in CeH_{2+c} the ordering, being a one-step 0 - b process for $c \leq 0.35$, at higher c (e.g. when $c \approx 0.40$) becomes a two-step process 0 - b - d, and finally, at $c \approx 0.45$ the latter transition is subdivided giving a *three-step* 0 - b - a - d process [6].

Trajectory modifications caused by the gradual variation of hydrogen concentration are shown in Figures 2, where are presented calculations performed for a hypothetical RH_{2+c} compound with $p = 0.9$. (A metallic lattice with the p-parameter located within the range $p(\mathrm{La}) < p < p(\mathrm{Ce})$ can be obtained in the corresponding alloys $La_x Ce_{1-x}$.)

The set of trajectories presented in Figures 1, 2 and a number of analogous calculations led to the conclusion that p and c parameters are the main characteristics of the system, which define the shape of the equilibrium trajectory in the (η_1,η_2)-plane.

3.2. ANOMALIES OF THERMAL PROPERTIES

Thermal properties of RH_{2+c} ordering compounds are represented by the internal energy function E(T) and the heat capacity function C(T). Internal energy function is determined by the expression [5]

$$E(\eta_1, \eta_2) = N\, k_B \{(1/2)\, [V_0 c^2 + V_1 (\gamma\eta_1)^2 + 2\, V_2 (\gamma\eta_2)^2] \qquad (10)$$

where the equilibrium values of order parameters $\eta_1(T)$ and $\eta_2(T)$ are inserted, $E(T) \equiv E(\eta_1(T), \eta_2(T))$. The heat capacity function is defined as the derivative of function E(T):

$$C(T) = \partial\, E(T) \,/\, \partial\, T \qquad (11)$$

The anomalies of both functions occur at phase transition temperatures. As examples, in Figures 3a and 3b are given the E(T) and C(T) functions calculated for $LaH_{2.25}$ ($c = 0.25$, $p = 0.7743$). Transition points in the corresponding curves are registered by the kinks, or finite discontinuities. The first-order, discontinuous transformation is associated with the internal energy discontinuity $\Delta E(T_{tr2})$ - the latent heat of transformation, while the continuous, second-order transition at T_{tr1} - with the kink of E(T) curve. Both transformations are indicated in the C(T) dependence as discontinuities of different strength $\Delta C(T_{tri})$. Results of calculations of E(T) and C(T) functions at a number of hydrogen concentrations for different p-values were published in [10] (for $p < 1$) and [11] (for $p > 1$).

Calculations of both functions in a wide concentration range $0 < c < 1$ for the given value of the energy parameter $p = 1.25$ (CeH_{2+c}-system) have shown that around the value $c = 0.5$ exists a concentration range $[c_0^{(1)} \approx 0.355,\ c_0^{(2)} \approx 0.645]$ where the disorder-order transition is continuous ($\Delta E(T_{tr}) = 0$), while outside of this region the same transformation is discontinuous ($\Delta E(T_{tr}) \neq 0$). It means that the disorder-order phase transition changes its type at $c = c_0^{(1)}$ and $c = c_0^{(2)}$. When $c = c_0^{(1)}$, or $c = c_0^{(2)}$, then the heat capacity discontinuity $\Delta C(T_{tr})$ becomes extremely large, as it does in the tricritical points of the ordering one-parameter systems. In Figure 4 is shown the concentration dependence of discontinuities of thermal functions, $\Delta E(T_{tr})$ and $\Delta C(T_{tr})$, calculated for the disorder-order transition in CeH_{2+c} ($c \leq 0.5$) [12].

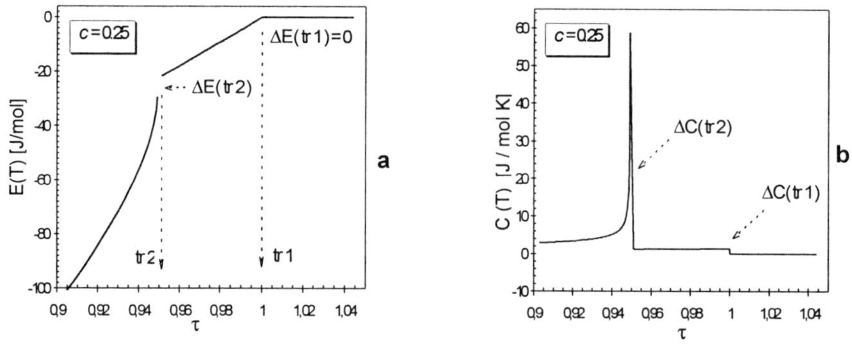

Figure 3. High-temperature parts of the internal energy E(T) and heat capacity C(T) functions in $LaH_{2.35}$. ($p = 0.7743$; $\tau \equiv T/T_{tr1}$) a - E(T) function, b - C(T) function.

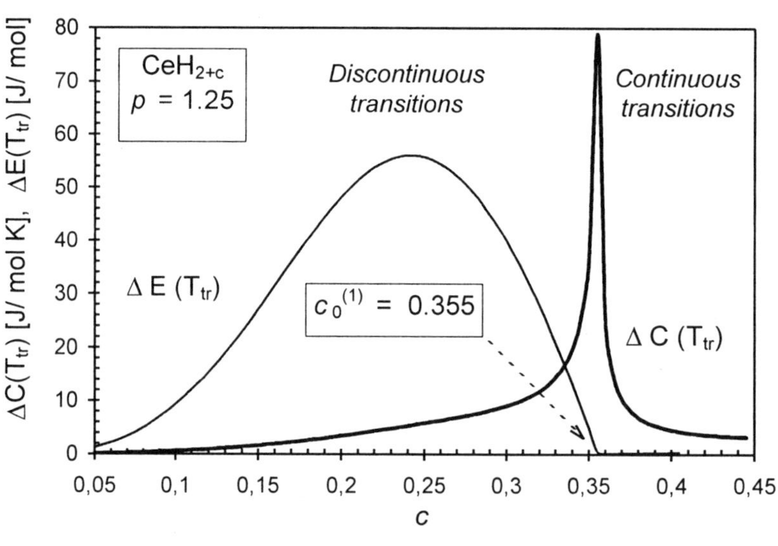

Figure 4. Concentration dependence of internal energy and heat capacity discontinuities, $\Delta E(T_{tr}; c)$ and $\Delta C(T_{tr}; c)$. Disorder-order transition in CeH_{2+c} . ($p = 1.25$).

Concentration values $c_0^{(1)}$ and $c_0^{(2)}$ are called *phase transition type change*, or *critical* concentrations.

Phase transition type change occurs in the order-order transformation as well. In the case of LaH_{2+c} ($p = 0.7743$) transformation $(\eta_1, 0) \leftrightarrow (\eta_1, \eta_2)$ is characterized by $\Delta E(T_{tr2}) \neq 0$ only at concentrations $c < c_0^{(1)} \approx 0.26$ and $c > c_0^{(2)} \approx 0.74$. Within the range $0.26 < c < 0.74$ the same transition becomes continuous [13].

3.3. PHASE TRANSITION TYPE CHANGE PHENOMENON

The mentioned above phase transition type changes in the disorder-order and order-order transformations for the first time were fixed in the calculations of equilibrium trajectories in CeH_{2+c} and LaH_{2+c} systems [6,7]. Particularly, in the case of CeH_{2+c} it was shown [6] that in the (η_1,η_2)-plane the distance between the disordered state and the high-temperature limit of the ordered state diminishes gradually at rising concentration, and at $c = c_0 = 0.355$ disappears [6]. Thus, it was established that at this concentration the discontinuous phase transformation, implying a jump between the states of different symmetries, transforms into a continuous transition, where all infinitely small values of order parameters are in equilibrium at corresponding temperatures. Simultaneously disappears the temperature interval ΔT determined as $\Delta T = T_{tr} - T_{min}$ (T_{tr} - disorder-order transition temperature, T_{min} - the lowest temperature until which can exist the metastable disordered state) [6, 9]. It must be noted that critical concentrations $c_0^{(i)}$ obtained in trajectory calculations [6] coincide with those determined from calculations of thermal functions E(T) and C(T) [12].

Analytical considerations [14,15] have shown that critical concentrations $c_0^{(i)}$ can be determined from special conditions imposed on the determinant (7). Particularly, for the disorder-order transition in compounds with $p > 1$ we have a system of equations [14]

$$\Delta(0, 0) = 0, \qquad \Delta(\eta_1, \eta_2) = 0, \qquad (12)$$

simultaneous solution of which gives the expressions for the critical concentrations [14]:

$$c_0^{(1)} = 0.5\ \{1 - [(p-1)/3]^{1/2}\}, \qquad c_0^{(2)} = 0.5\ \{1 + [(p-1)/3]^{1/2}\}. \qquad (13)$$

These expressions for $p = 1.25$ give the values : $c_0^{(1)} = 0.356$ and $c_0^{(2)} = 0.644$.

Existence of extremely large heat capacity anomalies in the phase transition type change points, i.e. at $c = c_0^{(i)}$, is explained mathematically in [16].

In the case of ordering systems with the energy parameter $p < 1$ (e.g. in LaH_{2+c}) we see an analogous phenomenon in the order-order transformation [7]. In this case the critical concentrations $c = c_0^{(i)}$ are determined from the conditions [15]

$$\Delta(\eta_1, 0) = 0, \qquad \Delta(\eta_1, \eta_2) = 0, \qquad (14)$$

which can not be solved analytically [15]. It have to be noted that in LaH_{2+c} compounds phase transition type change occurs as well in the disorder-order transition, but it is not accompanied by the heat capacity "anomalous anomaly" [13].

3.4. CRITICAL INDICES OF ORDER PARAMETERS

We consider the concentration region, $c_0^{(1)} \le c \le c_0^{(2)}$, where disorder-order (or order-order) transitions are continuous. In a one-parameter ordering system, in the vicinity of continuous transition point T_{tr} temperature dependence of the order parameter η is known to be presented by an approximate expression $\eta(T) = A\ (T_{tr} - T)^{\alpha}$, where α, called the *critical index*, equals $\alpha = 0.5$ for continuous transition point, and $\alpha = 0.25$ for $c = c_0^{(1)}$, or $c = c_0^{(2)}$, for the transition type change points ("tricritical points").

Our exact solutions of equations (8), $\eta_1(T)$ and $\eta_2(T)$, let us write as

$$\eta_1(T) = A_1\,(T_{tr} - T)^{\alpha_1}, \qquad \eta_2(T) = A_2\,(T_{tr} - T)^{\alpha_2}. \qquad (15)$$

If we assume that for "neighboring temperatures" T and (T + δT) critical indices are practically unchanged, $\alpha_i(T+\delta T) \cong \alpha_i(T)$, then from expressions (15) it can be deduced:

$$\alpha_i(T) = L_i(T) / L_0(T) \tag{16}$$

$$L_i(T) \equiv \ln\left[\eta_i(T_{tr} - (T+\delta T)) / \eta_i(T_{tr} - T)\right], \tag{17a}$$

$$L_0(T) \equiv \ln\left[(T_{tr} - (T+\delta T)) / (T_{tr} - T)\right]. \tag{17b}$$

Inserting in (16)-(17) the values of exact solutions $\eta_1(T)$ and $\eta_2(T)$ we determine the critical indices associated with the given temperature T, $\alpha_1(T)$ and $\alpha_2(T)$.

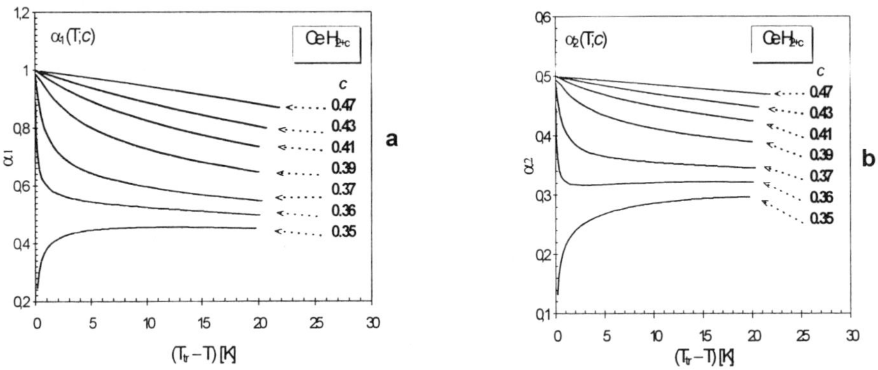

Figure 5. Critical indices α_i determined from solutions of eqs. (8). Disorder-order transition in CeH_{2+c}. ($p = 1.25$; $c_0^{(1)} = 0.356$). a - $\alpha_1(T; c)$; b - $\alpha_2(T; c)$.

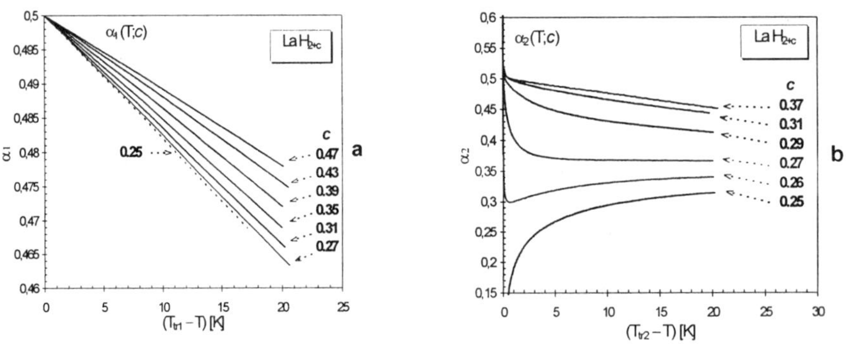

Figure 6. Critical indeces in LaH_{2+c}. ($p = 0.7743$; $c_0^{(1)} = 0.259$)
a - $\alpha_1(T; c)$ determined for the disorder-order transition $(0, 0) \leftrightarrow (\eta_1, 0)$;
b - $\alpha_2(T; c)$ determined for the order-order transition $(\eta_1, 0) \leftrightarrow (\eta_1, \eta_2)$.

In Figures 5, 6 are given critical indices $\alpha_1(T)$ and $\alpha_2(T)$ calculated for the disorder-order transition $(0,0) \leftrightarrow (\eta_1, \eta_2)$ in CeH_{2+c} ($p > 1$), and disorder-order $(0, 0) \leftrightarrow (\eta_1, 0)$ and order-order $(\eta_1, 0) \leftrightarrow (\eta_1, \eta_2)$ transitions in LaH_{2+c} ($p < 1$). In CeH_{2+c} at critical concentration $c \approx c_0^{(1)} \cong 0.356$ we have $\alpha_1(T) \approx 0.5$, $\alpha_2(T) \approx 0.3$, while as the difference $(c - c_0^{(1)})$ rises, $\alpha_1(T)$ tends to 1 and $\alpha_2(T)$ tends to 0.5. In LaH_{2+c} phase transition type changes at concentration $c_0^{(1)} = 0.259$ in the order-order transition $(\eta_1, 0) \leftrightarrow (\eta_1, \eta_2)$. Here $\alpha_2(T) \approx 0.3$ for $c = c_0^{(1)}$, and tends to 0.5, as the difference $(c - c_0^{(1)})$ rises.

4. Main conclusions

Peculiar thermodynamic properties of RH_{2+c} compounds caused by the existence of a pair of LRO parameters, η_1 and η_2, can be summarized as follows.

1. The ordering processes contain, generally, a disorder-order and a number of subsequent order-order transitions subdividing the total trajectory of the process into the steps consisting of continuous sequences of equilibrium ordered states. Equilibrium trajectories for compounds CeH_{2+c}, LaH_{2+c} and TbH_{2+c} differ by the number of steps, by the type of transitions between the steps, and by the shape of each given step. The total sequence of the equilibrium ordered states (the shape of the trajectory) is determined by the energy parameter p and concentration c. [5-7].

2. The disorder-order and order-order transitions reveal themselves by the discontinuities $\Delta E(T_{tr})$ and $\Delta C(T_{tr})$ of the internal energy function E(T) and the heat capacity function C(T). Discontinuities $\Delta E(T_{tr})$ and $\Delta C(T_{tr})$ are functions of the hydrogen concentration c and the energy parameter p [10, 11].

3. For a given RH_{2+c} system with a fixed energy parameter p there exist some critical concentrations $c_0^{(i)}$, where simultaneously $\Delta C(T_{tr})$ becomes extremely large and $\Delta E(T_{tr})$ (the latent heat of the transition) tends to zero, indicating thus the fact of phase transition type change [12, 13]. Phase transition type change phenomenon at $c = c_0^{(i)}$ was mathematically established: a) in the *disorder-order transition* for the systems with $p > 1$ [6, 13] and b) in the *order-order transition* for the systems with $p < 1$ [7, 12].

4. Equations defining the critical concentrations $c_0^{(i)}$ associated with the phase transition type changes in the disorder-order transitions (when $p > 1$) and in the order-order transitions (when $p < 1$) were formulated, respectively, in [14] and [15].

a) For compounds with $p > 1$ these equations were solved analytically and algebraic expressions $c_0^{(i)}(p)$ were written [14].

b) For compounds with $p < 1$ the dependences $c_0^{(i)}(p)$ were determined by numerical solutions of the corresponding equations [15]. It have to be noted that in this case besides the mentioned above phase transition type change in the order-order transition, an alike phenomenon exists in the disorder-order transition as well (at $c_1^{(i)} \neq c_0^{(i)}$), but it is not accompanied by a sharp anomaly of $\Delta C(T_{tr})$ [15].

5. For the first time we have discussed the critical indices of order parameters in the case of ordering systems with two LRO-parameters. Critical indices of η_1 and η_2 parameters, α_1 and α_2, were determined from the special representation of exact solutions $\eta_1(T)$ and $\eta_2(T)$, obtained on the basis of the undeveloped in power-series free-energy function.

a) In the RH_{2+c} compounds with $p > 1$ the disorder-order transition $(0, 0) \leftrightarrow (\eta_1, \eta_2)$ in the region of continuous transition (at concentrations $c \sim 0.5$) is characterized by the critical indices $\alpha_1 \cong 1$ and $\alpha_2 \cong 0.5$, while at concentrations tending to critical values $c_0^{(1)}$ or $c_0^{(2)}$ they gradually transform into the values $\alpha_1 \cong 0.5$ and $\alpha_2 \cong 0.3$.

b) In RH_{2+c} systems with $p < 1$ for continuous $(0, 0) \leftrightarrow (\eta_1, 0)$ disorder-order transitions we have $\alpha_1 \cong 0.5$, and for the order-order transition $(\eta_1, 0) \leftrightarrow (\eta_1, \eta_2)$ the critical index is $\alpha_2 \cong 0.5$ tending to $\alpha_2 \cong 0.3$ as c tends to $c_0^{(i)}$.

We hope that the enumerated details of the ordering processes in the systems with two LRO parameters, established mathematically in RH_{2+c} compounds, will stimulate the corresponding experimental investigations of these systems, especially in the neighborhood of the phase transition type change points.

5. References

1. Vajda, P. (1995) Hydrogen in rare-earth metals, includig RH_{2+c} phases, in L.Eyring, K. Gshneidner (eds.), *Handbook on the physics and chemistry of rare earths*. North-Holland, Amsterdam. v. **20**, pp. 207-291.
2. Fedotov, V.K., Fedotov, V.G., Kost, M.E., Ponyatovskii, E.G. (1982) Hydrogen ordering in the hydrogen solid solutions in rare-earth metals. *Sov. Phys. Solid State*, **24**, 1252 - 1257.
3. André, G., Blaschko, O., Schwarz, W., Daou, J.N., Vajda P. (1992) Experimental observation of hydrogen ordering in TbD_{2+x}. *Phys. Rev. B*, **46**, 8644-8647.
4. Udovic, T.J., Huang, Q., Rush, J.J., Schefer, J., Anderson, I.S. (1995) Neutron-powder-diffraction study of the long-range order in the octahedral sublattice of $LaD_{2.25}$. *Phys.Rev. B*, **51**, 12116-12126.
5. Ratishvili, I.G., Vajda, P. Ordering in the system β-TbH_{2+x}. (1993) *Phys.Rev. B*, **47**, 14062-14069.
6. Ratishvili, I.G., Vajda, P., Boukraa, A., Namoradze, N.Z. (1994) Ordering processes in monophase CeH_{2+x}. *Phys.Rev. B*, **49**, 15461-15469.
7. Ratishvili, I.G., Vajda, P. (1996) Hydrogen ordering in the superstoichiometric rare-earth dihydrides. The case of a system with an energy constanst ratio $p = V_2/V_1 < 1$. LaH_{2+x}. *Phys.Rev. B*, **53**, 581-587.
8. Khachaturyan, A.G. (1983) Theory of Structural Transformations in Solids. Wiley, New-York.
9. Ratishvili, I.G., Vajda, P. (1997) Phase diagrams of β-RH_{2+x} systems (R = La, Ce, Tb). Results of mean-field calculations. *J. All. Comp.*, **253-254**, 171-174.
10. Ratishvili, I.G., Vajda, P., Namoradze, N.Z. (1997) Heat capacity associated with hydrogen ordering in superstoichiometric rare-earth dihydrides : the case of LaH_{2+x}-like systems. *J. Phys. Chem. Sol.*, **58,** 59-62.
11. Ratishvili, I.G., Namoradze, N.Z. (1998) Heat capacity associated with hydrogen ordering in superstoichiometric rare-earth dihydrides : the case of CeH_{2+x} - like systems. *Bullet. of Georgian Acad. Sciences*, **158**, 229-232.
12. Ratishvili, I.G., Namoradze, N.Z. (2000) Heat capacity anomalies associated with phase transition type change in CeH_{2+x}-like superstoichiometric rare-earth dihydrides. *J. Phys. Chem. Sol.*, **61**, 1827-1838.
13. Ratishvili, I.G., Namoradze, N.Z. (2001) Heat capacity anomalies associated with phase transition type change in LaH_{2+x} -like superstoichiometric rare-earth dihydrides. *phys. stat. sol. (b)*, **225**, 1-14.
14. Ratishvili, I.G. (1997) Change of the disorder-order phase transition type in the two-component ordering alloys. *Bullet. of Georgian Acad. Sciences*, **155**, 198-201.
15. Ratishvili, I.G. (1999) Phase transition type change in order-order and disorder-order transformations of LaH_{2+x} - like superstoichiometric dihydrides. *phys. stat. sol. (b)*, **213**, 297-315.
16. Ratishvili, I.G., Namoradze, N.Z. (2000) Change in the order of the phase transition in ordering rare-earth hydrides: A heat capacity anomaly. *Physics of Metals and Metallography*, **89** (2), 145-149.

THE CONCENTRATION AND TEMPERATURE TRANSITION TO SUPERCONDUCTING STATE IN FULLERIDES OF ALKALI METALS

Z.A. MATYSINA, D.V. SCHUR, S.Yu. ZAGINAICHENKO
Dnepropetrovsk State University, 49000 Ukraine
Institute of Hydrogen and Solar Energy, P.O. Box 150, Kiev, 03150 Ukraine
Institute for Problems of Materials Science of Ukrainian Academy of Sciences, 3, Krzhizhanovsky str., Kiev, 03142 Ukraine

The statistical theory of phase transition in fullerides of alkali metals to superconducting state has been elaborated. The calculation of free energy of crystals has been performed on the basis of molecular-kinetic notion and the equilibrium equations have been examined. The critical temperature of superconductivity has been estimated, its concentration dependence has been ascertained. The calculation results have been compared with experimental data. The numerical estimation for energetic parameters of pairing of Cooper pairs electrons has been performed.

1. Introduction

It has been established experimentally that phase transformation to superconducting state takes place in fullerides crystals of alkali metals $A_{3-X}B_X\Phi$ (A, B = Li, Na, K, Rb, Cs; $\Phi = C_{60}$, x = 0, 1, 2, 3) with temperature decrease. The values of temperature T_C of phase transition for some crystals are listed in the Table 1 [1-4]. In caesium fullerides $Cs_3\Phi$ the superconducting state is not discovered, conceivably the critical temperature T_C for this crystal is close to absolute zero. With addition of rubidium atoms to $Cs_3\Phi$ crystals the $Cs_2Rb\Phi$ crystals are formed and superconductivity shows up in them with critical temperature $T_C = 33K$.

Table 1. The critical temperatures of superconductivity of alkali metals fullerides

Crystal	T_C, K
$Cs_3\Phi$	~ 0
$Cs_2Rb\Phi$	33
$CsRb_2\Phi$	31
$Rb_3\Phi$	29
$Rb_2K\Phi$	27
$RbK_2\Phi$	23
$K_3\Phi$	19
$Li_2Cs\Phi$	12
$Na_2Cs\Phi$	12
$Na_2Rb\Phi$	2,5
$Na_2K\Phi$	2,5

N. Veziroglu et al. (eds.), Hydrogen Materials Science and Chemistry of Metal Hydrides, 45–50.

The study of temperature and concentration phase transition to superconducting state of the fullerides crystals of alkali metals $A_{3-x}B_x\Phi$ (A, B = Li, Na, K, Rb, Cs; $\Phi = C_{60}$; x = 0, 1, 2, 3) is of interest for materials scientists. Such investigation for $Cs_2Rb\Phi$ crystals was performed in [5-7].

2. Theoretical results and discussion

In $A_{3-X}B_X\Phi$ crystals the $\Phi = C_{60}$ fullerenes form the frame of structure – face-centered cubic lattice (Fig.1). It is suggested that metals atoms A (or B) of smaller size are predominantly distributed over the tetrahedral interstitial sites, but of greater size occupy the octahedral interstitial sites and we shall take them respectively as interstitial sites of the first and second type. So tetrahedral interstitial sites are legal for A atoms, octahedral interstitial sites – for B atoms.

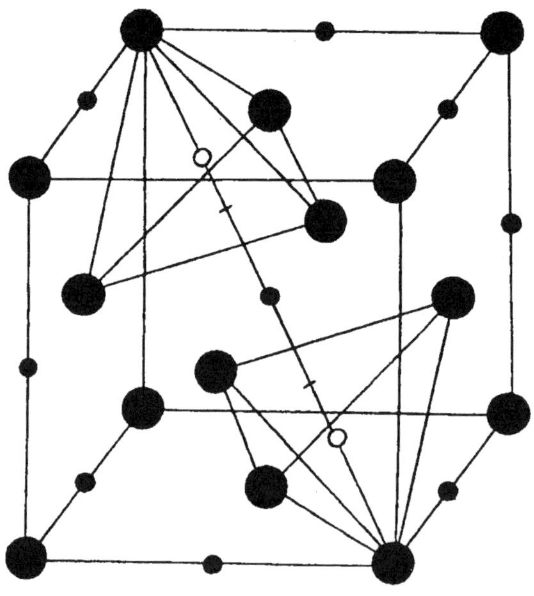

Fig. 1. Elementary cell of fulleride crystal of alkali metals

● - sites of face-centered lattice, in which fullerenes molecules C_{60} are distributed;

● ○ - octahedral and tetrahedral interstitial sites (not all are marked off), in which A and B metals atoms are located.

The potential field of molecular type [8-10] is entered into consideration for calculation of free energy. It is taken into account that A, B atoms have one unpaired (valence) electron. At high temperatures A, B atoms atoms are randomly distributed over the interstitial sites of first and second type and spins of electrons are disorientated. With lowering temperature the atomic order is found first of all on interstitial sites, later at more low temperature the spin order is established also in orientation of electrons moments of metal atoms A, B: the conductivity electrons with oppositely directed spins at the expense of interaction with crystal lattice are paired in Cooper bosons and their appearance should determine the superconductivity. We suppose the collinearity of electron spins and the primary their orientation is called as right r, the opposite orientation as left *l*.

The atomic and spin orders are characterized by parameters

$$\eta = \left(P_A^{(1)} - c_A\right)/\left(1 - \nu_1\right), \quad 0 \le \eta \le 1, \tag{1}$$

$$\sigma_1 = P_r^{(1)} - P_l^{(1)}, \ \sigma_2 = P_r^{(2)} - P_l^{(2)}, \ -1 \le \sigma_1, \sigma_2 \le 1, \tag{2}$$

where $P_A^{(1)}$ is probability of occupation of first type interstitial site by A atom, c_A is concentration of A atoms, $\nu_1 = 2/3$, $\nu_2 = 1/3$ are concentrations of interstitial sites of first and second type, $P_r^{(i)}, P_l^{(i)}$ are probabilities of the right r and left *l* orientations of spins of atomic electrons in interstitial sites of i = 1, 2 type.

The calculation of free energy per one lattice point is found in the form

$$\begin{aligned} f = {} & u - 8\upsilon\eta/3 - 16\omega\eta^2/9 + kT[2(c_A + \eta/3)\ln(c_A + \eta/3) + \\ & + 2(c_B - \eta/3)\ln(c_B - \eta/3) + (c_A - 2\eta/3)\ln(c_A - 2\eta/3) + (c_B + 2\eta/3)\ln(c_B + 2\eta/3)] - \\ & - J_o(4\sigma_1\sigma_2 + 3\sigma_1^2 + 3\sigma_2^2) - 2J'\eta(-2\sigma_1\sigma_2 + 3\sigma_1^2 - 6\sigma_2^2)/3 - \\ & - J\eta^2(8\sigma_1\sigma_2/3 - \sigma_1^2 - 4\sigma_2^2)/3 + \\ & + kT[(1+\sigma_1)\ln\frac{1+\sigma_1}{2} + (1-\sigma_1)\ln\frac{1-\sigma_1}{2} + \frac{1+\sigma_2}{2}\ln\frac{1+\sigma_2}{2} + \frac{1-\sigma_2}{2}\ln\frac{1-\sigma_2}{2}], \end{aligned} \tag{3}$$

where u, υ, ω are energetic parameters, expressed in terms of energies of electrochemical interaction of the nearest pairs $\alpha\beta$ = AB, $\alpha\Phi$ and of c_A, c_B concentrations of A, B atoms ($c_A + c_B = 1$). The energetic parameters J_o, J', J are expressed by energies $J_{\alpha\beta}$ of pairing of valence electrons of α, β atoms and c_A, c_B concentration by formulae

$$\begin{aligned} J_o &= c_A^2 J_{AA} + c_B^2 J_{BB} + 2c_A c_B J_{AB}, \\ J' &= c_A(J_{AA} - J_{AB}) - c_B(J_{BB} - J_{AB}), \\ J &= 2J_{AB} - J_{AA} - J_{BB}. \end{aligned} \tag{4}$$

The equilibrium values of spin order parameters are determined by expressions

$$\begin{aligned} kT\ln\frac{1+\sigma_1}{1-\sigma_1} &= 2J_o(3\sigma_1 + 2\sigma_2) + 4J'\eta(3\sigma_1 - \sigma_2)/3 + 2J\eta^2(-\sigma_1 + 4\sigma_2/3)/3, \\ kT\ln\frac{1+\sigma_2}{1-\sigma_2} &= 4J_o(2\sigma_1 + 3\sigma_2) - 8J'\eta(\sigma_1 + 6\sigma_2)/3 + 16J\eta^2(\sigma_1/3 - \sigma_2)/3. \end{aligned} \tag{5}$$

By assuming $\sigma_1, \sigma_2 \to 0$ in these formulae, expanding the logarithms into a series of σ_1, σ_2 and restricting to the linear terms of expansion, we receive the system of homogeneous equations for σ_1, σ_2. When determinant of this system is taken equal to zero, we find the equation for the critical temperature of spin ordering as follows:

$$\begin{aligned} & (kT_C - 3J_o - 2J'\eta_* + J\eta_*^2/3)(kT_C - 6J_o + 8J'\eta_* + 8J\eta_*^2/3) = \\ & = 8(J_o - J'\eta_*/3 + 2J\eta_*^2/9)^2, \quad \eta_* = \eta(T_C). \end{aligned} \tag{6}$$

As this formula shows, the critical temperature T_C is dependent on crystal composition

c_A, c_B, degree of long-range atomic order η and energies of pairing of valence electrons $J_{\alpha\beta}$.

In the specific case, when nearly all conductivity electrons are paired, we restrict our consideration to the interaction of spin electrons of Cooper pairs only for the nearest atoms. The calculation gives (instead of formulae (5)) more simple expressions as follows:

$$kT \ln \frac{1+\sigma_1}{1-\sigma_1} = 4(J_o - J'\eta/3 + 2J\eta^2/9)\sigma_2, \quad kT \ln \frac{1+\sigma_2}{1-\sigma_2} = 8(J_o - J'\eta/3 + 2J\eta^2/9)\sigma_1, \tag{7}$$

from which we find the critical temperature in the form

$$kT_C = 2\sqrt{2}(J_o - J'\eta_*/3 + 2J\eta_*^2/9), \quad \eta_* = \eta(T_C). \tag{8}$$

If it is remembered that at low temperatures, when superconductivity manifests itself, the atomic order is at it maximum and is defined by relations

$$\eta_m = \begin{cases} c_A/\nu_1 & \text{at} \quad c_A \le \nu_1, \\ c_B/\nu_2 & \text{at} \quad c_A \ge \nu_1. \end{cases} \tag{9}$$

So in terms of expression (4) we have the spin ordering temperature depending on concentration

$$kT_C = 2\sqrt{2}\begin{cases} J_{BB} + 3(J_{AB} - J_{BB})c_A/2 & \text{at} \quad c_A \le \frac{2}{3}, \\ J_{AA} + 3(J_{AB} - J_{AA})c_B & \text{at} \quad c_A \ge \frac{2}{3}. \end{cases} \tag{10}$$

As is obvious from this formulae, the $T_C = T_C(c_A)$ dependence is linear with a break in the point $c_A = 2/3$ at the distribution of A atoms in tetrahedral interstitial sites. At the primary A atoms location in octahedral interstitial sites the break of this curve is bound to be observed in the point $c_A = 1/3$.

Figure 2 illustrates the plots of spin ordering temperature T_C as a function of concentration c_A for $A_{3-x}B_x\Phi$ crystal, constructed by formulae (10) for different relationship between the pairing energies of electrons of A, B atoms.

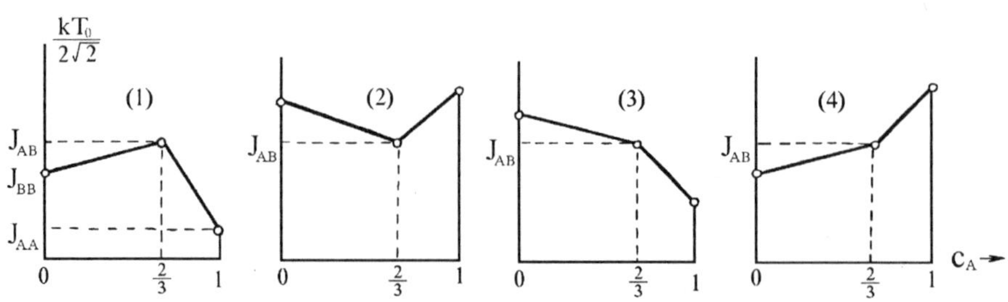

Fig. 2. The plots of possible concentration dependence of spin ordering temperature for $A_2B\Phi$ crystals. The construction of curves is performed by the equation (10) for different relationship between energies of pairing of valence electrons of A, B atoms.
1) $J_{AB} > J_{BB}$, $J_{AB} > J_{AA}$; 2) $J_{AB} < J_{BB}$, $J_{AB} < J_{AA}$;
3) $J_{BB} > J_{AB} > J_{AA}$; 4) $J_{BB} < J_{AB} < J_{AA}$.
The values of ordering temperatures are marked off by circles in the points $c_A = 0$; 2/3; 1, equal correspondingly to J_{BB}, J_{AB}, J_{AA}.

The experimental data of Table 1 allow the plots construction to be made of the concentration dependence of critical temperature of superconductivity for crystals $Rb_{3-x}Cs_x\Phi$, $Rb_{3-x}K_x\Phi$ ($x = 3c_A$). These plots are presented in Fig. 3. As figures have illustrated, the agreement between theory and experiment is excellent. The $T_C(c_A)$ dependence is found to be strictly linear with a break in the point $c_A = 2/3$ or $c_A = 1/3$. The experimental plot (a) of Fig. 3 corresponds to theoretical plot (1) of Fig. 2, but plot (b) of Fig. 3 is in agreement with theoretical (3) of Fig. 2. The first has a break in the point $c_A = 2/3$ ($x = 2$), the second – in the point $c_A = 1/3$ ($x = 1$). Such, indeed, it is bound to be, because the caesium atoms are greater, but the potassium atoms are smaller than the rubidium atoms.

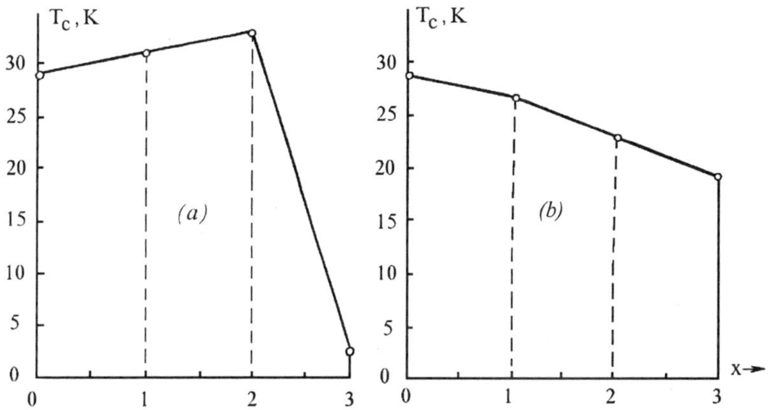

Fig. 3. Experimental plots of concentration dependence of critical temperature of superconductivity for $Rb_{3-x}Cs_x\Phi$ (a), $Rb_{3-x}K_x\Phi$ (b) crystals ($x = 3c_A$). The values of critical temperatures are marked off by circles for crystals:
(a) $Rb_3\Phi$, $Rb_2Cs\Phi$, $RbCs_2\Phi$, $Cs_3\Phi$; b) $Rb_3\Phi$, $Rb_2K\Phi$, $RbK_2\Phi$, $K_3\Phi$.

The formulae (10) permit the assessment of energetic parameters of electrons spin interaction, if critical temperatures of phase transition to the superconducting state are known. We take as our example the $Rb_{3-x}Cs_x\Phi$ crystals. For concentrations $c_{Cs} = 0, \frac{1}{3}, \frac{2}{3}, 1$ formulae (10) give the respective data

$$\begin{aligned}
&kT_C = 2\sqrt{2}J_{RbRb} && \text{at} \quad c_{Cs} = 0 && (Rb_3\Phi), \\
&kT_C = \sqrt{2}(J_{RbRb} + J_{CsRb}) && \text{at} \quad c_{Cs} = 1/3 && (Rb_2Cs\Phi), \\
&kT_C = 2\sqrt{2}J_{CsRb} && \text{at} \quad c_{Cs} = 2/3 && (RbCs_2\Phi), \\
&kT_C = 2\sqrt{2}J_{CsCs} && \text{at} \quad c_{Cs} = 1 && (Cs_3\Phi).
\end{aligned} \tag{11}$$

Substituting in equations (10) the numerical values of critical temperatures from Table 1, we find the pairing energies of valence electrons for CsCs, CsRb, RbRb atomic pairs and further the energy of spin ordering $\omega_C = kT_C/c_{Rb} = 6\sqrt{2}\ J_{CsRb}$

$$J_{CsCs} \approx 0, \quad J_{CsRb} = 1{,}00 \cdot 10^{-3} eV, \quad J_{RbRb} = 0{,}88 \cdot 10^{-3} eV,$$

$$\omega_C = 0{,}85 \cdot 10^{-2} eV. \tag{12}$$

The numerical values of energetic parameters (12) correspond to the case (1) of Fig. 2.

3. Conclusions

Thus, theoretical study of spin ordering in $A_{3-x}B_x\Phi$ crystals of fullerides of alkali metals allows the calculation of crystal free energy, the derivation of equations of equilibrium state, that determine the spin order parameters as a function of crystal temperature, its composition and ordering degree of A, B atoms. This calculation makes it possible to evaluate the critical temperature of spin ordering and its concentration dependence. The last is found to be linear with a break in the point $c_A = 1/3$ or $c_A = 2/3$. The results of calculations are compared with experimental data for $Rb_{3-x}Cs_x\Phi$, $Rb_{3-x}K_x\Phi$ crystals, in doing so for the first the numerical evaluation of energetic parameters of electrons pairing is fulfilled.

It should be pointed out that energy of atomic ordering for the majority of crystals is of the order one electron-volt in magnitude ($\omega \sim 1$ eV) and the critical temperature of atomic ordering is of the order of 10^3 K, i.e. two orders higher the critical temperature of spin ordering. Because of this, the energy of spin ordering $\omega_C(12)$ is found to be on two orders below the energy of atomic ordering and the assessment (12) is defined as realistic.

4. References

1. Haddon, R. (1992) *Account Chem. Res.* **25**, 127.
2. Haddon, R., Hebard, A., Rosseinsky, M. and Murphy, D. (1992) In: *Fullerenes* (Eds. Hammond, G.S. and Kuck, V.J.) ACS Symp. Ser. **481**, 71.
3. Sokolov, V.I. and Stankevich, I.V. (1993) *Uspekhi khimii* **62**(5), 455-473.
4. Trefilov, V.I., Schur, D.V., Tarasov, B.P., Shulga, Yu.M., Chernogorenko, A.V., Pishuk, V.K. and Zaginaichenko, S.Yu. (2001) *Fullereny – osnova materialov budushchego*, Kiev: Izd. ADEF – Ukraine, 148. (in Russian)
5. Matysina, Z.A., Schur, D.V. and Chumak, V.A. Kontsentratsionnyj i temperaturnyj perekhod v sverkhprovodyashchee sostoyanie tseziy-rubidievoy soli, M.: VINITI. 1999. 14 p. *Izv. vuzov. Fizika*. 2000. **2**, 100.
6. Matysina, Z.A., Chumak, V.A., Schur, D.V. and Zaginaichenko, S.Yu. (1999) *Metallofizika i noveish. tekhn.* **21**(9), 78-83.
7. Matysina, Z.A., Chumak, V.A., Schur, D.V. and Zaginaichenko, S.Yu. (2001) *Met. Phys. Adv. Techn.* **19**, 1245.
8. Smirnov, A.A. (1966) *Molekulyarno-kineticheskaya teoriya metallov*, M.: Nauka, 488. (in Russian)
9. Smirnov, A.A. (1986) *Obobshchennaya teoriya uporyadocheniya splavov*, Kiev: Nauk. dumka, 168. (in Russian)
10. Smirnov, A.A. (1992) *Teoriya fazovykh prevrashcheniy i razmeshcheniya atomov v splavakh vnedreniya*, Kiev: Nauk. dumka, 280. (in Russian)

HUMID ATMOSPHERE INDUCED PROCESSES OF HYDROGEN FORMATION AND EMBRITTLEMENT OF Sn-Al EUTECTIC

F. MUKTEPAVELA
Institute of Solid State Physics University of Latvia
8 Kengaraga Str, LV 1063, Riga, Latvia
M. VASYLYEV, V.G.KOSTYCHENKO
Institute of Metal Physics, Ukrainian Academy of Sciences
36 Vernadsky blvd., 03142, Kiev, Ukraine

1. Introduction

The environmental problems caused by the utilization of fossil fuels (petroleum, natural gas and coal) continue their escalation. Metal-hydrogen systems have aroused considerable interest as reliable, safe, ecological, and economic energy storage systems of the future. Hydrogen materials science will play an important role in hastening the transition to the hydrogen energy system. The problem of developing new economic methods for the production of hydrogen from water is actual. The present paper reports on a continuation of our earlier study of the embrittlement of the Sn-Al alloy in the humid air. In [1] we have found that this phenomenon is accompanied by the emission of hydrogen from bulk of alloys.

The aim of this work is to investigate the chemical composition together with the structural and mechanical properties of the Sn-Al alloy and Al/Sn phase boundaries. This study is important because of the application of the Sn-Al alloys for clad layers for sliding bearings in motor industry, and as catalysts in the reactionary systems [2-4]. In addition, the Sn-Al alloy can be applied for hydrogen storage in the case of the controllable emission of hydrogen.

2. Experimental

The investigation of the deformation behavior of interphase boundaries in the fine grained alloys of eutectics type causes significant difficulties. New opportunities are opened with use of the bimetallic joints, the interface in which is similar to a phase boundary in eutectic. Solid phases joints of Al/Sn with atomically clean interface were obtained by a special method of cold welding with deformation ~70%. [5,6]. The eutectic alloy (99.5 wt% Sn – 0.5 wt% Al) and bimetallic solid phase joints of tin and aluminum were prepared from 99.999% aluminum and 99.95% tin. The SIMS method

N. Veziroglu et al. (eds.), Hydrogen Materials Science and Chemistry of Metal Hydrides, 51–58.

was used for investigation of the chemical composition and hydrogen distribution processes. Analytical sensitivity of SIMS to hydrogen is rather high, the method allows to detect the appearance of the new chemical combinations of compounds (oxides, hydrides) and can perform the level-by-level depth analysis of the near-interface region [7,8]. Ar^+ ions with energy of 6 keV were used, the ion sputtering rate being 1.5 nm. min^{-1}. Chemical composition, fractographic, structural investigations and microhardness were carried out by use of X-ray diffractometer (DRON UM1), optical microscopes (Neophot 30), SEM (JEM-100, ASID-4D) and precision microhardness tester at loads 3-10mN.

3. Results and discussion

Fig.1 shows the decrease of Sn/Al phase boundary strength with the increase of exposure time in water vapor at 293K. The brittle fracture of deformed samples takes place after 1h. For the annealed sample the process of embrittlement also occurs, however with another rate. Thus, the process of embrittlement of Sn/Al phase boundaries is facilitated by the previous plastic deformation. As it was shown in [5], the plastic deformation in the Sn-Al alloy arises micropores and submicrocracks on the grain boundaries.

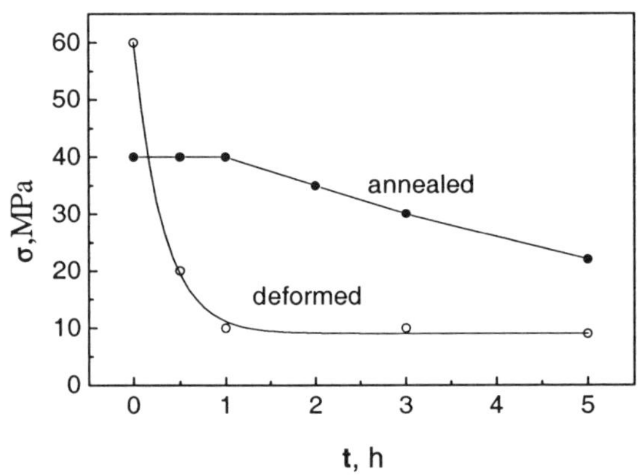

Figure. 1. Dependence of strength of the deformed at 293K and annealed at 423K Sn/Al joints on time of exposure to water vapor.

The results of SIMS analysis of the Sn-Al alloy (Fig.2) showed that the intensity of the Sn^+ -signal increases after exposure to water vapor. At the same time, this effect was very small for aluminum. The increase of the SIMS signal from metals always accompanies the formation of metal oxides. In our case the seven-fold increase of the Sn^+ signal is probably related to the formation of the oxidized layers of Sn. It is in agreement with to the results of the X-ray analysis. As concerns the aluminum phase, it is necessary to take into account the small content of aluminum in the alloy. For the Al content an intensity I^{+}_{Al} of ~ 0.11arb.units is expected, however, I_{Al}^{+} was even higher and reached the value of 6 arb.units. This relatively magnification was more significant for Al ($I_{Al}^{+}alloy/ I_{Al}^{+}pure=55$) than for Sn ($I_{Sn}^{+}alloy/I_{Sn}^{+}pure=7$), which is in agreement with the SIMS results for oxides of Al and Sn [8].

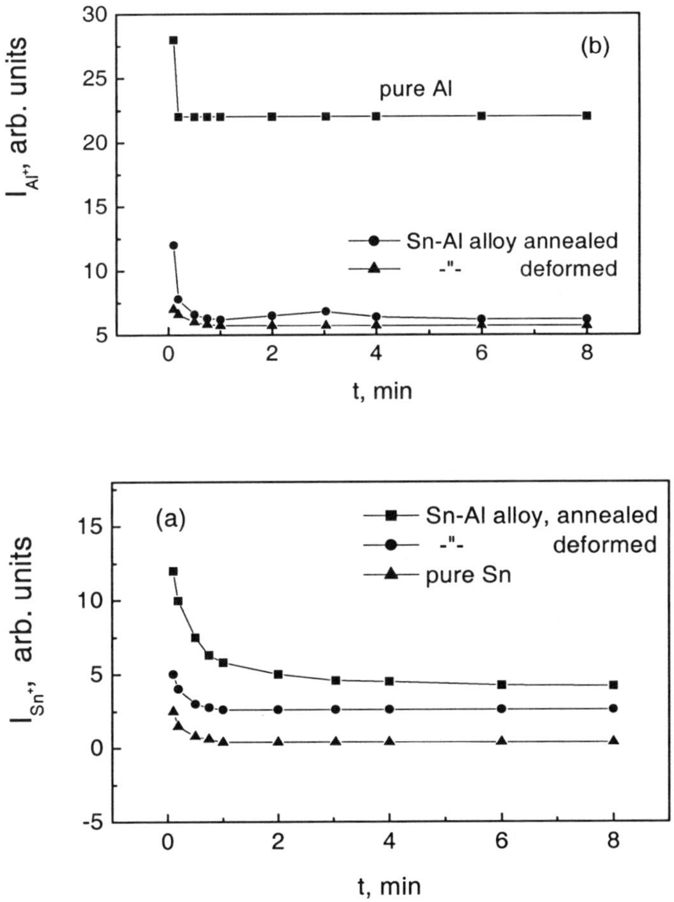

Figure2. Depth profiles of the of SIMS signal of Sn^{+} (a) and Al^{+} (b) of pure metals and the Sn-Al alloy after contact with water vapor during 1 month.

Less intensity of SIMS signals of alloys in deformed state is connected with the high density of cracks and macropores.

As shown in Fig.3, the SIMS depth profile analysis of Sn/Al phase boundary revealed that there was a considerable amount of hydrogen on the boundary. However, besides hydrogen an increased amount of oxygen compared with that distributed within the bulk was observed. This suggests that the water-induced change of the strength of phase boundaries in the eutectic Sn-Al alloy is caused by the influence of hydrogen and oxygen. Probably, oxidation process accompanies the penetration of hydrogen along the phase boundaries and leads to an irreversible failure of the phase boundary adhesion.

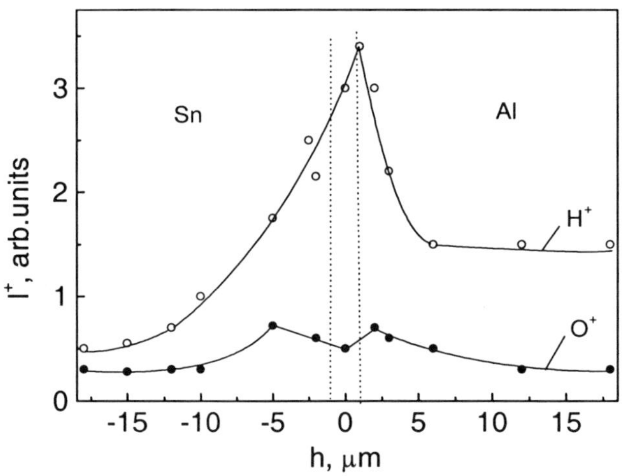

Figure 3. Depth profiles of the SIMS signal of hydrogen and oxygen on the Sn/Al phase boundary after contact with water vapor during 1 month. Position of "0" corresponds to phase boundary.

When considering the curve of hydrogen distribution near the boundary (Fig.3), other features are apparent, namely, hydrogen does not only accumulate on the boundary but is distributed in both the metals. The depth of penetration of hydrogen is larger for tin than for aluminum. The results of the electronic and optical microscopy and SIMS analysis have shown that eventually the process of hydrogen and oxygen interaction with tin expands into the inner tin grain and subgrain boundaries (Fig.4). The interaction of hydrogen mainly with the Sn surface is explained by the low energy of formation of gaseous SnH_4 (252 kJ mol^{-1} for AlH; 147 kJ mol^{-1} for SnH and -71,4 kJ mol^{-1} for SnH_4). It is the main cause of the penetration of atomic hydrogen into the bulk of an alloy. However, it is know [9], that formation of SnH_4 occurs only upon the fulfilment of two conditions: the presence of free bonds of tin and atomic hydrogen. The specific properties of the Sn/Al phase boundary satisfy these conditions.

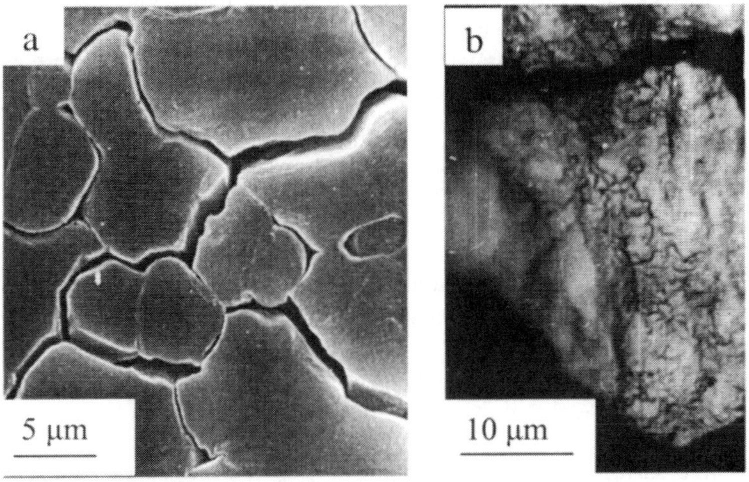

Figure 4. Grain boundaries brittleness of the Sn phase after exposure of the Sn-Al alloy to water vapor for 1-2 month (a), powder –like structure of the Sn-Al alloy after exposure to water vapor during 1year and testing at room temperature (b).

Let us consider those properties of the Sn/Al phase boundary. In [5] it was shown that the Sn/Al phase boundary has a higher value of energy (O.36 Jm^{-2}) than the energy of grain boundary for Sn and Al taken separately. Such relatively high values are due to the large positive energy of mixing (4.2kJ mol^{-1}), which, in turn, indicates a low energy of bonds of Sn with Al. Thereof there are always places of the broken contact and free bonds on the Sn/Al interface, that in particular was confirmed with the microscopic observations which have established the selective character of development of cracks. The appearance of atomic hydrogen on the Sn/Al interface is connected, first of all, to electrochemical processes which take place because of the high difference of standard electrode potentials (for Al $E^0_H= -1.4$V, for Sn $E^0_H =- 0.14$V [9]. A peculiarity of this phenomenon is that atomic hydrogen is not converted into molecular form on the surface, but penetrates along phase boundaries into the alloy. It is determined by the low energy of the formation of hydride of tin SnH_4 (-71.4 kJ mol^{-1}). However, it is also known that SnH_4 shows instability and decomposes at the room temperature [10]. The instability of tin hydride leads to accumulation of hydrogen in the bulk of the alloy and its conversion into the molecular form. The formation of molecular hydrogen results in stable cracks or in bubbles. Such phenomenon of hydrogen enrichments at grain boundaries was observed in [11,12] where it is shown, that as a result of interaction of water vapor with a surface of an Al-Zn-Mg alloy on internal grain boundaries are formed cracks and bubbles. Cracks open ways of an exit for hydrogen, at the same time

facilitating penetration of oxygen. These microscopic cracks of a hydrogen origin can render essential influence on the development of brittleness of Sn-Al alloy.
Let us consider the possibility of the appearance of a microscopic critical cracks on the Sn/Al phase boundary according to the Griffith criterion of brittleness

$$\sigma = \sqrt{\frac{EW}{\pi c}}$$

where σ is the tensile stress; W is an adhesion work, c – cracks size, E -Young modulus. From thermodynamics adhesion work is defined as

$$W = \gamma_0^{Al} + \gamma_0^{Sn} - \gamma_{pb}$$

where γ_0^{Al} ,γ_0^{Sn} the free surface energies of Al and Sn, γ_{pb} - Al/Sn phase boundary energy. Assuming that for Al/Sn phase boundary σ = 45 MPa [1], E=5.10^4 MPa, γ_0^{Al} = 0.7 Jm^{-2}, γ_0^{Sn} = 0.54 Jm^{-2} [9], γ_{pb} =0.36 Jm^{-2} [5], then the critical crack size $c \approx 7\mu$m. Optical microscopy investigations showed that cracks with sizes of 5 – 15μm appear on the surface of alloy in the places where the emission of hydrogen occurred. This is the evidence, that the stresses resulting in the formation of cracks are attributed to hydrogen. If hydrogen is enriched at the phase boundary region, local embrittlement of phase boundary takes place leading to intergranular crack initiation and propagation.

The investigations of process of emission of hydrogen from an alloy have shown, that within 1 month 1kg of alloy totally allocates up to 5-7 litres of hydrogen However, the process is non-uniform: it decelerates gradually in due course, but in the case of rupture of an alloy it renews sharply. We can assume, that emission of hydrogen is hampered because of oxidized layers near the external surface of alloy. For confirmation of this hypothesis we have carried out the investigations of the dependence of microhardness on penetration depth using the high precision microhardness method [13,14]. It was shown (Fig.5) that after long term contact with humid atmosphere (humidity ~80-90%) a soft flocculent layer is formed near surface of an alloy. This soft layer contains macrocracks and macropores, formed because of the hydrogen emission. A deeper layer has higher hardness (140 MPa) and is probably oxidized layer. Since depth 5-10 μm there is a decrease of hardness that apparently reflects a structural state of an alloy in the bulk. Fractographic investigations confirm these data.

Results show that the conditions of free emission of hydrogen out of the material may change during the exposure to the water vapor because of fast oxidation processes. After a long enough time, a dense oxidized layer formed along the sample perimeter may hamper hydrogen emission.
Besides it is established, that the kinetic of emission of hydrogen depends on the form and the state of an alloy. For example, chip-shaped or sawdust samples of the same

mass as the massive samples emit much smaller volume of the hydrogen, despite their strong deformation. It is connected with many factors, among them the increased role of external surfaces should be mentioned as hydrogen easy outlets. This creates a potential possibility of control of hydrogen accumulation in the bulk of the alloy and its emission.

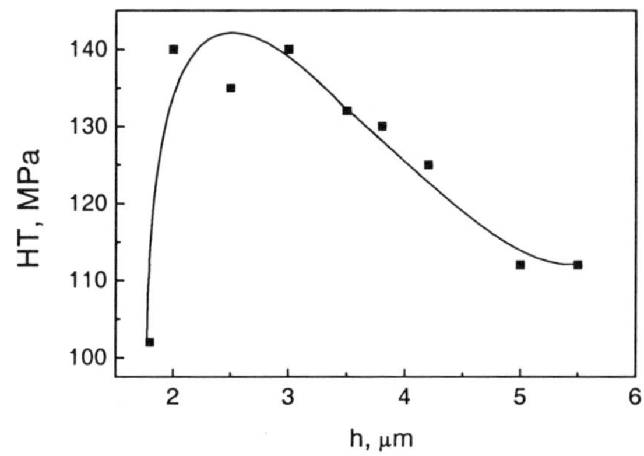

Figure 5. Microhardness as a function of indentation depth for Sn-Al alloy exposured in humid air during 2-3 month.

4. Conclusions

1. The irreversible brittleness of the Sn-Al alloy in water vapor is connected with the degradation of phase boundary adhesion due to accumulation of hydrogen and formation of oxidized layers on the phase boundaries.

2. The penetration of hydrogen and its accumulation in the bulk of alloy is because of the low energy of formation of unstable tin hydride and the high energy of phase boundary (0.36 Jm^{-2}). The processes are facilitated by preliminary plastic deformation due to formation of cracks.

3. It was found that after the long time exposure in the humid atmosphere the conditions of free emission of hydrogen out of alloy are changed. Results allow to assume an opportunity of the control of processes of accumulation and emission of hydrogen.

5. References

1. Muktepavel F, Upit G. (1984) The embrittlement of Sn-Al alloy in water vapor, *J.Mater.Sci.* **19** (2), 599-606.
2. Harris SJ, McCartney DG. Horlock A.J, Perrin C. (2000) Production of ultrafine microstructures in Al-Sn, Al-Sn-Cu alloys for use in tribological applications, *Mater. Sci. Forum,* **331**, 519-525.
3. Klyamkin SN, Hagstrom MT, Mescheryakova EV. (2000) Hysteresis in Ce-based metal hydrides, *J. Mater.Sci.* **35**, 133-137.
4. Yilbas BS, Coban A, Kahraman R, Khaled MM. (1998) Hydrogen embrittlement of Ti-6Al-4V alloy with surface modification by tin coating, *Int. J. Hydrogen Energy*, **23**, 483-489.
5. Muktepavela F, Maniks J. (1998) Mechanical properties and accommodation processes on metallic interfaces, *Nanostr. Mat.,* **10,** 479-484.
6. Maniks J. (1995) Methods of controllable internal rupture in solids and liquids for obtaining atomically clean surfaces and investigating their adhesion, *Vacuum*, **4,** 1439-1444.
7. Cherepin VT. (1978) Secondary ion emission of metals, *Adv. Mass Spektrom,* **1,** 776-781.
8. Vasylyev M.A., Panarin V.E, Deina A.A., Straumal B., Muktepavela F. (2000) Increase of the corrosion-resistance after low energy ion treatment, *Czech. J.Phys*, **50**, 1-5.
9. Samsonov G.V. (1976) *Handbook of the properties of elements. Part II.*: Metallurgy, Moscow, (in Russian).
10. Kolachev B.A. (1966) *Hydrogen embrittlement of metals,* Metallurgy, Moscow (in Russian)
11. Scamans G.M. (1978) Hydrogen bubbles in embrittled Al-Zn-Mg alloys, *J. Mater. Sci.* **13**, 27-36.
12. Wunderlich W., Niegel A., Gudladt H. (1990) Influence of hydrogen on the interface structure and properties of fatigued Al-Zn-Mg –bicrystals, *J.de Physique (Paris),* **C1,** 709-714.
13. Varcheniya S.A., Muktepavela F.O, Upit G.P. (1970) The effect of the indentation depth on the value of hardness and dislocation mobility near the surface of LiF crystals, *Phys.Stat.So.l (a),* **1**, K165-K167.
14. Upit G.P and Varchenja S.A., (1973) The size effect in the hardness of single crystals, in J.H. Westbrook and. H. Conrad (eds) *The Science of Hardness Testing and Its Research Applications*, American Society For Metals, Metals Park, Ohio, 135 -.

CALORIMETRIC INVESTIGATIONS OF METAL-HYDROGEN SYSTEMS FOR INTERMETALLIC COMPOUNDS OF LAVES PHASE TYPE

E.YU. ANIKINA, V.N. VERBETSKY
Moscow State University, Chemical Department
Russia, Moscow, 119992 Leninskiye Gory

Abstract

In the present work the review of the literature data and the results obtained in our laboratory, attributed to the direct calorimetric investigations of hydrogen interactoin with Laves phase intermetallic compounds, is carried out.

1. Introduction

The intermetallic compounds (IMC) on the base zirconium (titanium) with Laves phase structure C14 and C15 are rather perspective materials for the development of the energy storage systems and the hydrogen accumulation.

The effectiveness of the intermetallic hydrides for these and other technological purposes rests on their thermodynamic properties and, thus knowledge and understanding these properties have the paramount importance. There are a lot of the works, in which hydrogen interaction with IMC on the base Laves phases C14 and C15 have been investigated by means of the measure of P –X (X =H/IMC) isotherms. However, the investigations, in which these systems were studied by the calorimetric method, in literature are significantly less, though this method permits to appreciate the thermodynamic parameters of hydrogenation-dehydrogenation reaction more precisely, to measure the enthalpy of the reaction directly and to investigate of the mechanism of hydrogen interaction with IMC more carefully.

2. Results and discussion

2.1. ZrV_2.

A calorimetric investigation of the hydrogen interaction with ZrV_2 was carried out in [1,2] at 433 K. The plots of the values Peq. versus X and $\Delta H - X$ have been obtained for the hydrogen absorption from 10^{-4} to 1 MPa.

It is known that ZrV_2 [3] does not form the hydride phase at temperature $T > 280$ K, however, in ref. [1] the authors have got the interesting dependence $\Delta H - X$. In the initial region the $|\Delta H|$ values sharply decrease with the increase X until they reach a constant value of $|\Delta H \text{ plat.}|$ at X = 1.2 then in the region $1.2 < X < 2.7$ there are two well-definite regions with the ΔH constant values namely $|\Delta H\text{plat.}| = 115$ kJ/mole H_2

N. Veziroglu et al. (eds.), Hydrogen Materials Science and Chemistry of Metal Hydrides, 59–68.

and |Δ Hplat.| = 85 kJ/mole H_2. The first plateau length is 0.7 X and the second plateau length – 0.64 X. The authors did not study the higher concentrations region. The presence of two plateaux on the plot of Δ H versus X authors [1] explained as ZrV_2 consists of two hydride forming elements and the hydrogen atoms dissolved in metal matrix can occupy two types of tetrahedral interstitial sites Zr_2V_2 and ZrV_3.

2.2. $ZrCr_2$.

The calorimetric investigations of the hydrogen interaction with $ZrCr_2$ are presented in a few works [1,2, 4, 5]. Pedziwiatr and co-work. [4] has given the |Δ Hplat.abs.| and |ΔHplat.des.| values at 23°C equal 39.2kJ/mole H_2 and 36.4 kJ/mole H_2, respectively. The authors have pointed out that these data, obtained by calorimetric method, differ significantly from the data, that Shaltiel [6] obtained by the use of van't Hoff expression for pressure-composition isotherms (Δ H = -46kJ/mole H_2

Pedziwiatr et.al. [4] explained this difference that using of van't Hoff equation assumes the reversible behaviour of this system, however, the IMC – H_2 systems exhibit hysteresis effect in the pressure-composition isotherms and hence they are not behaving reversibly.

In ref. [2] it has been obtained the values Δ H at 433 K: |ΔHα| = 33.0 kJ/mole H_2 and |ΔHplat.abs.|=38.0 kJ/mole H_2 ($0.8 < X < 2.6$), that in practice coincides with the data [4].

In Magomedbekov works [1, 2, 5] the study of the $ZrCr_2$ – H_2 system has been carried out in the temperature interval from 330 to 530 K. The authors pointed out that the |ΔHabs.| values significantly decreased on the range corresponding α - phase and reached the constant value at X = 1.1 (T = 380 K, α→β transition). This fact testified to non-ideal system behaviour and the deviation from Civert's law.

Though Pebler and Gulbransen [3] did not observed a hysteresis in the dependencies ΔH from the hydrogen concentration in the $ZrCr_2$- H_2 system, whereas Magomedbekov et.al.[5] found out the express enthalpy hysteresis at T < 435 K, moreover the differences in the enthalpy values mainly showed on the phase boundaries (when β-phase forms or decomposes), where |Δ H des.|>|Δ Habs.|. At 330 K the hysteresis value is equal 11.4 kJ/mole H_2 and then this value changes to 0 when experimental temperature arises up to T > 435 K. Hysteresis of the Δ G values is not much (markedly lower), only 0.9 kJ/mole H_2 at T = 330 K, that testifies in accordance with authors about significant compensate influence of the entropy factor. In the range of α→β transition|Δ H abs.|>|Δ Hdes.| from 2kJ/moleH_2 at 330 K to 0 at 435 K. The authors pointed out that the systems irreversibility disappears at T>480 K, i.e. the hydrogen desorption at high temperatures puts into restoration of initial structure.

2.3. $ZrMn_2$.

The hydrogen interaction with $ZrMn_2$ has been studied by calorimetric method a number of researchers [2, 4, 7, 8, 9, 10]. In the paper [4] the values |Δ Hplat.abs.| and |Δ Hplat. des.| have been determined at 23 °C (40.4 kJ/mole H_2 and 38 kJ/mole H_2, respectively). The authors marked that obtained values of |Δ H| differ from the |Δ H|

values, about which reported in [6] (|Δ Habs.| = 53 kJ/moleH_2). The last values have been obtained by use of van't Hoff expression for pressure-composition isotherm. The authors [4] explained existing difference this way. The use of van't Hoff expression entails the assumption that the system behaviour is reversibly, whereas the system invariably exhibits hysteresis effect in the P –X isotherms and hence they are not reversible.

In ref. [8] the phase boundaries were determined for the $ZrMn_2$ – H_2 system: for α - phase 0.06<X< 0.1 and for β - phase 2.44 < X < 2.73. Magomedbekov et. al. [2, 8] remarked, that in the α- phase the Δ H values decrease sharply with increasing X, that is a peculiarity a lot of the IMC – hydrogen systems. In the range, corresponding to α→β transition (0.1< X < 2.4) the Δ H values are constant, moreover the authors pointed out [8], that |Δ Hdes.| > |Δ Habs.| (ΔHα→β = -42.5 kJ/mole H_2).

The hydrogen solution in the β-phase (X > 2.7) is accompanied by some decreasing |Δ H| with arising X.

The authors pointed out [8], that there is a significant difference between the |Δ Habs.| and |ΔHdes.| values on the range 2.44 < X < 2.73, which correspondents to α+β/β phase boundaries. This difference equals ~ 10 kJ/mole H_2 (|Δ Hdes.| > |Δ Habs.|). The workers suppose that this phenomenon is connected with the peculiarity of crystal structure of the intermetallic hydrides, namely the crystal lattice still saves the parameters of β-phase when hydrogen desorbes from it.

The $ZrMn_2$ – H_2 system in ref. [7] has been studied by a calorimetric method at a series of temperatures from 273 to 473 K. A plot of Δ H as a function of a hydrogen concentration in the metal matrix was drown on the base of calorimetric determined partial enthalpies for hydride formation from which one can see that the Δ H values decrease with increasing X across plateau. These results do not coincide with the data [2, 8] in which the authors marked a permanence of the partial Δ H values in the region of α→β transition. On the base of the absorption data obtained from the P –X isotherm measurements from 273 to 473 K the partial enthalpies values were calculated from van't Hoff equation at various values of X along the plateaux. The enthalpies showed a decrease in the values with increasing X. Flanagan [7] pointed out, that the absolute values for the enthalpies, determined from the van't Hoff plots for the formation plateaux were about 2 J/moleH_2 smaller, then the values determined calorimetrically. In [7] hysteresis has been determined as 1/2 RTln(Pf/Pd) and it was equal 3.59kJ/moleH_2.

Zhang and Wallace in [9] investigated the hydrogen interaction with $ZrMn_2$ in the temperature range 22 - 200°C by both P –X isotherm methods and a calorimetric method. The calorimetric values of ΔHabs. and Δ Hdes. for $ZrMn_2H_{3.6}$ were 43.3 kJ/mole H_2 and 45.4 kJ/mole H_2, respectively.

2.4. $ZrMn_{2+Y}$.

The hydriding characteristics of $ZrMn_{2+Y}$ has been studied by a number of workers. The intermetallic compounds can exist over a wide range of nonstoichiometry, i.e., from y= - 0.2 to 1.8. The plateau pressures increases markedly with y. Another feature in behaviour of this system is that their plateau slope and a degree of slope increase with increasing y.

TABLE 1. Calorimetric data of hydrogen interaction with intermetallic compounds.

IMC	X, (H/IMC)	T, K	|ΔHα|, kJ/moleH2	|ΔHplat.| a., kJ/moleH2	|ΔHplat.| d., kJ/moleH2	Ref.
ZrV_2	1.2-1.9	433		115		1, 2
ZrV_2	2.06–2.7	433		85		1
$ZrCr_2$*		296		39.2	36.4	4
$ZrCr_2$	0.4-0.8	433	33			1, 2
$ZrCr_2$	0.8–2.6	433		38		1, 2
$ZrCr_2$	1.1–2.3	330		52		5
$ZrCr_2$	1.1–2.3	380		44.2		5
$ZrCr_2$	1.1–2.3	530		36		5
$ZrV_{0.2}Cr_{1.8}$		433		48		2
$ZrV_{0.2}Cr_{1.8}$	0.9–2.8	433		47		1
$ZrV_{0.5}Cr_{1.5}$	1.1–3	433		57 – 52		1
$ZrV_{0.5}Cr_{1.5}$		433		57		2
ZrCrFe	3.3	298		34.1	35.7	9
$ZrMn_2$*		296		40.4	38.0	4
$ZrMn_2$	0.1–2.44	336		42.5		8
$ZrMn_2$		336	25.0			2
$ZrMn_2$	3.6	298		43.3	45.4	9
$Zr_{0.8}Mn_{0.2}Mn_2$	0.08–2.06	333		27.5		2, 10
$Zr_{0.8}Mn_{0.2}Mn_2$		333	22.7			2, 10
$Zr_{0.7}Mn_{0.3}Mn_2$	0.09–1.72	333		27.5		2, 10
$Zr_{0.7}Mn_{0.3}Mn_2$		333	19.1			2, 10
$Zr_{0.79}Mn_{0.2}Mn_2$		300		32.9		11
$Zr_{0.62}Mn_{0.38}Mn_2$		300		29.8		11
$ZrMn_2Mn_{0.8}$*		296		32.9	29.9	4
$ZrMn_2Mn_{1.8}$*		296		29.8		4
$ZrCr_{0.6}Fe_{1.4}$*		296		29.1	28.5	4
$Zr_{0.75}Ti_{0.25}Mn_{1.1}Fe_{0.9}$*		296		30.2	29.4	4
ZrVMn		433	55.0	70.0		2
ZrCrMn	0.27–2.52	333		43.0		2, 10
ZrCrMn		333	27.5			2, 10
$ZrMn_2Fe_{0.8}$*		296		25.9	25.0	4
$ZrMn_2Co_{0.8}$*		296		23.8	20.9	4
$ZrMn_2Ni_{0.8}$*		296		26.4	25.0	4
$ZrMn_2Cu_{0.8}$*		296		35.9	31.6	4
$Ti_{0.8}Zr_{0.2}Fe_{0.1}V_{0.4}Mn_{1.5}$	0.5-2.5	298		15.5**		21
$Ti_{0.98}Zr_{0.02}Mn_{1.5}V_{0.43}Cr_{0.5}Fe_{0.09}$		306			22.8±0.2	23
$Ti_{0.9}Zr_{0.1}Mn_{1.5}V_{0.25}Cr_{0.1}Fe_{0.05}Co_{0.05}Ni_{0.05}$		306			22.4±0.2	23
$Ti_{0.9}Zr_{0.1}Mn_{1.5}V_{0.25}Cr_{0.1}Fe_{0.05}Co_{0.05}Ni_{0.05}$	0.2-0.6	306	21.9±0.3			23
$Zr_{0.8}Ti_{0.2}CrFe$	0.1-0.6	318	28.0±0.2			24
$Zr_{0.8}Ti_{0.2}CrFe$	0.6-2.2	318		30.7±0.3		24
$Zr_{0.8}Ti_{0.2}CrFe$	0.2-1.0	461	20.5			24
$Zr_{0.8}Ti_{0.2}CrFe$	1.0-2.0	461			31.7±0.7	24
$Zr_{0.8}Ti_{0.2}CrFe$	0.2-1.2	488	20.6±0.6			24
$Zr_{0.8}Ti_{0.2}CrFe$	1.2-1.9	488	26.7±0.4			24

* - integral Δ H ** - kJ/mole H

Flanagan and co-work. [7] have been investigated the $ZrMn_{2+Y}$ – H_2 system at temperatures from 273 to 473 K. Absorption isotherms for these system with y = -0.2, 0, 0.5 1.0 and 1.5 have been measured and the dependencies Δ Habs./des. – X in the plateau region have been obtained by a calorimetric method too. On the base of these data the authors concluded, that values of calorimetric determined enthalpies for hydride formation decrease with X across the plateaux, moreover the extent of the sloping increased with the degree of nonstoichiometry, i.e. with $|y|$.
The authors [7] set a task to find out which thermodynamic parameter is responsible for the observed sloping. For these purposes it has been calculated the plateau entropies from equation (1):

$$\Delta S(plat) = \Delta H(plat.)(f,d)/T - 1/2\, R\ln P(f,d) \qquad (1)$$

It is appears, that the plateau entropies are unaffected by hydrogen concentration. This way Flanagan and co-work. established that the marked increase in Pf and Pd with increasing y is due to energetic reasons and this may be related to the decrease in the unit cell size with the increase of y.
Values for hysteresis at a fixed value X are given as: 2.33 kJ/mole H (y = - 0.2), 3.59 kJ/mole H (y=0), 3.16 kJ/mole H (y = 0.5), 2.35 kJ/mole H (y = 1.0).
The calorimetric studies of the hydrogen interaction with nonstoichiometric IMC on the $ZrMn_{2+Y}$-base were carried out in the works [2, 11]. Results are presented in Table 1.
As [12] Sinha and Wallace showed that a deviation from stoichiometry in $ZrMn_{2+Y}$ conditioned by partial substitution of Zr by Mn, so the intermetallic compounds $ZrMn_{2+Y}$ can be presented as $Zr_{1-y}Mn_yMn_2$. The authors [11] remarked, that Δ H values of hydride formation, obtained by a calorimetric method, are lower one for $ZrMn_2$-hydride. The similar results were obtained in [4] (see Table 1).
In the papers [2, 10] plots of a concentration dependence Δ H from X for $Zr_{0.7}Mn_{0.3}Mn_2$ and $Zr_{0.8}Mn_{0.2}Mn_2$ are given, their shapes do not differ from similar dependencies Δ H – X for the IMC – H_2 system.
In the ref. [10] the phase boundaries $\alpha/(\alpha+\beta)$ and $(\alpha+\beta)/\beta$ for the $Zr_{0.7}Mn_{0.3}Mn_2$ and $Zr_{0.8}Mn_{0.2}Mn_2$ systems, namely for α-phase – 0.06< X < 0.09 and 0.05< X < 0.08 and for β-phase – 1.72< X < 2.03 and 2.06< X < 2.31, respectively. The $Zr_{0.7}Mn_{0.3}Mn_2$ and $Zr_{0.8}Mn_{0.2}Mn_2$ systems show increasing the $|\Delta H|$ values with increasing X from 28.9 to 25.5 kJ/ mole H_2 and from 30.8 to 27.5 kJ/mole H_2, respectively. In the region of the $(\alpha+\beta)/\beta$ phase boundaries the |Δ Hdes.| values are significantly more than the |ΔHabs.| values. These differences are equal to 12 kJ/mole H_2 for $Zr_{0.8}Mn_{0.2}Mn_2$ and 9 kJ/mole H_2 for $Zr_{0.8}Mn_{0.2}Mn_2$. The similar results have been obtained by Dantzer [13] for the $LaNi_5$-H_2 system.
In the ref. [4] Pedziwiatr and co-work. have published the results of calorimetric investigations the systems on the $ZrMn_2T_{0.8}$ – based (T = Fe, Co, Ni, Cu. Mn) at 23°C (see Table 1). These data permit to make conclusion, that the values of |Δ Habs.| and |Δ Hdes.| are minimal for $ZrMn_2Co_{0.8}$ .This result agrees with the fact, that the hydride of this alloy exhibits the highest vapour pressure. The author suggested that Co has a particular effect on the band structure in $ZrMn_2$ when it enters the lattice by partial replacement of Mn and/or Zr.

2.5. $Zr(Fe_{0.75}Cr_{0.25})_2$

In 1990 Grant et.al. published the work devoted to the calorimetric investigation of $Zr(Fe_{0.75}Cr_{0.25})_2 - H_2$ system [14]. Previously Nortwood et.al. [15] have researched this system by means of the P –X isotherm method at different temperatures (from –5 to 40°C) and the values of ΔH, ΔS and ΔG have been calculated from van't Hoff plots in the usual manner. There is a broad region of α→β transition on the plots P – X absorption and desorption isotherms ($0.2 < X < 2$), and these P –X isotherms have shown quite pronounced hysteresis and sloping plateau.
In ref.[14] it has been obtained the dependencies P –X and ΔH – X at 273 K. The authors [14] have pointed out, that data for this hydride/ dehydride sequence were obtained over a period of 21 days, this fact was a feature of this study, and steady state conditions were obtained after each gas aliquot.
The hysteresis in the region $0.15 < X < 2.0$ is large (ln Pabs./Pdes. = 3.45). A notable feature of the $Zr(Fe_{0.75}Cr_{0.25})_2 - H_2$ system is the persistence of the pressure hysteresis well beyond the obvious marked change in slop in P – X curve (at X ≅ 2.0 in absorption), which is taken as a boundary between a two-phase coexistence region and a single-phase region in different researches. In this region where $X > 2.0$ (in absorption) P –X reversibility only occurs for $X > 2.5$. Such a behaviour of pressure hysteresis has not been observed in the true steady state P –X diagrams available for other hydride systems such as $LaNi_5 - H_2$ [16] and $CaNi_5 - H_2$ [17, 18]. The researches suggested [14] that composition $2<X<2.5$ (in absorption) is also a two-phase coexistence region, i.e. there are two compound hydrides, β and γ, in the $Zr(Fe_{0.75}Cr_{0.25})_2 - H_2$ system at 273 K with values of X of 1.8 – 2.05 and $X > 2.33$, respectively. The small compositional difference between these phases could yield the high slope in the P –X data in their coexistence region.

2.6. ZrCrFe.

A calorimetric study of the hydrogen interaction with ZrCrFe was carried out in ref. [19] in the temperature range 320 – 563 K and $0< P < 6$ MPa. The P –X isotherms and the dependencies Δ H – X were obtained for the absorption and desorption process. The authors pointed out the presence of pressure hysteresis on the P –X isotherms at 320 – 470 K, however, there is no enthalpy hysteresis in this system. The authors marked interesting feature of this system, which shows P -X isotherms. At 320 K there is the sharp change of the slope on the P –X dependence at X = 2, and P – X reversibility in this system appears only at $X \geq 2.7$. To this phenomenon corresponds to the stepwise change of the $|\Delta H_{abs.(des.)}|$ value from 35.6 ± 0.4 kJ/mole H_2 in the range $0.3< X < 2.06$ to 42.7±0.9 kJ/moleH_2 in the range $2.0 < X < 2.7$.
To compare dependencies P –X with Δ H – X in ref. [19] authors came to the conclusion that at 320 K there are two hydrides in the ZrCrFe – H_2 system: ZrCrFe – $H_{2-\delta}$ (β - hydride) and ZrCrFe – $H_{3-\delta}$ (γ - hydride).
The α-solution region for ZrCrFe – H_2 system is narrow ($0 < X < 0.2$). The plateau length reduces with rising experimental temperature, and the α-solution region widen from X = 0.2 at 320 K to X = 1.2 at T=563 K while the analogous shift in the region of

hydride phases existence in more noticeable ($\Delta X \cong 2$). The authors remarked that the temperature $T \cong 477$ K is the critical temperature, above which in the ZrCrFe – H_2 system a hydrogen solution (or solutions) should exist in metal matrix. At T > 430 K the pressure hysteresis was not found and the absorption/ desorption P –X isotherms in practice coincide.

In ref. [9] for the ZrCrFe – H_2 system it has been obtained P –X isotherms and measured the integral heat formation and decomposition of $ZrCrFeH_{3.3}$. The desorption isotherms were obtained at t=22, 100, 150 and 200°C. The P –X isotherms data in ref. [9] coincide with the data from [19]. In the research [9] the ΔHdes. values has been calculated using the van't Hoff equation for the concentration range 0.5<X<2.0 (T = 298 K). Comparing the results from [9] and [19] for theΔ Hdes. values one can conclude that these values of Δ Hdes. are close tj each other, however, for the ZrCrFe – H_2 system the method P –X isotherm measurement is not enough informative and it does not permit to reveal more precise mechanism of the hydrogen interaction with ZrCrFe. The values of the integral heat formation and decomposition $ZrCrFeH_{3.3}$ obtained by the calorimetric method in ref. [9] also are close to the values ,obtained in ref. [19], but they do not show the way, on which the ZrCrFe – H_2 system moved from X = 0 to X = 3.3 and return back.

2.7. $ZrCrFe_{1.2}$.

The hydrogen interaction with IMC on the base Laves phase C14 $ZrCrFe_{1.2}$ has been studied by a calorimetric and P –X isotherm methods in the work [20]. The $ZrCrFe_{1.2}$.-H_2 system has been investigated in the temperature range 315 – 523 K. For all temperatures the desorption isotherms are presented and for 315 K also the absorption isotherm. The authors pointed out that experimental results obtained at 315 K showed a certain pressure hysteresis as well as an enthalpy hysteresis.

The worker marked the difference between of the Δ H = *f*(X) for absorption and desorption process at 315 K. There is one region with constant values Δ H on the plot of the dependence Δ H – X in absorption while in desorption process it could be noted two distinct region with constant values of ΔH: the first Δ Hdes. = 33.6 ± 0.4 kJ/mole H_2 in the range 0.4< X < 1.8 and the second Δ Hdes. = 36.5±0.5kJ/mole H_2 in the range 1.9 < X < 2.3. The researchers suppose that two region of constant enthalpy values are possibly due to existence of two hydride phases $ZrCrFe_{1.2}$- $H_{2-\delta}$ and $ZrCrFe_{1.2}$- $H_{3-\delta}$ corresponding to β- and γ-hydrides.

In the region of hydrogen α-solution in IMC the $ZrCrFe_{1.2}$-H_2 system behaves similar to all AB_2 –H_2 system which have been mentioned above. In the range 0 < X < 0.4 the values of |Δ Habs.(des.)| sharply decrease and further they increase up to the plateau values, moreover the quantitative data coincide in full as for absorption and desorption, In the region of hydrogen solution in hydride the dependence |Δ Habs.(des.)| - X has a maximum at X = 2.5, and |Δ Habs.|<|Δ Hdes.| in the region 2.2< X < 2.5, which agrees with the data obtained in [5, 10, 21].

It should bee noted, that at rising the experimental temperature from 315 K to 353 K the enthalpies ΔHα→β and Δ Hβ→γ transition change: the β-hydride dissociation enthalpy value increases negligibly, the γ-hydride dissociation enthalpy value reaches 41.2 ± 0.3 kJ/mole H_2 at 338 K and practically does not change up to 353 K. However rising of the experimental temperature up to 393 K leads to drastic decreasing of values Δ H, corresponding $\alpha \rightarrow \beta$ transition to 32.2 ± 0.4 kJ/mole H_2 and disappearance of the plateau, corresponding $\beta \rightarrow \gamma$ transition. The authors conclude that there is a critical temperature in the range of 353 – 393 K above which γ- hydride does not form. At higher temperatures, besides the hydride phase, two types of hydrogen solution in IMC form with increasing ΔH values which decrease with increasing experimental temperature up to 523 K.

2.8. $Ti_{0.9}Zr_{0.1}Mn_{1.1}V_{0.1}$

The hydrogen interaction with $Ti_{0.9}Zr_{0.1}Mn_{1.1}V_{0.1}$ has been studied in [22] at 72, 88, 103 and 116 °C and a hydrogen pressure 0< P < 50 atm by a calorimetric and P-X isotherms measurements methods. For all temperatures desorption isotherms and the |Δ Hdes.| - X dependencies are presented. The desorption isotherms have significant plateau sloping. Another feature of this system is an anomalous large α-region for all measured P-X isotherms.

In the plot of |Δ Hdes.| versus X the enthalpy data show two distinct region with a constant value of desorption enthalpy in the range from 72 to 103 °C. The authors [22] proposed this is due to the existence of two hydride phases, $Ti_{0.9}Zr_{0.1}Mn_{1.1}V_{0.1}H_{\sim 1.0}$ and $Ti_{0.9}Zr_{0.1}Mn_{1.1}V_{0.1}H_{\sim 2.0}$, corresponding to β1- and β2-hydrides. At moving from 72 to 103°C the β1- hydride domain decreases, whereas the β2 –hydride region expends at the same time from 1.1< X < 1.67 at 72°C to 0.9< X < 1.5 at 103°C. At 116°C there is drastic change in the $Ti_{0.9}Zr_{0.1}Mn_{1.1}V_{0.1}$ –H_2 system from the results the authors[22] assumed that 103°C is close to the critical temperature for β1-hydride, above which temperature it does not exist. At 116°C the authors [22] assign three regions with constant values of desorption enthalpy, namely in a range 0.25<X< 0.50 hydrogen dissolution is accompanied by significant large enthalpy values (|ΔHdes.|= 54.17 ± 0.53kJ/moleH_2), then in the range 0.58< X < 1.20 enthalpy values are less exothermic then the values correlated for β2-hydride, while in the range X > 1.20 at 116°C there is no the constancy of concentration dependence of |Δ Hdes.| values. The authors explained this complex character of hydrogen interaction in the $Ti_{0.9}Zr_{0.1}Mn_{1.1}V_{0.1}$-$H_2$ system by odering hydrogen in the tetrahedral holes formed by different metals, as well as possible structural changes in the metal matrix.

2.9. $Ti_{0.8}Zr_{0.2}Fe_{0.1}V_{0.4}Mn_{1.5}$.

The hydrogen interaction with the technologically important multicomponent intermetallic compounds has been studied by a calorimetric method in [21]. Flanagan and co-work. have found that the |ΔHplat.| values decrease with increasing X along plateau. This decrease is more apparent on the range 0.5< X < 1.0 than in the last range of plateau. The average value |Δ Hplat.|= 15.5 kJ/mole H; The values decrease about 0.5kJ/mole H across plateau. The authors do not know as to explain this fact. The decrease is too small for that to explain the entire increase of 1/2RTln(Pf). The enthalpic factors are the most probable cause of the increase in Pplat.

3. Conclusions

The review of the literature date carried out in this paper let us make some conclusions.

1. Using the calorimetric method it has been found that there is the temperature dependence of thermodynamic characteristics in a number of IMC systems.
2. A comparison of the results, obtained by P – X measurements, with the calorimetric data of the concentration dependence Δ H – X permits us to determine the location of the phase boundaries, phase transitions and also the IMC – H_2 system behaviour on these boundaries more precisely.
3. The existence of two hydride phases has been found in a number of the IMC – H_2 system using the calorimetric obtained dependencies Δ H = f(X).

Acknowledgements

This work was supported in part by RFBR Grants .99-02-16383 and 00-15-97-457.

4. References

1. Krupenchenko, A.V., Magomedbekov, E.P. and Andreev, B.M. (1991) Concentration dependence of the enthalpy of hydride formation in the $Zr(V_yCr_{1-y})_2$ – hydrogen system, Russ. Zh. Neorganichekoy khimii 36,(6), 1595-1603.
2. Krupenchenko, A.V. and Magomedbekov, E.P. (1992) Study of stability of intrmetallic compounds ZrB_2–type (B= V, Cr, Mn) at hydrogen interaction by equilibrium-calorimetric method, Russ. Zh. Neorganicheskoy khimii 37 (1), 174-181.
3. Pebler, A. and Gulbransen, E.A. (1967) Equilibrium studies on the systems $ZrCr_2$–H_2, ZrV_2–H_2 and $ZrMo_2$–H_2 between 0°and 900°C. Trans. Met. Soc. AIME 239 (10), 1593-1600.
4. Pedziwiatr, A.T., Craig, R.S., Wallace, W.E. and Pourarian, F. (1983) Calorimetric enthalpies of formation and decomposition of hydrides of $ZrMn_2$, $ZrCr_2$ and related system, J. of Solid State Chemistry 46, 336-341.
5. Krupenchenko, A.V., Magomedbekov, E.P. and Vedernikova, I.I. (1990) Calorimetric investigation of thermodynamic properties of the $ZrCr_2$-hydrogen system, Russ. Zh. fizicheskoi khimii 64 (11), 2897-2908.
6. Shaltiel, D., Jacob, I. and Davidov, D. (1977) Hydrogen absorption and desorption properties of AB_2 Laves-phase pseudebinary compounds, J. Less-Common Met. 53, 117-131.
7. Luo, W., Majorowski, S., Clewley, J.D. and Flanagan T.B. (1989) The thermodynamics of the $ZrMn_{2+X}$ – hydrogen system, Z. Phys. Chem. 163, 81-86
8. Krupenchenko, A.V., Magomedbekov, E.P., Andreev, B.M. and Dyatlova, O.V. (1991) Calorimetry of hysteresis transition in the $ZrMn_2$ – hydrogen system, Russ. Zh. Neorganicheskoi khimii 36, (1), 175-181.
9. Zhang, L.Y. and Wallace, W.E. (1988) Hydrogenation characteristics of the isoelectronic alloys $ZrMn_2$, ZrCrFe and ZrVCo, J. of Solid State Chemistry 74, 132-137.
10. Krupenchenko, A.V. and Magomedbekov, E.P. (1994) Hysteresis in the IMC system AB_2- type. Influence of IMC composition, hydrogen concentration in the solid phase and annealing, Russ. Zh. Neorganocheskoy khimii 39 (!), 151-155.
11. Pourarian, F., Sinnha, V.K., Wallace, W.E., Pedziwiatr, A.T. and Craig, R.S. (1982) Hydrogen entropies of the $ZrMn_{2+y}$ system. In Proceeding of a NATO International Symposium on the Electronic Structure and Properties of Hydrogen in Metals, held March 4 –6, in Richmond, 385-390.
12. Sinha, V.K. and Wallace, W.E. (1982) $Zr_{07}Ti_{0.3}Mn_2Fe_{0.8}$ as a material for hydrogen storage. J. of the Less-Common Met. 87, 297-303.
13. Dantzer, P., Orgaz, E. and Sinha, V.K. (1989) Influence of Hysteresis on the Thermodinamic Properties of the $LaNi_5$-H_2 System, Z. Phys. Chem. 163, 141-148.
14. Grant, D.M., Murray, J.J. and Post M.L. (1990) The thermodynamics of the system $Zr(Fe_{0.75}Cr_{0.25})_2$ + H_2 using heat conduction calorimetry, J. of Solid State Chemistry 87, 415-422.

15. Ivey, D.G. and Northwood, D.O. (1984) Hydrogen absorption-desorption characteristics of $Zr(Fe_xCr_{1-x})_2$. In Proceeding of the 5-th World Hydrogen Energy Conference, Toronto, Canada, 15-20 July 1984, Ed. Veziroglu T.N., Taylor J.B.:1395-1404.
16. Murray, J.J., Post, M.L. and Taylor J.B. (1981) Thermodynamics of the $LaNi_5$ – H_2 system by differential heat flow calorimerty II: α and β single phase regions, J. of the Less-Common Met. 80, 211-219.
17. Murray, J.J., Post, M.L. and Taylor, J.B. (1983) Thermodynamics of the system $CaNi_5$ – H_2 using differential heat conduction calorimetry, J. of the Less-Common Met. 90, 65-73.
18. Grant, D.M., Murray, J.J. and Post, M.L. (1987) Thermodynamics of the γ- and δ-hydrides and deuterides of $CaNi_5$. J. Chem. Thermodyn, 19, 1217-1226.
19. Sirotina, R.A., Umerenko, E.A. and Verbetsky, V.N. (1996) Calorimetric investigation of hydrogen interaction with intermetallic compound ZrCrFe, Russ. Neorganicheskie materially 32(6), 710-714.
20. Ivanova T.V. and Verbetsky, V.N. (1999) Calorimetric investigation of the hydrogen interaction with $ZrCrFe_{1.2}$, J. of Alloys and Compounds 293-295, 34-37.
21. Flanagan, T.B., Luo, W. and Clewley, J.D. (1993) Characterization of multi-component metal hydrides using reaction calorimetry, Z. Phys. Chem. 179, 35-44.
22. Anikina, E.Yu. and Verbetsky, V.N. (2002) Calorimetric investigation of the hydrogen interaction with $Ti_{0.9}Zr_{0.1}Mn_{1.1}V_{0.1}$, J. Alloys and Compounds 330-332, 45-47
23. Sirotina, R.A., Mitrohin, S.V., Zakharova, M.A. and Verbetsky, V.N. (1993) Calorimetric investigation of multicomponent Laves phase interaction with hydrogen and deuterium, J. of Alloys and Compounds 202, 41-45.
24. Sirotina, R.A. and Verbetsky, V.N. (1994) Calorimetric investigation of interaction in the $Zr_{0.8}Ti_{0.2}ZrFe$-H_2, Russ. Izvestiya RAN. Neorganicheskie materialy 30 (2), 197-200.

THE INFLUENCE OF HYDROGEN ON STRUCTURE AND PROPERTIES OF TITANIUM POWDERS

O.M. SHAPOVALOVA., E.P. BABENKO, J.V. BABENKO
Dnepropetrovsk National University
ul. Nauchnaya, 13, Dnepropetrovsk, 49081, Ukraine

Abstract

It is established, that structure and properties of titanium powders depend on methods of their productions and degree of pollution by impurity. Nature of formation of titanium hydrides by processing of various concentration of hydrochloric acid is investigated.

Titanium hydrides can be received from titanium powders, which in Ukraine can be done by different ways: iodide, electrolytic, electrolytic purifying, magnesium-heat, sodium-heat, chemicomechanical [1,2,3].

As it has appeared, different nature of titanium powders essentially influences on:

- Interaction of particles of titanium powders with gas impurity including hydrogen, that determines their unequal chemical structure;
- Density and porosity of a material;
- The periods of the crystal lattice of titanium;
- Propensity to absorption of hydrogen and hydrides formation;
- Form and sizes of particles of powders;
- Macro-, micro- and thin structure of crystals;
- Characteristics of metal strength;
- Technological properties: bulk weight, compressibility and corrosion stability.

Pollution of particles of titanium powders by gas impurity, first of all by hydrogen, is caused by a way of their manufacture and subsequent interaction with an environment. In figure 1 are submitted histograms of change of the contents of hydrogen (a), oxygen (b) and nitrogen (c) depending on a nature of powders in a half-logarithmic scale.

It follows from the submitted data that under the contents of all impurity the purest are the powders of iodide and purified titanium. The amount of impurity increases in titanium powders of different ways of manufacture in a sequence: iodide, purified, magnesium-heat, electrolytic, sodium-heat, chemicomechanical.

Titanium powders are sated with gas impurity during their manufacture and subsequent interaction with an environment.

We have investigated 65 sets of titanium powders purified by electrolytic way. As influence on microhardness and hardness can render not only gas impurity, but also other elements, their role is estimated on size of factors of pair correlation.

N. Veziroglu et al. (eds.), Hydrogen Materials Science and Chemistry of Metal Hydrides, 69–76.

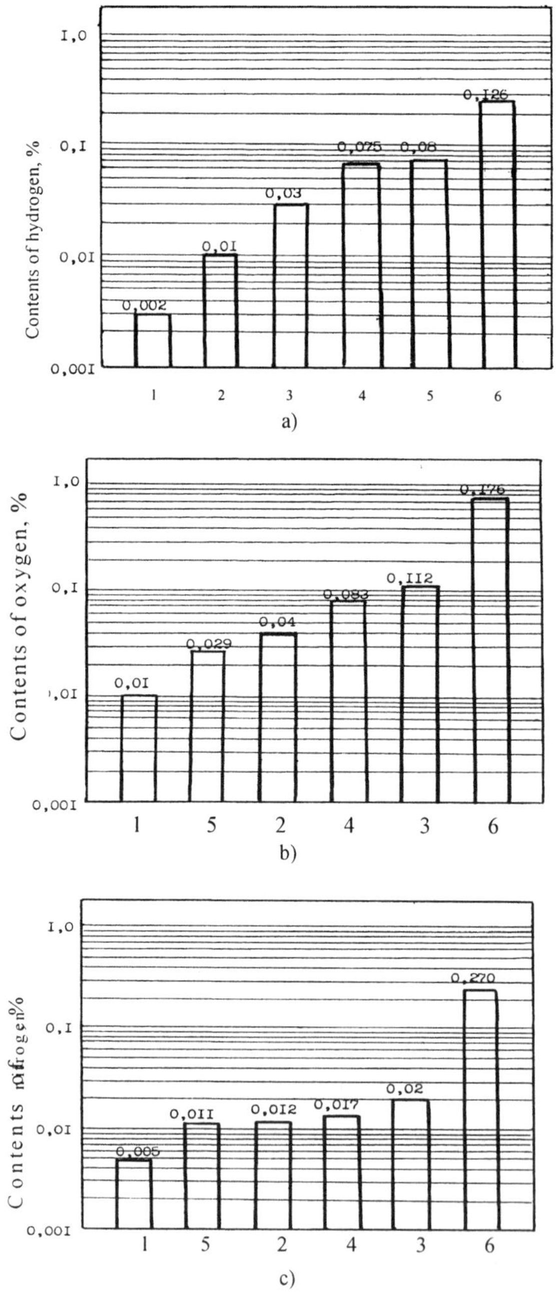

1 - iodide, 2 - purified, 3 - magnesium- heat, 4 - electrolytic, 5 - sodium-heat, 6 – chemicomechanical

Figure 1. Histogram of change of the contents of hydrogen, oxygen and nitrogen in titanium powders depending on a way of their manufacture

Stronger influence of gas impurity is caused not only by interaction with titanium, but also by contents of impurity adsorbing by powders. (Table 1)

Table 1. Middle chemical composition of purified titanium powders

Material	HB, MPa	Impurity, % mas.												
		H_2	O_2	N_2	Cl_2	C	Fe	Si	Mn	Ni	Al	V	Cr	Σ imp.
Purified powders	86,7	0,0108	0,029	0,012	0,042	0,005	0,011	0,003	0,003	0,003	0,005	0,003	0,002	0,1384

It follows that the greatest influence on hardness is rendered by hydrogen, oxygen and iron. It is connected to the fact that hydrogen in purified titanium was a decimal order more, than in iodide titanium. The oxygen and nitrogen, that usually strengthen titanium (0,1 % of oxygen strengthens titanium by 150 MPa, 0,1 % of nitrogen - by 200 MPa) were in the bound status as oxide and nitride, therefore they rendered smaller, than it is possible to expect influence on hardness of purified powders. As on the data [4] solubility of hydrogen in α-titanium at 0 °C makes 0,0009, at 78 °C $-7x10^{-8}$, at 196 °C – $4x10^{-23}$ % of weight, other hydrogen forms titanium hydrides. Hence in unalloyed titanium of the marks VT1-0 and VT1-00, in which the contents of hydrogen 0,01 and 0,008 % to mass respectively is supposed, the hydrogen is not only in the dissolved status in a crystal lattice of titanium, but it is necessary that as titanium hydrides. Do metallographic analysis of structure of particles of purified titanium from 0,012 % H_2 we found out large enough amount of apportionments of titanium hydrides of plastic form located on planes of sliding with certain crystallographic orientation determined under angles 60, 90 and 120° to each other (Figure 2). There are dark color pores in a particle of a powder.

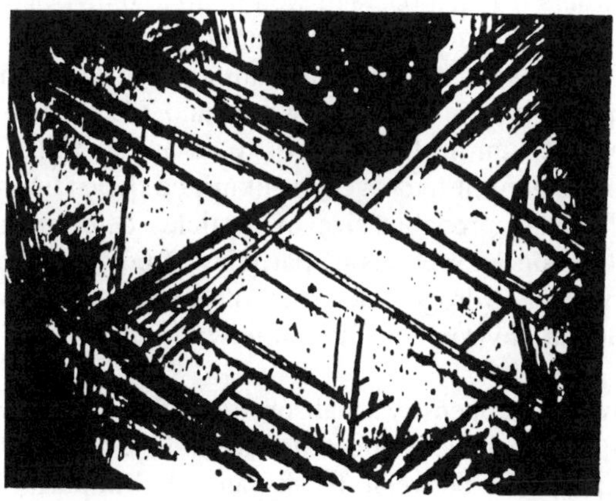

Figure 2. Crystallographic focused needles of titanium hydride in a particle of purified titanium powder, x300

Iron and manganese also belong to a family of strong reinforces of titanium (1 % of iron causes hardening of titanium $\Delta\sigma_B$=65 MPa, 1 % of manganese - on 75 MPa), however their influence is considerably smaller, than that of the gas impurity. Besides amount of metal impurity in purified powders is very low. So, titanium powder iron present in purified strengthens titanium on 0,65 MPa, while oxygen - on 43,5 MPa.

Appreciable influence of hydrogen on durability of super purified titanium (in it the total impurity makes 0,140 % of weight, against 0,181 % of weight in iodide) is connected to specific volume of titanium hydrides in larger by 13,2 % comparison with titanium. The dissolved atoms are more likely placed in octahedral pores of a crystal lattice of titanium, as their radiuses are well integrated (respectively 0,041nm and 0,062 nm). The presence of atoms of hydrogen in tetrahedral emptiness because of their small radius (0,034 nm) is poorly probable. Hydrides of titanium, having large specific volume in comparison with specific volume of a matrix, are placed in places with low density of packing of atoms and with the increased potential energy. Such places are the planes of sliding, twinning {1010} and {1011}, boundary of grains (Figure 3). We have found allocations of titanium hydrides on the surface of particles (Figure 4), which act above a plane of a crystal, keeping, however, form of focused plates.

Increase of the contents of hydrogen in titanium powders in different ways of manufacture, while of the same size, correlates with growth of microhardness (Figure 5).

Especially significant there was hardening of the purest material – electrolitically purified powder (Figure 6).

The revealed natural hardening of titanium powders with increase of the contents of hydrogen can be connected only with significant increase of amount of titanium hydrides, which specific volume of which is much greater than the one of titanium. It was confirmed by metallographic, electron microscope and radiographic analysis of powders of different ways of manufacture: the amount of hydride apportionments consistently grew according to increase of microhardness. In figure 3 the change of amount of hydrides and their morphology with increase of the contents of hydrogen in purified titanium powders from 0,003 % of weight up to 0,024 % of weight is shown.

Amount, structure, the sizes and form apportionments of titanium hydrides are also connected to a nature and character of manufacture of titanium powders. So, mass quantity of titanium hydrides in iodide powders and received by chemicomechanical manufacturing of magnesium-heat wastes differed in 60 times. On picture 7, the diagram of change of amount of hydrogen is submitted depending on the most probable size of particles. The most probable size of particles determined using a method of Saltykov [5] on 21 tests on 11 sections of particles of powders received by electrolytic method. As the amount of hydrogen absorbed on a course of technological process of manufacture and from an environment is visible with reduction of the sizes of particles of powders grew.

In titanium powders propensity to absorption of hydrogen, density, bulk weight, corrosion stability in aggressive environments depend on the way of their manufacture. On a figure 8 the dependence between density and total impurity in industrial powders of electrolytic, sodium-heat and magnesium-heat is submitted, the degree of pollution which grew according to the specified sequence.

Density was sharply reduced at increase of the contents of the total impurity (ΣI from 0,16 up to 0,3 % of weight). The further increase of ΣI rendered considerably smaller influence. It is connected with change of the periods of the crystal lattice,

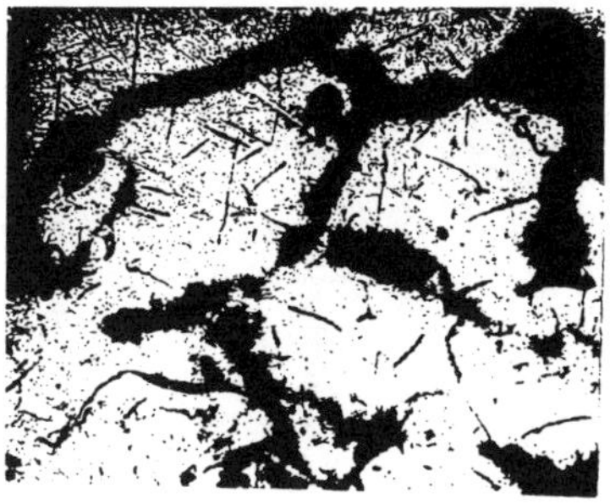

0,003 %

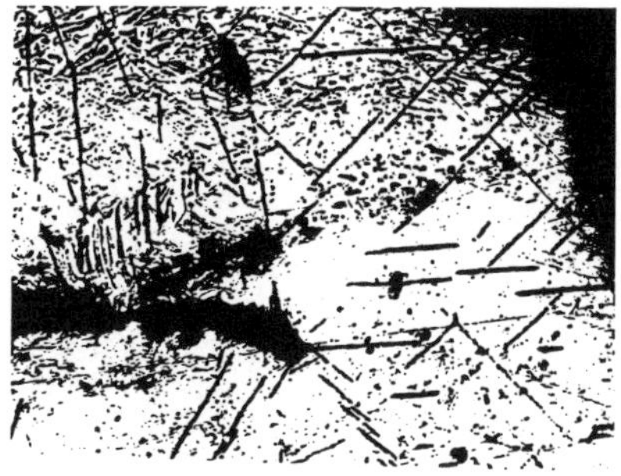

0,006 %

0,008 %

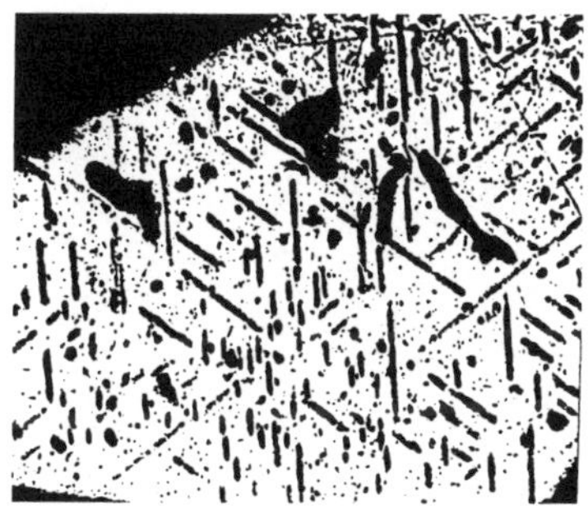

0,012 %

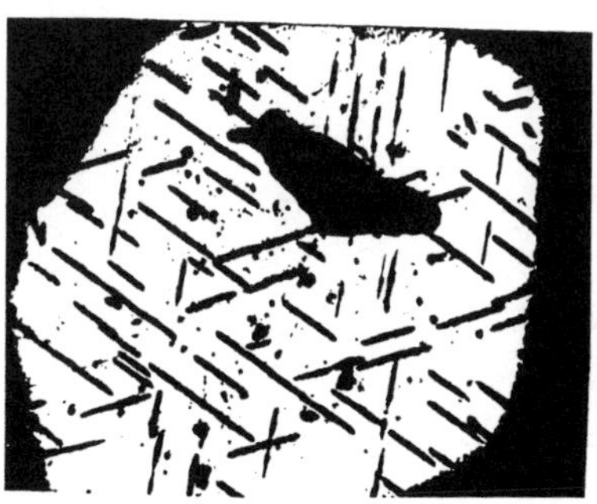

0,016 %

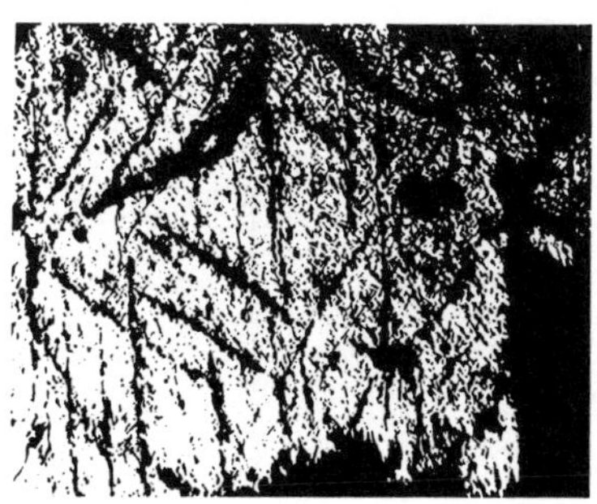

0,024 %

Figure3. Titanium hydrides in particles of purified titanium powders with increasing contents of hydrogen, x 300

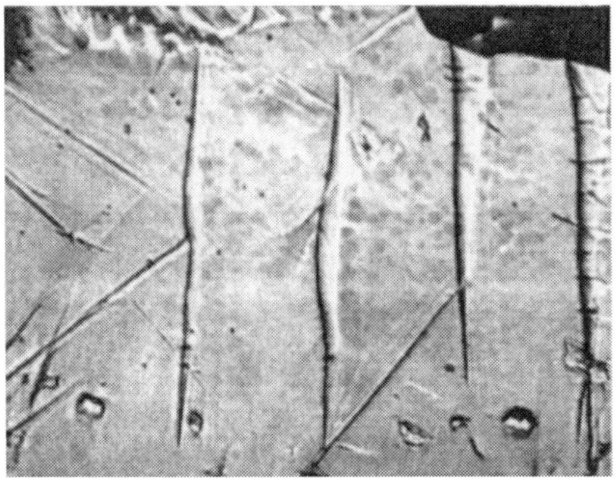

Fig 4. Convex crystallographic orientation titanium hydrides on the surface of particle of electrolytic powder, x 300

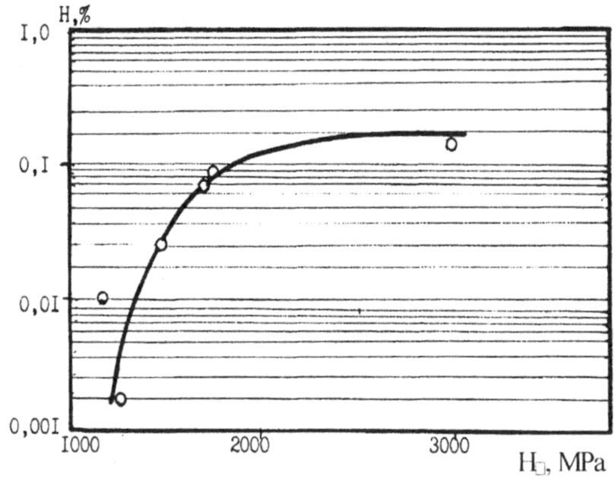

Fig.5. Changing of microhardness and contents of hydrogen in purified titanium powders

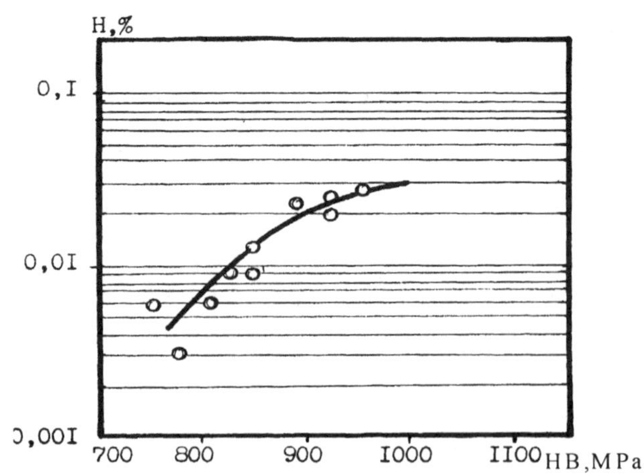

Figure 6. Strengthening of electrolytic purified powder by hydrogen

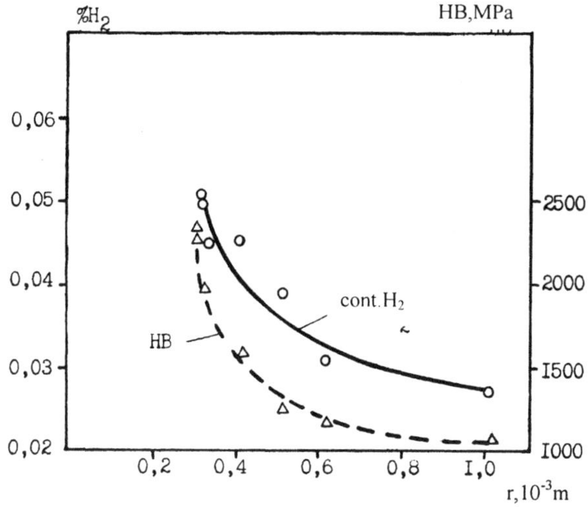

Figure 7. Changing of the contents of hydrogen and hardness of electrolytic titanium powders depending on the most probable size of particles

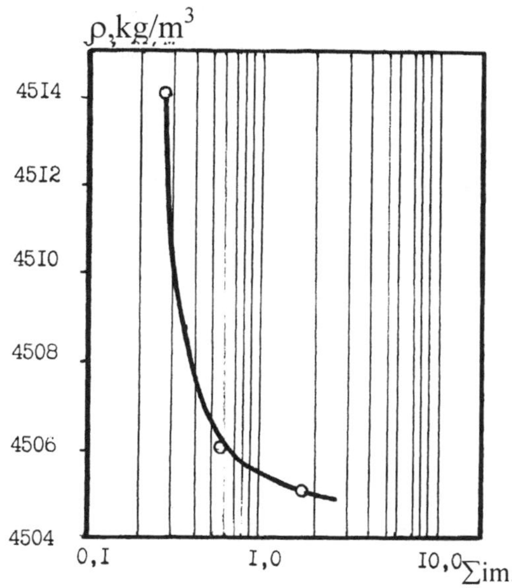

Figure 8. Dependence on density of titanium powder of variety method of it production from degree of pollution (Σimp.)

quantity of hydrides, impurity and porosity of particles of powders. To this, the influence of impurity on bulk weight was close: it was reduced the more, the higher were the contents of hydrogen in electrolytic purified titanium powder (figure 9).

The corrosion stability of powders depends on a degree of their pollution. By numerous tests in 5, 10,15, 20, 30 and 37-per sent solutions of HCl the proved [1] that the speed of corrosion grew at increase of a degree of pollution of powders.

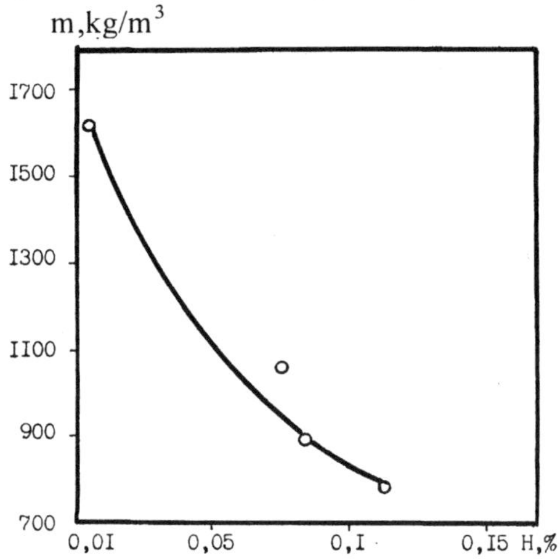

Figure 9. Change of microhardness and contents of hydrogen in purified titanium powders.

It has appeared that kinetics, mechanism of growth and morphology of hydride inclusions were in direct dependence on a nature of powders.

By results of the carried out researches it is possible to conclude:

1) Structure and the properties of titanium powders essentially depend on a way of their manufacture and degree of pollution by impurity;

2) Structure, kinetics, mechanism of growth and morphology of hydride inclusions are determined by a nature of titanium powders and can considerably differ in materials of different ways of manufacture.

3) At use of titanium powders for obtaining hydrides it is necessary to choose purest titanium powders, for example, received by electrolytic refinement.

References

1. Shapovalova, O.M., Babenko, E.P. Ivchenko, T.I. (2000) Working out of methods of control of hydrogen in titanium alloys, *IX-th Intern. Sciense-techn. Conf.*, Zaporozhye, p.p. 75-77.
2. Shapovalova, O.M., Babenko, E.P. (1985) Corrosion-resisting of titanium powders in hydrochloric acid solutions, *RJ Metallurgy,* 1363.
3. Shapovalova, O.M., Babenko, E.P. Ivchenko, T.I. (1999) Hydrogen in titanium and technologies, *Collection works NATO intern. conf. " Hydrogen description of materials and chemistry of hydrides metals "*, Yalta, p. 209.
4. Mackwillan, A.D., Mackwillan, M.K. (1958) *Titanium*, Metallurgy, M.
5. Saltykov, S.A. (1976) *Stereometric metallography,* Metallurgy, M.

ACTIVATION ANALYSIS OF THE STRUCTURAL CLUSTERING TRANSFORMATIONS IN METAL HYDRIDE - FORMING SYSTEMS

V.G. TKACHENKO, I.I. SHULJAK, A.M. STRUTINSKY*,
V.V. FRIEZEL, O.V. KRIVIK

International Center for Electronic Materials Science and Applied Problems of Aerospace Technology,
Krzhyzhanovsky Str 3, Kiev-142, 03680, Ukraine
**Taras Shevchenko National University,*
Academician Glushkov Prospect 6, Kiev-22, 03601, Ukraine
Faculty of Physics, M.V. Lomonosov Moscow State University, Vorobevy Gory, Moscow, 119992 Russia.
[2] *International Laboratory of High Magnetic Fields and Low Temperatures, 95 Gajowicka str, 53-421 Wroclaw, Poland.*
[3] *Faculty of Chemistry, M.V. Lomonosov Moscow State University, Vorobevy Gory, Moscow, 119992 Russia.*

Abstract

In search of the generalized approach the results of investigations of the structural transformations in decaying hydrogenated solid solutions on the base of HCP Mg, Mg-Ba, FCC Al, Al-Li are summarized and analyzed in terms of hydrogen-induced diffusion concept and of clustering a structure model with the estimates of the rate of relaxation and the activation parameters of diffusion at the temperature range of 293-900K by means of structure-sensitive methods of measurements. It is found that observed accelerated (shot-range) diffusion is realized in the interstitial and substitutial solid solutions with cluster-forming structure responsible for the high rate of decay and for the transience of the hydride nucleation at low (room) temperatures. In accordance with it mixed (excess hydrogen vacancy-solutes) diffusion attributed by participation of the mobile diffusive cluster is considered to be one of the most probable rate-controlling mechanism which causes the anomalous deviation of hydrogen diffusivity from classic (Arrhenius) behavior at low (room) temperatures.

1. Introduction

Metal (Me) hydride - forming systems attract the increasing attention owing to their importance in hydrogen and thermal storage devices applications. Neverthe-

N. Veziroglu et al. (eds.), Hydrogen Materials Science and Chemistry of Metal Hydrides, 77–95.

less, analysis of the world trends indicates the presence of problem for hydrogen absorption in light metal hydrides for its using as an automobile fuel. During the investigations of thermodynamic properties and hydrides formation kinetics in Me (Ti, V, Mg, Al)-H systems it has been established [1, 2] that slow kinetics of hydrogenation (for example, at 923 K and under pressure to 150 bar) and unstable conditions of hydride decomposition appear to be the main lacks of these alloys. Necessity of the further search is discussed in periodical literature in order to formulate the concept for new alloys with much more active kinetics of hydrogenation [3, 4]. These results are also of great interest for modern technologies, in particularly, for the development of the high-current UHF electronics on the base of high quantum efficiency materials, first of all, light metal systems Mg-Ba and Al-Li which operate at the technical vacuum [5] as well as for the use them in dynamic application where low moment of inertia, high resistance to deformation and good stiffness-to-weight are essential.

In the framework of statistical-thermodynamical approach [2, 6] it has been established that hydrogen theoretically may be the source of large quantity of the excess (node, structural, non-intrinsic) vacancies. To give a rigorous test of this concept the first direct experimental evidence are obtained of what excessive hydrogen vacancies concentration exceeds that value for the thermally equilibrium ones more then two orders (of the amounts) [4, 7]. The proposed physical model is consistent with these results. But up to now no rearrangement mechanism of atomic structure containing excessive concentrations of point defects, as well as no any regularity of the influence of these point defects on kinetics of decay and hydride nucleation. This situation should take account of the working-out a concept for active hydrogenation and for search of alloys with active kinetics of hydrogenation. Moreover, as is expected theoretically [2, 6] and has been shown experimentally [4, 6, 8] information about the early stages of solid solutions' decay in Me-H-bearing systems is needed for problem decision of the hydrogen storage in hydrides of light metals. At any case this is still a matter of the further investigation.

An interpretation of the decay kinetics at the early stages for interstitial solid solutions is also hampered by lack of experimental information, especially by the values of frequency factor and activation energy. Therefore analysis of regularities and peculiarities of decomposition of hydride-forming systems becomes the object of intensive investigations, have greatly advanced the knowledge of this area, but basic information on the physical nature of phenomena is still scarce. In connection with it knowledge of the rate and activation energy for the diffusion of hydrogen and H-like interstitial solutes such as carbon, oxygen and nitrogen are probably the most important.

This work summaries results of original investigations and computer calculations for some hydrogen alloys to analyze the effects of H and H-like solutes in metals with different Pieirls' barriers and to identify possible physical mechanisms which introduce the prevail contributes to properties of FCC and HCP metals using structure-sensitive technique of measurements.

Further measurements have been continued, mainly on non-ideal close - packed crystals, in particular H- free and H - electrochemically charged HCP Mg and Mg-Ba-alloys with hydrogen content control in the ingots by time - of-flight mass spectrometry. The main aim is to reveal the rate - controlling mechanism operative at the earliest stage of hydride formation using the available data of internal friction, electroresistivity, X-ray photoelectron -and Auger electron spectroscopy. The another aim is to estimate the influence of alloying element on the structural and phase transformations in hydrogenated solid solutions basing on the Young's modulus (E) and the internal friction (Q^{-1}) measurements in the wide temperature range.

2. Experimental

Alloying elements and interstitial solutes are added to the alloys by previously established procedure which has appeared in our foreign publications [5, 7]. All the decrement damping, elastic modulus, resistivity and microyielding measurements reported here are also made using the same procedure which has been described in detail earlier [2]. The part of the technique is described in the captions as well.

Hydrogenation accomplished by hydrogen charges of solid solutions is successively placed in the samples, at compositions up to saturation and super saturation. In this study, alloys of Mg-Ba and Mg-Ba-H systems have been investigated using the electrochemical introducing the hydrogen into the cast ingots produced by vacuum/argon induction melting. Besides, due to high absorbability of Ba the HCP Mg-Ba alloys can absorb a considerable amount of additional hydrogen (0.6...1.4%) even from argon environment during induction melting.

A complete set of temperature dependent internal friction, elastic modulus and resistivity data for hydrogen-free, hydrogen - and H-like doped metals with different Peierls' barriers as well as for a series of hydrogen-charged and hydrogen-free light alloys of Mg-Ba and Al-Li systems is obtained in the temperature range up to 900K.

Measurements of Q^{-1} and E were carried out in the KHz frequency range using the pulse echo technique (fi=2 kHz). Errors in Q^{-1} measurements were about of 4% (for background 5×10^{-3}) and 15% (for higher level background). Values of E(T) were measured to an accuracy of 0.2%. Activation energy (a.e.) values of the obtained maximums of $Q^{-1}(T)$ are determined by the familiar method of Wert-Marx and by the peak half-width measurement as well. Frequency factor τ_0 was calculated from equation for the relaxation time.

The observation of an internal friction peak as a function of temperature T_{rp} implies that the relaxation time τ is temperature sensitive according to the relation

$$\tau^{-1} = \tau_0^{-1} \exp(-E_H/RT_{rp}),$$

where τ_0^{-1} - frequency factor (equal to stress-induced jump atom frequency ν); E_H - activation energy for interstitial hydrogen and solutes diffusion.

From measurements of the peak at different frequencies the values of parameters T_{rp} are obtained for the frequency factor τ_0^{-1} and the a.e. E_H. Thus interstitial and substitutional mobilities during these relaxations are deduced from the temperature dependence of their respective relaxation times.

The effects of absorption of elastic energy in the BCC, FCC and HCP crystals after chemical and electrolytic hydrogenation of various intensity are also measured with use of semi-automatic relaxometer having sufficient precision to measure the internal friction (3-5%) and the modulus (up to 1%).

In the present paper we also summarize some measurements of the stress-induced jump atom frequencies $\nu=\tau_0^{-1}$ for substitutional and interstitial atoms in solid solutions using inelastic techniques which permit to study pair relaxation, the diffusion motion of these clusters and hydrogen-solute interaction. A series of such measurements are made for a variety of hydrogen concentrations at various temperatures in order to completely specify the' relaxation properties of solutions of this type and to deduce the nature of the observed interactions.

3. Results

3.1. STRUCTURAL CLUSTERING AND RELAXATION EFFECTS IN HYDROGENOUS SOLID SOLUTIONS

Collected data of diffusion parameters calculated from Arrhenius relation for hydride forming and non-hydride forming systems are presented in Tables 1,2 and 3. Relaxation peaks associated with pair defect - relaxation in solid solutions are identified on the base of the change in their heights (amplitudes) and temperature position at different alloy compositions and frequencies.Hydrogenation followed by heat treatment is accompanied by clustering a structure at the expense of the excess vacancies (V) generated in the stress field of mobile hydrogen. As a source of non-intrinsic point defects it stimulates formation of the Ba-V clusters at higher temperatures (Fig. 1), while at higher concentrations it destroys Ba-V clustering structure in saturated Me-H solid solutions giving rise to development of Ba-H clusterization

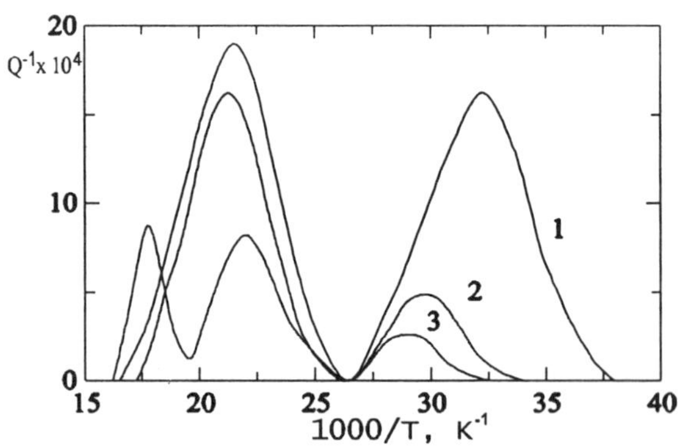

Figure 1. The discrete temperature spectrum of internal friction $Q^{-1}(T)$ for Mg–2.1%Ba–1.1%H alloy firstly revealing atom Ba-V clusters relaxation maximums of vacancy origin with a.e. of 0.8eV at 310K; grain-boundary relaxation maximum with a.e. of 1.1-1.3 eV at 495K and impurity (segregation) - induced relaxation maximum to be caused by Ba-H cluster - segregation at grain boundaries (1- quenching; 2- annealing at 673K for 1 hrs; 3- the same (slow cooing).

TABLE 1. Measured parameters of inelastic pair (cluster) relaxation in interstitial and substitutional solid solutions of hydrogeneous alloys.

Compositions of alloys	a.e. E_H, eV	τ_0^{-1}, sec^{-1} ($\approx\nu$)	T, K	Relaxation mechanism
		Hydride forming systems:		
FCC Ni-H	0.41	2.8×10^{13}	400-95	Quantum hydrogen diffusion [11]
	0.23	1.0×10^{6}	100-230	
BCC Nb-H	-	10^{10}	-	[3]
HCP Mg	1.02	2×10^{13}	471	Oxygen relaxation peak
HCP Mg-1%Ba	0.91	1.8×10^{13}	420	
HCP Mg-5%Ba	1.06	2.1×10^{13}	480	Pair (cluster) relaxation "exess V - atom of Ba" (f_2=2kHz)
HCP Mg-5%Ba-H (low concentrations of H)	0.87	5.8×10^{13}	473	Pair (cluster) relaxation "exess V - atom of Ba" cluster (f_2=2kHz)
HCP Mg-5%Ba-H (high concentrations of H)	0.64	6.8×10^{12}	321	Pair relaxation "atom of H - atom of Ba" cluster (f_2=2kHz)
HCP Ti-0.4...0.6%H	0.5		77-293	"excess V - atom of H" (f_1=1Hz)
		Non-hydride forming systems (f_2=2kHz):		
				Pair (cluster) relaxation:
FCC Al-2%Li	0.96	1.6×10^{13}	440	"excess V - atom of Li»
FCC Al-H	0.63	7.0×10^{12}	338	"excess V - atom of H"
FCC Al-2%Li-H	0.88	9.6×10^{12}	463	«excess V - atom of H"

TABLE 2. Values of activation energies for diffusion in HCP Be-0.1%Fe-0.1%C, H solid solution.

Measured effect	Controlling mechanism	a.e. E_H, eV
Isotopic method	Self diffusion	~3.5
Isothermals $Q_i^{-1}(\tau^*)$ and $G(\tau^*)$	Volume (long-range) diffusion of isolated carbon atoms	~0.6
Zener pair relaxation (Fe-Fe)	Shot-range substitutional diffusion of Fe	~2.1
Pair (cluster) relaxation	Shot-range rotation by cluster of Fe-C	~1.1
Pair (cluster) relaxation	Shot-range rotation by cluster of Fe-H	<0.6

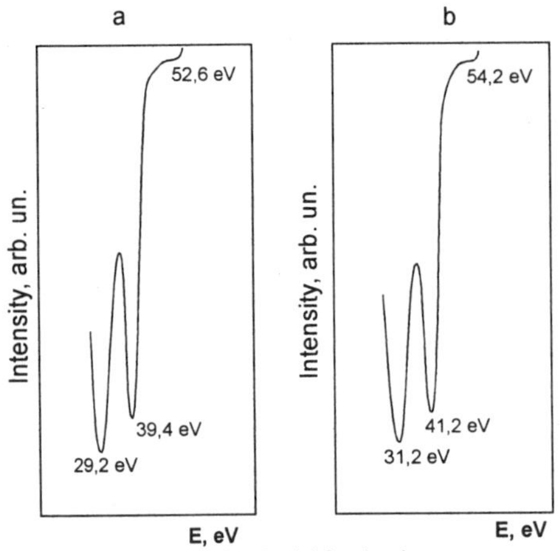

Figure 2. Chemical shifts in low – energy region of the Auger – electron spectrum (LVV – transition) for Magnesium (a) ang Mg-1%Ba-1.1%H alloy (b)

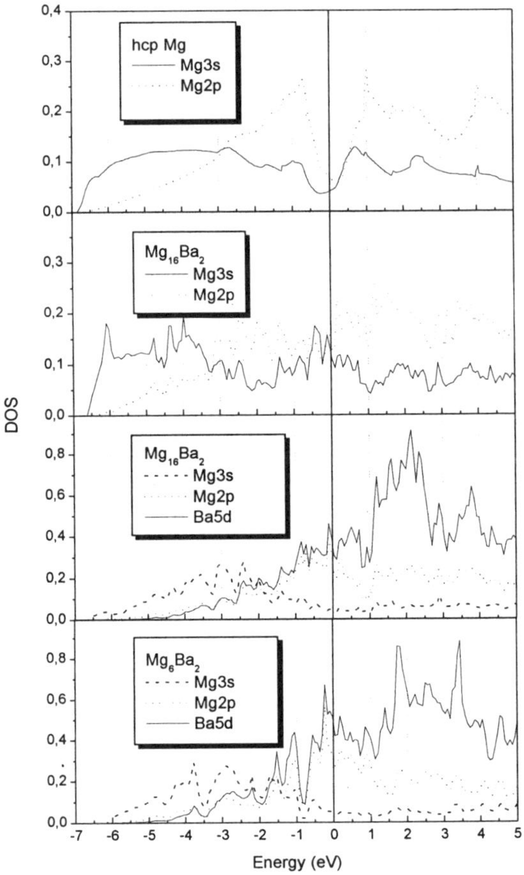

Figure 3. Partial density of electronic 3s-, 2p-states for magnesium atom and 5d- state for barium atom in HCP Mg.

at lower (room) temperatures due to the high absorbility of the barium to the hydrogen (Fig. 1, Table 1). Diffusion mobility of the cluster is enhanced by mixed diffusion mechanism (partially - on excess vacancies, partially - on interstitial sites) followed by long- range segregation of Ba-H clusters to grain boundaries (Fig. 1). Hydride - like segregations (Ba, H) and (Mg, Ba, H) types with nanostructure to be occurred at dislocations and grain boundaries differ essentially from well-known Cottrell atmospheres due to their chemical composition close to corresponding compound.

Clustering a structure in hexagonal crystals is confirmed by X-ray photoelectron and Auger electron spectroscopy revealing the chemical shifts [9] and s, p, d - hybridization of chemical bonds [10] in cluster-formed structure (Fig. 2 and 3). Contrary to the thermoemission approach X-ray as well as UV-photoelectron spectroscopy can be used as advanced short-impulse methods not only for the thorough physical-chemical analysis of the equilibrium phase diagrams but also for the detail investigations of pre-phase (GP-zone) transformations. The variations concern valence band transition in the low-energy regime below 150eV i.e. for Auger electrons from valence bond [9]. Auger transitions involving valence electrons of solid compound (crystalline HCP Mg and Mg-Ba-H alloy) reveal the energy (chemical) shifts to be caused by changing the binding energy shifts of each electron as well as the relaxation energy. Photoelectron spectra excited by X-ray (XPS) also yield valuable information about the new local environment change in Mg-Ba-H system due to changing the chemical bond as a result of microalloying [10]. Valence level participating in the Auger transition (Fig. 2) and assuming s, p, d – hybridization (Fig. 3) for a cluster-hydride-like segregation or GP-zone (Mg, Ba, H) is connected with the hydrogen bonds according to a self-consistent-field calculation [10].

Thus, the barium-bonded hydrogen effect is that the excess V- concentration generated by absorbed hydrogen provides for the high rate of clustering on the base of accelerated diffusion of Ba in substitutional subsystem and facilitates formation of the mobile diffusive cluster Ba-H operative at low (room) temperatures due to more strong interaction these solutes. Such a cluster relaxes by Finkelstain-Rozin mechanism and distinguishes the bonding energy higher than that for Mg-H and Ba-V (0.2eV) [2, 4]. Moreover, it becomes integral component of the structure capable to relaxation and responsible for high diffusive mobility of Ba at low (room) temperatures with e.a. close to e.a. for hydrogen migration in the HCP lattice of Mg (0.6eV).

Observed behavior of Mg-Ba-H alloys resembles the behavior of Be -Fe-C, H alloys with HCP lattice [2] revealing Finkelstain-Rozin (Fe-C, H) relaxation and Zener (Fe-Fe) pair relaxation respectively (Table 1 and Table 2).

According to activation analysis (Table 1, 3) hydrogen absorption rate in Mg-Ba alloys at low temperatures is much higher than that of Mg at low temperatures due to additional accelerated Ba-H-cluster segregation at the structural defects (dislocations and grain boundaries) by mixed mechanism (partially - on vacancies, partially - on interstitial sites) for cluster compound (Fig. 3). Creation of easily - mobile clusters of Ba-H reduces critical temperature of initiation of the pre-phase (GP - zones) Mg, Ba, H - precipitates on grain boundaries up to room temperatures.

The most typical results of the $Q^{-1}(T)$ and E(T) measurements for the Mg-Ba and Al-Li systems are presented in Fig. 4 and Fig. 5. It can be safely concluded that the presence of the frequency and temperature dependences, as well as of modulus relaxation are indicatives of their same inelastic origin (Fig. 4, 5 and Table 1). This will be taken as sufficient identification of the relaxation origin of these peaks.

According to Table 1 there is no apparent difference in the values of τ_0^{-1} and E at change of the concentration of Ba up to 5% pointing out the same inelastic relaxation mechanism, very likely, to be attributed to stress-induced reorientation of cluster "vacancy - atom of Ba" [4]. Relaxation nature of the $Q^{-1}(T)$ maximums is also confirmed by the occurrence of the folds associated with the modulus relaxation to be observed on the E(T) curves (Figs.4 and 5).

It is significant that at high frequencies ($f_2 \approx 2$ kHz) sharp intensification of Q^{-1} background becomes noticeable at essentially lower temperatures in comparison with low- frequency internal friction data ($f_1 \approx 1$Hz). By reason of this it is impossible to obtain the meaning results as to grain boundary and segregation relaxation in Mg-Ba alloys. These data are unobtainable because of the background high level for all investigated states of substitutional solid solutions, while the discrete temperature spectrum of Q^{-1} for interstitial solid solutions is registered fully. These data for Al-Li alloys are plotted in Fig. 5 and presented in Table 1.

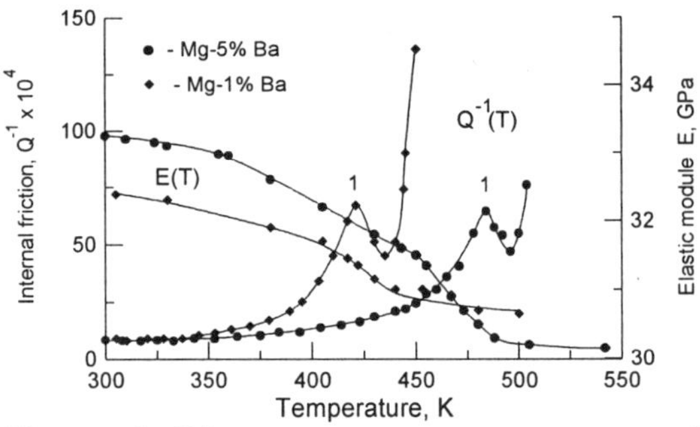

Figure 4. Discrete temperature spectrum of internal friction $Q^{-1}(T)$ with measured damping component (above background) and elastic modulus changes E(T) for Mg-Ba-1.1%H alloys (f=2kHz): 1 - cluster (Ba, excess vacancy) - induced pair relaxation peak.

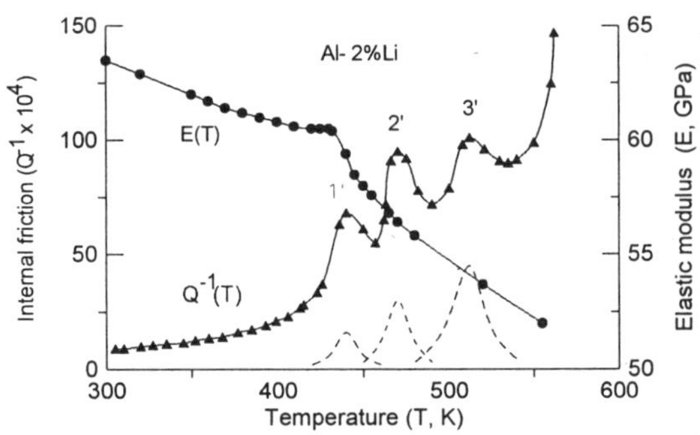

Figure 5. Discrete temperature spectrum of internal friction $Q^{-1}(T)$ with measured damping component (above background) and elastic modulus changes E(T) for Al-2%Li alloys (f=2kHz):1' - cluster (Li, excess vacancy) - induced pair relaxation peak; 2' – grain boundary relaxation peak; 3' - Li segregation-induced impurity relaxation peak.

Obtained values of τ_0^{-1} and E_H appear to be lower than that for the volume diffusion of Li in FCC AL and of Ba in HCP Mg [2] testifying to the acceleration of diffusion in cluster-forming structures of this type. Formation of mobile diffusive clusters and alloying element segregations at structural defects (grain boundaries) should classify as the important peculiarity of these alloys which are of practical interest for recovery of photoemissive properties of surfaces (single crystals). Hence clustering a structure at the most early stages of decay, by its very nature, is common characteristics for the interstitial and substitutional solid solutions to be independent on the lattice position of alloying element and crystal lattice type. Data for Mg-Ba alloys containing up to 5%Ba are plotted in Fig. 4 and Fig. 6 and also well presented in Table 1. The temperature activated damping peak (labeled peak 1) is located on $Q^{-1}(T)$ curves at approximately 420K and 480K representing the original data, corrected for background damping. Its temperature position depends on the Ba concentration in substitutional solid solutions. Similar results are obtained for Al-Li interstitial solid solution (Fig. 5) In accordance with the calculation results (Table 1) these peaks may be caused by relaxation of vacancy-Ba atom pairs in the HCP lattice of Mg and by relaxation of vacancy-Li atom pairs in the FCC lattice of Al.

3.2. ANELASTIC EFFECTS IN HYDROGEN - CHARGED MAGNESIUM ALLOYS WITH CLUSTER-FORMING STRUCTURE

The absorbed and barium-bonded hydrogen accelerates diffusion mobility of Ba at lower temperatures by mixed diffusion mechanism, with the participation of the excess hydrogen vacancies.

Isolated atom or a node monovacancy in the anisotrpic HCP lattice could not be a relaxation center, because it saves crystal's symmetry. In saturated HCP solid solution, the formation of vacancies occurs in the stress field of solved mobile atoms, and the initial stages of decay demonstrate the enhanced willing to clusterize finishing with the formation low-energy configuration and mobile diffusion clusters «vacancy - solutes».

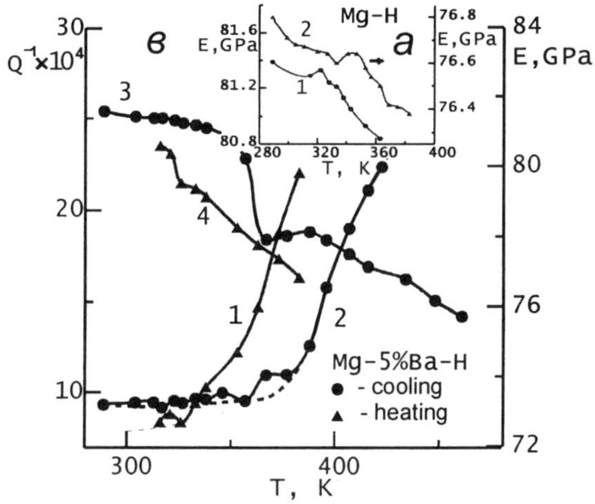

Figure 6. Temperature dependencies of the elastic modulus for the hydrogen-doped Mg crystals: 1-for concentrated solid solution; 2- for the saturated solid solution (a). Temperature dependencies of the internal friction (1, 2) and of the elastic modulus (3, 4) for Mg-5%Ba≥1.4%H alloy after the intensive hydrogenation and saturation of the solid solution (b).

So, according to $Q^{-1}(T)$ measurement the observed maximum (Fig. 6) is originated from pair relaxation caused by stress-induced reorientation of the «vacancy - atom Ba» clusters which form anisotropic fields of internal deformation with simmetry different from that of the HCP Mg crystal.

Indeed, in saturated Mg-H solid solution of the Mg-5%Ba≥1.4%H alloy (Fig. 6) cluster (Ba, H) – induced pair relaxation peak is observed at about 300K. The folds and extremum points are also detected on cooling curves of E(T) at about 300 and 400K indicating the origin of the phase transformations both in bulk Mg matrix (MgH_x) and grain boundaries ($MgBaH_x$) during measurements. Diffusive mobility of Ba-H clusters at low (room) temperatures is appear to be much higher then that for Ba-V clusters at higher temperatures. Clustering a structure in Mg-Ba-H alloys is accompanied by segregation of (Ba, H)-clusters at the grain boundaries (Table 3 and Fig. 6).The calculation data for the creep behavior at constant grain size show two distinctly different stress dependence's for Mg-7%Ba≤1.4%H alloy (Fig. 7a). The stress exponent n increases from 0.8 to 3.25 at higher stresses. A transition (threshold stress) from diffusional creep regime to power law creep indicates the activation of grain boundary strengthening mechanism corresponding to Cobble creep typical for grain boundary diffusion (Fig. 7a). According to an Arrenius dependence where the logarithm of the steady state strain rate is plotted against the reciprocal of temperature (Fig. 7.b) the Mg-7%Ba≤1.4%H alloy exhibits a pronounced diffusion creep regime with a.e. for creep deformation ~0.48eV. The a.e. is much lower then that for bulk (lattice) self-diffusion in HCP Mg (1.46-1.38eV ≈136.5 kJ/mole) [2]. At the same time the a.e. increases to a value exceeding that of grain boundary self-diffusion in HCP Mg (~0.3eV) due to presence of thermally stable pair defects «atom Ba – V» → «atom Ba – atom H» clusters.

The barium makes more active the kinetics of hydrogenation and thus increases additionally the hydrogen storage capacity of HCP Mg at the expense of the grain boundary Ba, H-segregation (Figs. 6, 7, Table 3). As a result, being possessed to high absorbability to hydrogen, barium increases the hydrogen absorption rate, and therefore the hydrogen absorption capacity in Mg-Ba alloys with cluster-formed structure. Nevertheless, it is difficult to avoid the surface oxidation which will cause a considerable loss of hydrogen storage capacity in the alloy powder metallurgy route.

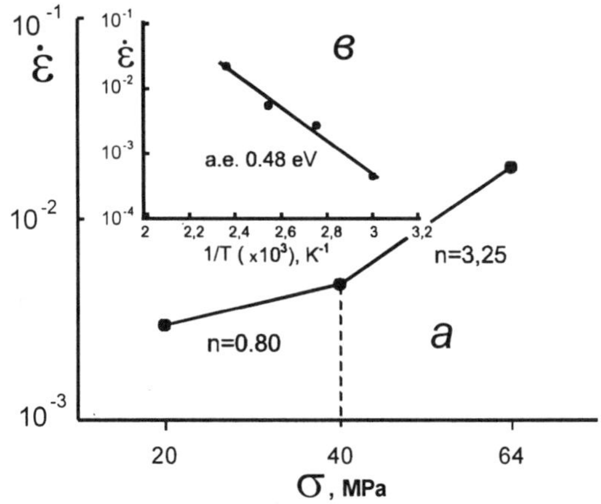

Figure 7. Typical plot of the stress σ vs strain rate (h^{-1}) for Mg–7%Ba–0.6…1.4%H alloy on a log/log scale at constant creep temperature of 150°C (a) with the estimate of a.e. for creep microdeformation at constant creep strength of 64 MPa (b).

It is interesting to compare the variations in modulus changes in studied temperature range (Fig. 6). It is shown, in particularly, that temperature interval 423-433K divides the curves in two regions corresponding to the absence (below 423K) and to presence (above 433K) of the modulus relaxation. It means that low temperature range characterizes the phase transformations (changes) and that at higher temperatures one indicates the effects of relaxation origin.

The following sequence of non-relaxation transformations is observed for concentrated and saturated Mg-H solid solutions:

- dislocational strengthening of HCP Mg lattice caused by formation of domains, coherently bonded with the host metal and associated with the appearance of abnormal curves fitting of dependence E(T) on temperature, at heating and at cooling;

- phase transformation with hydride formation at 320K and at 340K (cooling) by analogy in BCC Nb-H [12, 13].

- sharp decreasing of E at higher temperatures because of coherent interface destruction and formation of non-coherent hydride (Fig. 6).

It is significant that these results coincide fully with the data obtained for Ti-H saturated solid solutions [4]. However, in the present case the cooling and corresponding heating curves are not similar. The observed thermal hysteresis is of course a very common phenomenon (Fig. 6) associated with phase transformations in metal-hydrogen systems during measurement [14]. At the same time in this unstable conditions it is practically impossible to obtain damping data for hydrogen HCP Mg, because of intensive hydride formation and dislocational strengthening. In connection with it only elastic modulus anomalies on the temperature curves are shown in Figs.4 and 6. Repeated measurements of E(T) reveal the further growth of the values. Temperature position and height of E-anomalies («maximums») responsible for dislocation strengthening of Mg matrix by hydrides are a function of the hydrogen concentration.

Saturated hydrogenation of Mg-5%Ba produces a series of additional internal friction peaks which form a complex pattern of the discrete damping spectra. In the present case it appears that data can be resolved by comparison of the measurable spectra for Mg-5%Ba, Mg-H and Mg-5%Ba-H at constant concentration of hydrogen (Figs.4, 6, 8 and Table1).

It is also seen (Fig. 8) that essential addition of H to Mg practically suppresses the amplitude of 483K-peak which is observed in Mg-5%Ba and is associated with solute cluster "excess vacancy - atom Ba". Whether this is a true effect of hydrogen is not know with certainty at this time, but the reduction may be caused the solubility limit change in substitutional system Mg-Ba and formation of hydride (Ba, H). Using inelastic technique it is impossible to establish even approximate composition of the hydride. It is also found that the amplitude of the oxygen peak attributed by pair relaxation of O-O atoms retains its meaning with increasing hydrogen concentration.

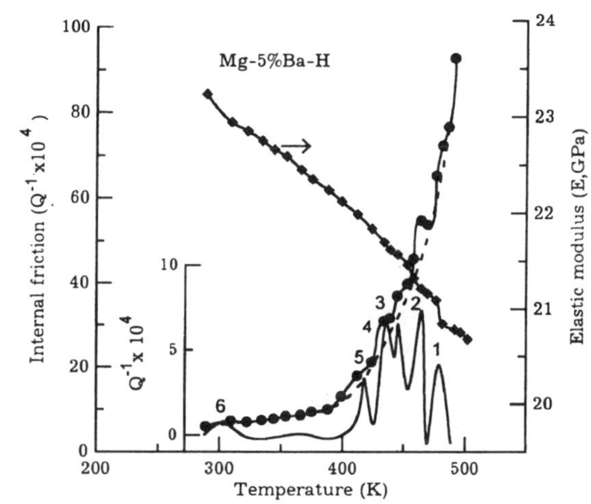

Figure 8. Temperature dependencies of the internal friction and of the elastic modulus for Mg-5%Ba hydrogen doped alloy (concentrated solid solution). Low temperature scales - discrete temperature spectrum of $Q^{-1}(T)$ after substracting from the background. 1 - cluster (Ba,V)-induced pair relaxation peak; 2 - oxygen relaxation peak stimulated by forming O-O pair defects; 3, 4- splitting relaxation peak caused by pair defects of H-O and of H-V; 5 - in all probability Snoek-Koster effect associated with segregations of H-Ba at dislocations; 6- cluster (Ba,H)-induced pair relaxation peak.

It should be emphasized unusual ageing behavior of the hydrogen-charged Mg crystals during low temperature measurements. The hysteretic effects are introduced into the decrement and modulus data by low temperature ageing. The hysteretic phenomena are observed to occur at the various hydrogen concentrations during heating and cooling. Observed effects are not similar to the behavior of H atoms (up to 1 at. %) in dilute solid solution of Nb single monocrystals [13] where the hydrogenation causes activation of two relaxation mechanisms associated with Nb hydride formation. The nature of these inelastic relaxation has not yet been established.

4. Discussion

4.1. HYDROGEN-INDUCED DIFFUSION

It has been point out [6] that during diffusion the mobile hydrogen atoms induce the formation of excess node (non-intrinsic, structural) vacancies and in such a way accelerate the self-diffusion in the local stress field. Hydrogen-induced diffusion development depends on the parameters of interatomic interactions, the character of hydrogen spatial distribution in concrete systems and accompanies by the formation of pair defects, so named clusters "excess vacancy - atom H".

This physical interpretation is supported by theoretical estimates of the binding energies which are in a good agreement with the direct experimental data obtained by electron-positron annihilation technique and other precise structure-sensitive methods for hydride-forming and hydride-non-forming alloys systems [4, 7]. Statistical-thermodynamical analysis of dissolved hydrogen behavior carried out in the framework of the interstitial solutes' deformation interaction theory [6] predicts the effect of essential increasing of self-diffusivity for FCC Fe ions due to the introduction and the rise of excessive concentration of structural vacancies generated by absorbed hydrogen. The regularities of hydrogen-induced diffusion for FCC Fe-H (austenite-H) system are of common character inherent the other concrete hydrogen-bearing systems which differ mainly with values of free energies for alloys and of Furier-energy components for the effective interaction Me-Me and H-H. It is also seen, that in the enough wide range of values the concentration of excess vacancies (C_H) exceeds at the least two orders of that for thermally equilibrium ones typical for melting temperature of pure metals ($10^{-3}...10^{-4}$). Theoretical investigations of Me-H solid solution are carried out on only model objects FCC Fe-H because of often insuperable methodological and technical difficulties in receiving austenite states for pure Fe matrix. But at this time data obtained for analysis of the deformation ageing of austenite alloys Fe-Ni-C [15] one should consider as the indirect confirmation of the validity of our approach (approximation) [2, 6].

Relaxation methods can be used to investigate the presence and concentration of pair defects, because clusters of atoms have crystal symmetries lower than that of the host crystals and cause mechanical relaxation in solid solutions [16]. One type of cluster which has not so long been identified is an association of excess vacancy with hydrogen dissolved interstitially in solvent metals [4, 6]. In the previous investigations no detailed information of any of these clusters has been obtained. A new physical model [4] is used for a tentative identification of the relaxation peaks associated with stress-induced reorientation of the clusters.

4.2. DECAYING ALLOYS OF Mg-Ba-H SYSTEM WITH CLUSTER-FORMING STRUCTURE

Being possessed of the high absorbility to the hydrogen dissolved Ba suppresses the high temperature relaxation peak 1 due to cluster relaxation of "vacancy - Ba atom" (Fig. 4 and Fig. 8) and stimulates the activation of relaxation mechanism at 308 K that is confirmed by the modulus relaxation in this area and by the appearance of small thermally activated $Q^{-1}(T)$ peaks which are detected at much more low (room) temperatures. (Fig. 8 and Table 1). Moreover, alloying addition eliminates the nucleation stage of hydrides at 321K and 300K (Fig. 6 and Fig.8, peak 6). Hence these effects are accompanied by the destruction of the cluster-forming structure at higher temperatures and initiation development of the corresponding cluster-forming structure at low (room) temperatures. Indeed, after hydrogenation 479K-peak 1 caused by the "vacancy - Ba" cluster relaxation significantly decreases at low hydrogen concentrations and then vanishes at higher ones. At the same time new small peak 6 occurring in the temperature range 300-330K accommodating to the region of low temperature phase transformations (Fig. 8) is cluster relaxation origin. Its temperature position corresponds to an a.e. of 0.6eV. Thus, the hydrogen effect is that the excess concentration of vacancies generated by absorbed hydrogen provides for the high rate of clustering on the base of accelerated diffusion of Ba in substitutional sub-system and facilitates formation of mobile cluster Ba-H on the base of strong interaction between these solutes. Such a cluster relaxes by Finkelstain-Rozin mechanism and distinguishes bonding energy (0.6eV) higher than that for Ba-V (0.2eV [4]), H-H, H-V and Me-H. Indeed, peak associated with pair relaxation of H-O and of H-V are only detected at low concentration of hydrogen (Fig. 8). Atoms of Ba locate at the excess vacancies generated by dissolved hydrogen inhibiting the stress-induced H-H ordering (at octahedral and tetrahedral interstices) to be need for hydride initiation, especially at structural defects (dislocations, grain boundaries). Thus, the strong cluster Ba-H becomes integral component of the structure capable to relaxation and responsible for high diffusive mobility of Ba at low (room) temperatures to occur by the Ba-H pair rotation with a.e. equal to a.e. for hydrogen migration in the HCP lattice of Mg (0.6eV).
In this case diffusive mobility Ba-H cluster at low (room) temperature appear to be much higher than of that for Ba-V cluster at high temperatures (Table 1).

After hydrogenation of Mg-Ba alloy the temperature position of 479K-peak induced by cluster relaxation of Ba-vacancy is not exchanged retaining values of τ_0^{-1} be constant (Fig. 8, Table 1). However, amplitude of this peak decreases sharply with increasing the hydrogen concentration. This leads one to conclusion that hydrogen eliminates cluster-forming structure of Ba-V substitutional solid solution at higher temperatures and gives rise to creation in Mg-Ba-H alloys of Ba-H cluster-forming structure more effective for the diffusion acceleration at low (room) temperatures due to the presence of excess vacancies generated by dissolved hydrogen. High diffusive mobility of Ba-H cluster is realized by Finkelstain-Rozin mechanism which is associated with rapid diffusion and location of interstitial atoms (H) around substitutional alloying elements (Ba). Observed behavior of Mg-Ba-H alloys resemble the behavior of Be-Fe-C alloys causing Finkelstain-Rozin (Fe-C) pair relaxation and Zeener (Fe-Fe) pair relaxation respectively [2].

The peaks occurring at 430K and 440K (Fig. 8, splitting peak 3) are as yet not completely understood due to its absence in other experiments. The peak 5 could possible be to the Snoek-Koster type, but this remains for further work to decide.

4.3. SHORT-RANGE AND LONG-RANGE DIFFUSION IN HCP SOLID SOLUTIONS

Following the familiar Snoek theory the isolated interstitial solutes rearrange (on interatomic distances) themselves with high rapidity in ordered regions around dislocation by Snoek ordering mechanism of locking [17]. Next they segregates on long-range distances and form impurity atmospheres around steady-state or moving dislocations by Cottrell mechanism of locking. From analysis of kinetic curves (isotherms) obtained by means of the dislocation damping and the recovery of amplitude-independent internal friction Q^{-1} measurements it does follow that time dependencies of the increase in $Q_i^{-1}(\tau) = 1/(1+\beta\tau_1^{2/3})$ in the theoretical Cottrell-Bilby coordinates obey a $\tau_1^{2/3}$ law (indicative of Cottrell dislocation locking) valid for all investigated alloys being to be independent upon the nature of concrete metals [2, 4, 7]. Hence the first stage of decay for the interstitial solid solutions could be controlled by the volume (long-range) diffusion of dissolved atoms to dislocations in order to form the impurity atmospheres by Cottrell-Bilby mechanism. As has been shown this process is completed by phase transformations at dislocations for saturated solid solutions with BCC and HCP lattice [2]. However, taking account of the different rates of ageing for these metals at low temperatures it is unlikely that Cottrell mechanism should be considered as controlling mechanism responsible for the decay kinetics at the most early stages. High rapidity of the process would rather attributed by Snoek dislocation-solute locking which depends on the rate of production of vacancies [18].

Moreover, Cottrell-Bilby mechanism is not able to discover the rapid nature of deformation ageing using concept of isolated atoms. It is very difficult to test these notions experimentally because of the low lattice symmetry, particularly, in HCP crystals where isolated interstitial atom can not be relaxation center. Meanwhile at the present case Snoek mechanism of stress-induced ordering only manifests as the phenomenon of relaxation by pair defects which raise the lattice symmetry. Indeed, in Ti, Mg and Be-rich solid solutions are found to be pair (cluster) relaxation effects including Snoek damping caused by ordering of interstitial atoms in HCP lattice to be distorted by substitutional elements, as well as Zeener pair relaxation, Snoek-Koster type relaxation and pair relaxation "substitutional atom - interstitional atom" by Finkelstain-Rozin mechanism [2, 6]. Simultaneous appearance and disappearance of pair (cluster) relaxations "substitutional atom - substitutional atom" and "interstitional atom - substitutional atom" in this systems indicates interaction of the decaying substitutional and interstitional solid solutions. For example, Snoek solute peak of Be-C and Finkelstain-Rozin peak of Fe-C are simultaneously detected in HCP Be lattice distorted by Fe atoms [2]. It should be important emphasized that formed cluster Fe-C possesses higher diffusive mobility and lower a.e. which is approximately three times less than that for the self diffusion and two times less than that for the substitutional (vacancy) diffusion of Fe atoms (Table 2). Hence these data available appear consistent with a accelerated (stimulated) diffusion concept and with a mechanism of stress-induced ordering within the local (short-range) zone about relaxation center.

High rate of early stage of the ageing for HCP and BCC alloys suggests that the dislocations are being locked by a Snoek ordering mechanism rather than either Cottrell mechanism. Following to our publications [6, 17] the hydrogen lowers the height of a Snoek carbon peak and Snoek-Koster peak in BCC Fe of various purity due to hydrogen hardening which is accompanied by the occurrence of some density of the excessive dislocations. The latter are the traps for a part of the hydrogen atoms excluding them from the relaxation process and redistributing them between BCC lattice and generated dislocations. The obtained data confirm the most probable model of the β-relaxation as is offered associated with thermally activated unpinning of screw component of dislocations which are hold up by clusters "interstitial atom - vacancy". Indeed, intensity and kinetics of hydrogenation essentially influence on the width of a β-relaxation maximum due to the change of relaxation times' distribution function.

Taking account of that point defects (excess vacancy or interstitials) are generated during straining by the non-conservative motion of jogs on mobile screw dislocations they are to be dependent upon the nature of the jog. Following to modified versions of Cottrell model for the Portevin-Le Chatelier effect [2, 18] repeat (serrated) yielding is caused by short-range diffusion of clusters "excess vacancy - solute" which provide for the dislocation locking with the participation of the strain-produced vacancies. Indeed, the disappearance (annealing) of excess vacancies at appropriate sinks and vacancy annihilation give rise to deviations of experimental data from Cottrell model [18]. Obtained results give some evidence of excess vacancy stimulated diffusion concept.

4.4. HYDROGEN-STIMULATED DIFFUSION MECHANISM IN HCP SOLID SOLUTIONS

The ageing reactions, decay stages of solid solutions, hydride formation sequences and hydride precipitation defy synonymous interpretation because of unprecise and unsettled notions about diffusion mechanism. Since then, the extremely high diffusive mobility of interstitial hydrogen in BCC, FCC and HCP solid solutions has been the subject of extensive investigations striking by variety of finding due to hydrogen effects as well as by existence of the discrepancy in estimates and of the uncertainty in the diffusion activation energy values. Indeed, hydrogen trapping effects appear to cause an apparent increase in the measured a.e. for H diffusion, while a decrease in that is connected with quantum tunneling effects [3, 19]. In connection with it their treatment [3, 19, 20] illustrates the theoretical possibility of a deviation from classical (Arrhenius) temperature dependence for the H diffusive mobility expressed in terms of the relaxation time

$$\tau = \tau_0 \exp (E_H / kT_{rp}) \qquad (1)$$

and the diffusivity

$$D = D_0 \exp (- E_H/RT). \qquad (2)$$

Hydrogen diffusion anomaly induced by quantum diffusion of H [11] is really discovered in FCC Ni below 300 K. Hydrogen diffusivity in solid solutions BCC Nb -H as measured at T<250 K by the Gorsky effect technique also exhibits a deviation from a simple Arrhenius temperature dependence, i.e. apparent a.e. for H diffusion decreases as the measurement temperature decreases [3, 21]. It means that H diffusivity are larger than expected by extrapolation from measurements made at higher temperature. Analysis of the available data indicates however, that in the temperature range 300 - 20 K the hydrogen relaxations in high purity BCC Nb and Nb-O (N) alloys are shown to obey a classical Arrhenius temperature dependence at high temperatures and to exhibit enhanced relaxation rates at low temperatures [3]. It is also well established that in Nb this intermediate temperature region of non -classical behavior of H is much lower than the temperatures at which deviations from the Arrhenius correlation are noted in the Gorsky effect measurement. This effect is not explicable on the basis of solute trapping of the isolated H atoms would be to cause an apparent increase in a.e. [3]. Moreover, analysis of the jerky flow (Portevin - Le Schatelier effect) in solid solutions (Fe-Ni-H,C and so on) indicates enough high rate of the low temperature ageing at 293 K. But up to now neither theory, for example, Cottrell (elastic correlation) or Suzuki (chemical correlation) can explain high rate of impurity locking at room temperatures using Maxwell-Boltzman equation and classical Arrhenius law. These effects are also not explicable on the basis of the long-range diffusion of isolated atoms (hydrogen) after Gorsky mechanism. Calculation of the τ to be needed for isolated interstitial atom (carbon) in austenite to perform (occur) single jump at 293 K may be carried out using Maxwell-Boltzman relation [15]. According to this τ constitutes 10^{10} sec (i.e. 300 years) for single atom of carbon in austenite (FCC Fe). This value is less some orders of that for single atom of hydrogen, but relaxation rate doesn't become enough high. Consequently in these cases one should to attract different (other) controlling mechanisms for explaining accelerated diffusion of hydrogen.

According to Table 1 in Mg-Ba-H alloys the stress-induced atom jump frequency ($\nu = \tau_0^{-1}$) is increased up to 5.8×10^{13} sec^{-1} at low hydrogen concentration and then is decreased up to 6.8×10^{12} sec^{-1} at high hydrogen concentration. At the same time at low (room) temperatures the calculated values of τ_0^{-1} appear to be much more near to Debye frequency ($\nu_0 \sim 10^{13}$ sec^{-1}) than of that typical for quantum hydrogen diffusion (Table 1). Al-H and Al-Li-H alloys reveal the similar result. Atoms of H reduce essentially the values of activation energy for the short-range diffusion a low (room) temperatures (approximately 0.6eV at 338K) independently from the lattice type and from the level of Pieirls' barrier (FCC Al or HCP Mg). The observed effect may be explained by development of the mixed (interstice-vacancy) diffusion mechanism [7] or its (hydrogen-substitutional atom) modification responsible for deviation from classical (Arrhenius) behavior typical at low temperatures for some H-like light interstitial, even He in FCC Cu [22].

The development of the mixed diffusion mechanism in HCP solid solutions are actually accompanied by important deviation of $\ln D_{ett}=f(l/T)$ dependence from Arrhenius linear relation in the temperature range. So, effective diffusivity D_{ff} for solutes which migrate through interstices partially and through vacancy sites partially in equilibrium condition can be determined by expression:

$$D_{eff} = n_i D_i + n_v D_v \tag{3}$$

where D_i and D_v are self diffusivities of isolated solutes through interstices and through vacancy sites respectively; n_i, n_v - are their volume fractions. In this case $\ln D_i = f(1/T)$ and $\ln D_v = f(1/T)$ dependencies obey the linear low.

5. Conclusions

1. Observed components of Q^{-1} temperature spectra demonstrate that transient contributions to the internal friction are deduced to be vacancy origin. It is suggested that these contributions occur by a stress-induced rearrangement (rotation) of mobile diffusive clusters, in particularly, substitutional atom of Ba - interstitial atom of H. Clustering develops at a rate greatly accelerated by the presence of the excess hydrogen vacancies even at a low (room) temperatures. In this case a relaxation rate τ^{-1} is governed by an activation energy and frequency factor τ_0^{-1} of 10^{12}–10^{13} sec^{-1}. Thus vacancy lattice sites and clusters become integral components of the structure of the solid solution (interstitial and substitutional types). The observed sequence of the relaxation and phase peaks indicates that relaxation centers are responsible for both decay kinetics and first stage of hydride initiation. Hence transience of inelastic relaxation may be explained by hydrogen high diffusive mobility of its clusters causing the anomalous deviation of diffusivity from classic behavior by Arrhenius mechanism.

2. The hydrogen effect is that the excess concentration of non-intrintic vacancies generated by absorbed hydrogen provides the high rate of clustering on the base of accelerated diffusion of Ba in substitutions subsystem and in this way facilitates formation of mobile cluster Ba-H on the base of strong interaction between these solutes. Ba-H nanocluster becomes the integral component of the structure responsible for high diffusive mobility of the Barium of lower temperatures.

3. Accelerated (short-range) diffusion is observed in interstitial and substitu-

tional solid solutions with cluster-forming structure. It becomes much more effective in hydrogenated solid solutions to occur at rate greatly accelerated by the presence excess hydrogen vacancies.

4. Short-range diffusion accelerated by activation of hydrogen vacancy-solutes clusters is one of the most probable rate-controlling mechanism of the decay of the hydrogenated solid solutions with cluster-forming structure responsible for the anomalous deviation of hydrogen diffusivity from classic (Arrhenius) behavior at lower (room) temperatures. The high rate of the short-range diffusion is provided for by the mixed (hydrogen vacancy-solutes) mechanism of migration. The mobile diffusive clusters and hydride-like segregations are responsible for transience of chemical reactions and rapidity of ageing at low (room) temperatures. The concept of hydrogen-induced diffusion and the model of cluster-forming structure are formulated for describing of rapid decomposition of hydrogenated solid solution at the earliest stage of hydride nucleation i.e. hydride-like segregation and pre-phase (GP-zones) precipitates.

It this time hydrogenous effects classification is being composed in account of the clustering a structure for choice of the light and precise hydrogenated alloys with more active kinetics of hydrogenation. Another possible elucidation of the data obtained is also under way, as some additional experimental results will have to be obtained elsewhere.

6. References

1. Nolhier-Guillerm, Ch. (1979) Stockage de l'hydro gene les hydrures de metaux legers, Paris: Univ. Paris.
2. Tkachenko, V.G. (1996) Physical fundamentals of forming and recovery of mechanical properties for rare metals, light and precise alloys, Kyiv: Naukova Dumka.
3. Baker, C. and Birnbaum, H.K. (1973) Anelastic studies of hydrogen diffusion in niobium, Acta Metall. **21,** 865-872.
4. Tkachenko, V.G., Strongin, B.G., Maksimchuk, I.N., Friezel, V.V., Grechko V.P., and Likhtarovich S.P. (1996) Peculiarities of solid solution decay of hydride-forming and non-hydride-forming systems with close-packed crystal structure, Int. J. Hydrogen Energy **21,** 1091-1096.
5. Tkachenko, V.G., Maksimchuk, I.N., and Shklover, V.V. (1995) Thermodynamical features of hydrogen diffusivity in metal hydride systems as the sources of plasma forming matter for laser spark in vacuum, Thesis of IV NATO Int. Conf. "Hydrogen Materials Science and Chemistry of Metal Hydrides", pp. 200-201.
6. Smirnov, A.A., Tkachenko, V.G., Maksimchuk, I.N., Shklover, V.V., Bugajev, V.N. and Tatarenko, V.A. (1993) Hydrogen-indused reconstructive, dimensional and diffusive effects in non-stoichiometric compounds and substitutional alloys. Scientific Report No. 02940000293, reg. No. 0193U028810. National Committee for Science and Technology of Ukraine, project No. 2/700, Kyiv.

7. Tkachenko, V.G., Kolesnik, L.I., Maksimchuk, I.N. and Sklover, V.V. (1996) Dislocation mechanism of hydrogen embrittlerhent of metals and alloys with HCP and FCC crystal structure, Int. J. Hydrogen Energy **21,** 1105-1113.
8. Berry, B.S. (1959) The role of vacancies in an unusual anelastic phenomenon, Acta Metall. **7**, 741-745.
9. Tkachenko, V.G., Kondrashev, A.I., et al. (1999) Electron Structure of metal alloys with increased quantum yield of photoemission, VI Int. Contf. "Hydrogen Material Science and Chemistry of Metal Hydrides". Katsiveli, Crimea, Ukraine, p. 77.
10. Tkachenko, V.G., Timoshevski, A.M., Korduban A.M., Maksimchuk, I.N. (2000) "High efficiency materials for massive high-current photocatodes", Scient. Rep. No. 0198U000369 Ntl Committee for Science and Technology of Ukraine Kyiv.
11. Combetee, P., Renard, M., Grilhe, I. (1972) Interaction between dislocations and hydrogen atoms in nickel, Ber Kernforschungsanlage Julich, Conf 6, **2,** pp. 821-832.
12. Baker, C. and Birnbaum, H.K. (1973) Anelastic studies of hydrogen diffusion in niobium, Acta Metall. **21**, 865-872.
13. Buck, O., Thomson, D.O., Wert, C.A. (1971) The effect of hydrogen on the low temperature internal friction of a niobium single crystal, J. Phys. Chem. Solids **32**, 2331-2344.
14. Kofstad, P. and Butera, R.A. (1963) Internal Friction in the Tantalum-Hydrogen System, J. Appl. Phys. **34,** 1517-1520.
15. Rose, K.S.B., and Glover, S.G. (1966) A study of strain-ageing in austenite, Acta Metall. **14**, 1505-1516.
16. Nowick, A.S. (1953) Progress in Metal Physics, London, Pergamon Press, p. 357.
17. Strongin, B.G., Kopilov, V.I., Varvus, I.N., Tkachenko, V.G. (1997) Low temperature β-relaxation of hydrogen iron with plasma spraying coating of molybdenium, 5th Int Conf "Hydrogen Materials Science and Chemistry of Metal Hydrides". Katsiveli, Yalta, Ukraine. pp. 114-115.
18. Murty, K., Mohamed, F.A. and Dorn, I.E. (1971) Effect of vacancy sinks and sources on serrated yielding due to solute locking, Scr. Metall. **5**, 1087-1092.
19. Buck, O., Thomson, D.O., Wert, C.A. (1971) The effect of hydrogen on the low temperature internal friction of a niobium single crystal, J. Phys. and Chem. Solids **32**, 2331-2344.
20. Oriani, R.A. (1970) The diffusion and trapping of hydrogen in steel, Acta Metall. **18**, 147-157.
21. Schaumaun, G., Volke, J. and Alefeld, G. (1970) The diffusion coefficient of hydrogen and deuterium in vanadium, niobium and tantalum by Gorsky-effect measurements, Phys. Status Solidi **42**, 401-413.
22. Wilson, W.D., Bisson, C.L. (1971) Inert cases in solids: interatomic potentials and their influence on rare-gas mobility, Phys. Rev. B **3,** 3984-3992.

SIMULATION OF HYDRIDE HEAT PUMP OPERATION WITH REFERENCE TO VEHICLE REFRIGERATING DEVICES

YU.I. SHANIN
Research Institute of Scientific and Industrial Association Luch
24 Zheleznodorozhnaya, Podolsk, Moscow Region 142105, Russia

1. Introduction

Hydride heat pumps (HHP) continue to be a subject of investigations because of their potential abilities. The problem of hydride devices commercialization as before depends on material science and chemistry of hydrides [1]. Therefore the basic researches of hydrides are concentrated in the field of materials. There is also experience in the world on creation of experimental samples of hydride heat devices. The substantiation of a material composition, design and modes of operations are the necessary part of investigations during development of HHP. The mathematical models and set of the programs for computer simulation of HHP processes were developed earlier [2]. Examples of calculations in the first, "dot" approximation were presented there. In these calculations sorption processes were described by average parameters of hydride reactors (sorbers) - average temperature, average content of hydrogen, average pressure. The results of calculations made with use of more complex second approximation, in which the final thermal conductivity of hydride layers was taken into account, are considered in this article. The calculations were carried out for individual modules of a conditional HHP, the results were compared with experimental data for the same modules of the real HHP - components of installation simulating operation of the vehicle conditioner [3]. Heat of cooling liquid of the internal combustion engine (with a level of maximum temperature 100 – 110°C) is supposed to be used as a source of heat for HHP operation.

2. Mathematical model

Development of the mathematical model and programs for its realization were carried out by Fedorov E.M. [2]. At the first stage of application the developed model allows to make a preliminary choice of hydrides pairs based on the specified intervals of operation temperatures and pressure of hydrogen taking into account the design features of hydride sorbers [3]. At this stage the main decision on ability of the HHP with selected hydrides to work is taken (i.e., does the thermodynamic cycle operate) in the specified ranges of working parameters.

N. Veziroglu et al. (eds.), Hydrogen Materials Science and Chemistry of Metal Hydrides, 97–106.

The second stage presents the mathematical model of working process in hydride sorbers and gives an opportunity of optimization of sizes and modes of HHP operations.
The mathematical model accounting the basic features of hydride sorbers design and processes occurring in it: method of heating (by electric power or by heat exchange) and heat removal, thermal conductivity of hydride sorbers and elements of design, of hydrogen transport with the appropriate heat release and absorption.
The following initial information is necessary for model:

- Diagram of hydride sorbers with identification of material and sizes;
- Thermophysical and transport properties of materials;
- Way of control of operation installation (time, temperature, pressure).

Calculations tested in experiments can find out the answers to the following questions:

- What allowable geometrical parameters of some elements within the framework of the chosen construction diagram (for example, sizes of hydride sorbers, diameters of channels for coolant, walls thickness, number and shape of various inserts with high thermal conductivity etc.);
- How to control the work of hydride sorbers receiving the maximum cold productivity, maximum coefficient of energy performance etc. (for example, what optimum relationship between power and time of heating, what should be temperature difference between sorbers and external coolant, what should be effective factors of thermal conductivity of hydrides etc.).

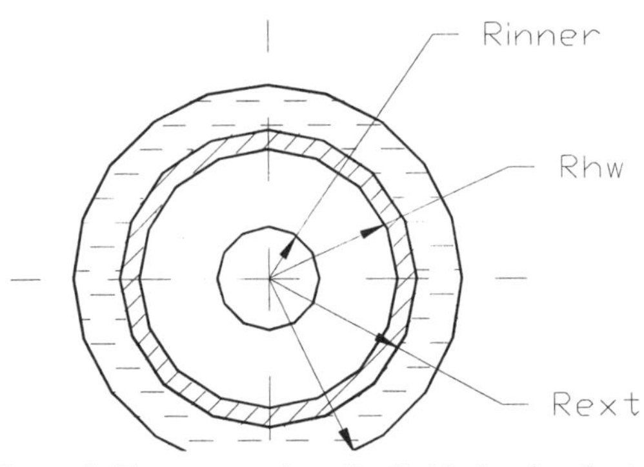

Figure 1. The cross section of cylindrical sorber (central collector of hydrogen is not shown): R_{inner} - hydride layer internal radius, R_{hw} - internal radius of a pipe (or hydride layer external radius), R_{ext} - external radius of a pipe, R_{out} - external radius of coolant layer.

The mathematical model was developed for one-dimensional non-stationary case. It took into account the real distribution of physical variables – fields of temperature, pressure, concentration of hydrogen in volume of a hydride layer of HHP and their change in time. The change of physical values only along one radial coordinate of cylindrical sorbers was taken into account within the framework of the one-dimensional model, Fig. 1.
The cause for application of such approach is a small ratio of sorber tubes radius to their length. The central channel with small hydraulic resistance for gas transport (hydrogen supply and removal to/from a hydride layer) promotes pressure equalization in various sections of sorbers. The factor breaking the constancy of parameters along sorber length can be change of temperatures of coolant in external paths during longitudinal flow of sorbers. At the relatively small coolant rates these changes can be significant. It will cause the large deviation from assumptions of the model

and will result in necessity of deriving of more rigid model (taking into account axial transfer of heat and hydrogen along sorber length). This factor becomes a major factor causing the large discrepancies of calculated and experimental results.
Equations set describing a thermal mode of HHP consisting of two sorbers with high-temperature (HTH) and low-temperature (LTH) hydrides can be written as follows:

$$\overline{\rho C_p}\left(\frac{\partial T}{\partial t} + \frac{\rho}{V} gradT\right) = div\ \overline{\lambda}\ gradT + \overline{Q}_v\ , \qquad (1)$$

$$x = x_{eq}(P,T)\ , \qquad (2)$$

$$N_h^0 = N_{hydr}^1 + N_{hydr}^2 + 2N_{H_2}^1 + 2N_{H_2}^2\ , \qquad (3)$$

where ρ -density, C_p - heat capacity of a sorber, T -temperature, V-volume of a hydride bed, λ_w -thermal conductivity of material, Q- energy, X- concentration of hydrogen, N- quantity of hydrogen in a sorber, N_{H2}- quantity of hydrogen in a gas phase in the appropriate sorbers, N_{hydr}- quantity of hydrogen in hydride.
The first equation is the equation of thermal conductivity (which is solved for each hydride layer and body shells; in the latter the members accounting absorption or desorption of hydrogen are absent). The second member in the right part of equation takes into account heat release or absorption during hydrogen sorption:

$$\overline{Q}_v = 0.5\Delta H_f n_{Me}\frac{dx}{dt}\ ,$$

where dx/dt is the rate of change of a hydride phase composition, and ΔH_f is specific heat of hydrogen dissolution, n_{Me} - density of metal sublattice of hydride, which determines a solid phase composition $x = nH/n_{Me}$.
The composition of hydrides in sorbers is calculated according to values of hydrogen pressure and temperature of hydrides on equilibrium isotherms.
The equation (1) is written down in the assumption of absence of temperature jumps between gas and solid phases in a hydride layer. The equation (2) shows that the hydrogen concentration in each point of a calculated region is determined from conditions of equilibrium. The equation (3) expresses a condition of constant hydrogen amount in a heat pump. The upper index *0* refers to the whole system, indexes 1 and 2- to various sorbers. Such approach is applicable as to the systems with a horizontal plateau on isotherms pressure -concentration, and to the systems with isotherms having slope in all area of homogeneity.
The initial hydrogen pressure and hydrides temperature in the system can be given as the initial data. The hydrogen content in sorbers at the initial moment is found out of equilibrium conditions. Pressure in sorbers during operation was assumed to be the same, this, in turn, assumes minimum hydraulic losses of flowing hydrogen.

The boundary conditions are defined by type of heat exchange. In our experimental case hydride sorbers exchange heat with an external coolant. The condition of heat transfer on their outside surface is set by Newton law:

$$-\lambda_w \, grad \;\; T = \alpha (T_w - T_c), \tag{4}$$

where λ_w- thermal conductivity of wall material, α- heat transfer characteristic, T_c, T_w - temperature of a coolant and wall.

The thermophysical and transport properties of materials, including parameters of equilibrium experimental isotherms in the metal – hydrogen systems, are stored as library of files that allows to carry out fast search of various combinations of materials during computations.

At the initial stage of development of a modular sorber design the model gives ample opportunities on alternative computations by change of design embodiment and material composition of the module. For the developed module the model allows to optimize modes of its operation by alternative change of parameters (initial hydrogen pressure, temperature and flow of coolant, cycle time).

3. Results and discussion

3.1. EXAMPLE OF CALCULATION

The results of computations of HHP design based on modules are presented as an illustration. The design of HHP was described in detail earlier [3, 4] (experimental results on this installation are received by Frolov V.P.). The HHP works in a refrigerator mode on the basis of hydride pair $LaNi_5$ - $MmNi_{4.15}Fe_{0.85}$.

The results of calculations on one-dimensional model for the module mentioned in work [3] will be presented below. The module is made out of stainless steel tubes with an external diameter R_{ext}=28 mm and wall 1 mm in thickness. Length of the module containing high-temperature hydride ($LaNi_5$) is 740 mm, weight of hydride in it - 1.07 kg. Length of the module containing low-temperature hydride ($MmNi_{4.15}Fe_{0.85}$) is 530 mm, weight of hydride - 0.714 kg. An insert made as aluminium foil corrugation 0.15 mm in thickness is placed inside tubes to increase thermal conductivity of hydride layer (it occupies about 5 % of total amount of a hydride layer and increases its effective thermal conductivity by several times). The central part of the module is occupied by a collector for hydrogen transport (is not shown in Fig. 1) serving also as a filter. The external part of the module tubes is washed by water as in heat exchanger "pipe in pipe", tube spacing is 2 mm.

Isotherms for the specified materials were measured experimentally and were used in calculations after their computer processing. Temperature of a hot coolant – water - changed in a range from 65 to 105 °C, temperature of cooling water is 10 °C, initial temperature of cooled water 10 °C. The calculation was carried out for a case of continuous feed of water with constant input temperature in half-cycle of cooling (open-ended contour). Typical curves of change of averaged integral temperature of hydride layer (along sorber tube radius) during periodic HHP operation are shown in Fig. 2. It can be seen that HTH temperature (T_2) is lowered to approximately minus 10 °C.

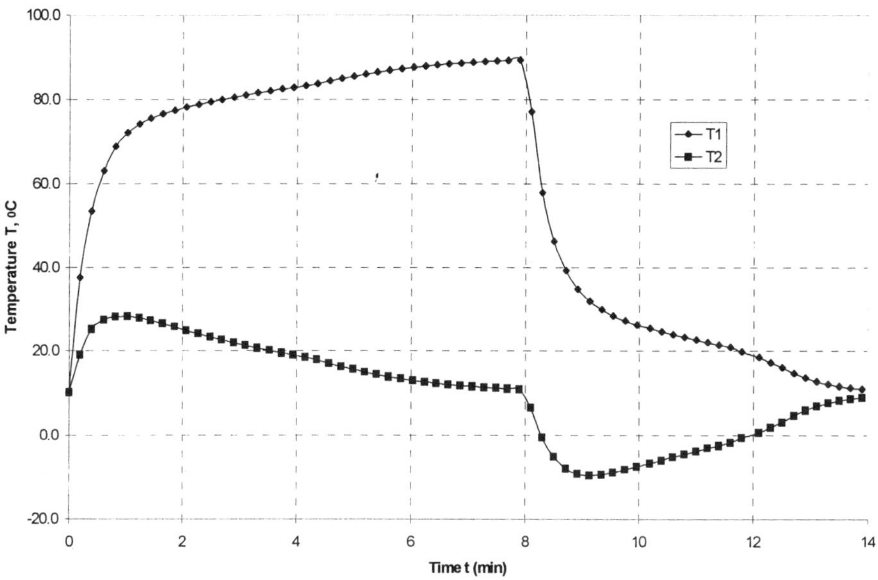

Figure 2. Cyclogram of module parameters change in a cycle: T1 - high-temperature sorber, T2 - low-temperature sorber.

Low thermal conductivity of hydride filling, high rates of heating and the cooling of sorber can cause significant differences of temperatures. The process can be very complicated with absorption of hydrogen in one parts and desorption in other of parts of the same sorber volume. Naturally, it is resulted in decrease of an overall performance of HHP in a whole. Typical radial temperature distributions in sorbers show that temperature and hydrogen concentration gradients appear at the initial stages of non-stationary processes of sorber heating and cooling resulting in the maximum inhomogeneities of radial hydride layer temperature field.

A phase point movement in coordinate pressure - concentration is shown in Fig. 3. It can be seen that the significant part of the phase diagram of two-phase region "works" so providing the HHP high designed module parameters.

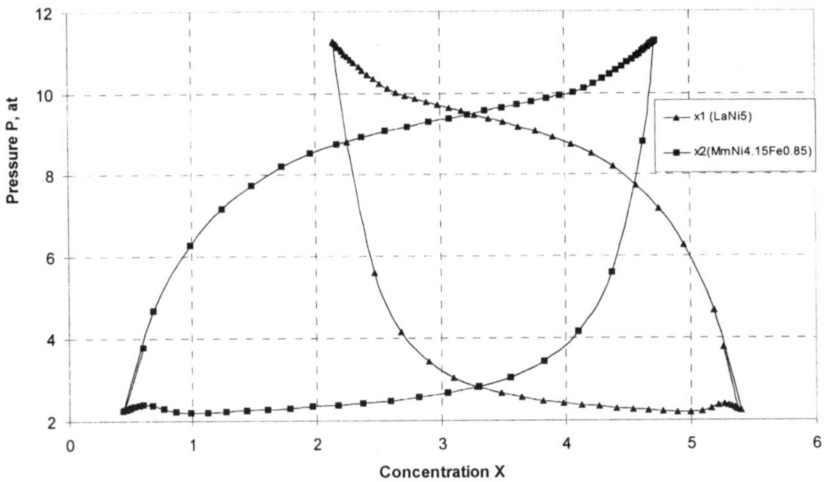

Figure 3. Cycle dependence of pressure and concentration of hydrogen in hydrides.

Taking into account the pressure - concentration curves hysteresis (during hydrogen absorption and desorption) and final rates of heterogeneous reactions of hydrogen sorption essentially increases the complexity of computations. Therefore, the influence of these factors was estimated only for limited number of variants. Summarizing the results of these estimations, it is possible to tell that the hysteresis (at the ratio of sorption pressure to desorption pressure at a level of 1.2-1.3) reduces an overall performance of HHP approximately by 30-50 %. This reduction is caused by two reasons: 1) shift of isotherms ***P-X*** (pressure - concentration) in the adverse part, i.e. lowering of desorption pressure and increase of absorption pressure of hydrides, 2) some part of running cycle now can be without change of hydrides composition and, respectively, without release and absorption of heat of hydrogen dissolution.

After that the influence of some parameters without ***P-X*** curves hysteresis on the change of pump characteristics was examined.

3.2. INVESTIGATION OF INFLUENCE OF PARAMETERS ON THE CHARACTERISTICS OF THE MODULE-TYPE HHP

3.2.1. *Initial pressure of hydrogen in sorbers*

The change of charging pressure has an essential effect on work of the module. The dependences of averaged on time of heating q_h and cooling power q_c vs. level of initial pressure above hydride (at temperature 10°C) are presented in Fig. 4. When pressure is lower than $1 \cdot 10^5$ Pa, hydrogen content in hydride corresponds to a linear part of the phase diagram. Power is low, because there is small amount of hydrogen in an alloy. There is a flat extremum of identified characteristics in a range of 1.5 to $3 \cdot 10^5$ Pa. With the further growth of charging pressure, the amount of hydrogen traveling between sorbers and, respectively, power characteristics falls. The appreciable change of characteristics begins, when charging pressure exceeds $(4\text{-}5) * 10^5$ Pa. The preliminary calculations have shown that the overall HHP performance can be increased by different amount of hydride alloys in sorbers. In particular, in our case weight of HTH purposely exceeded approximately 1.5 times weight of LTH.

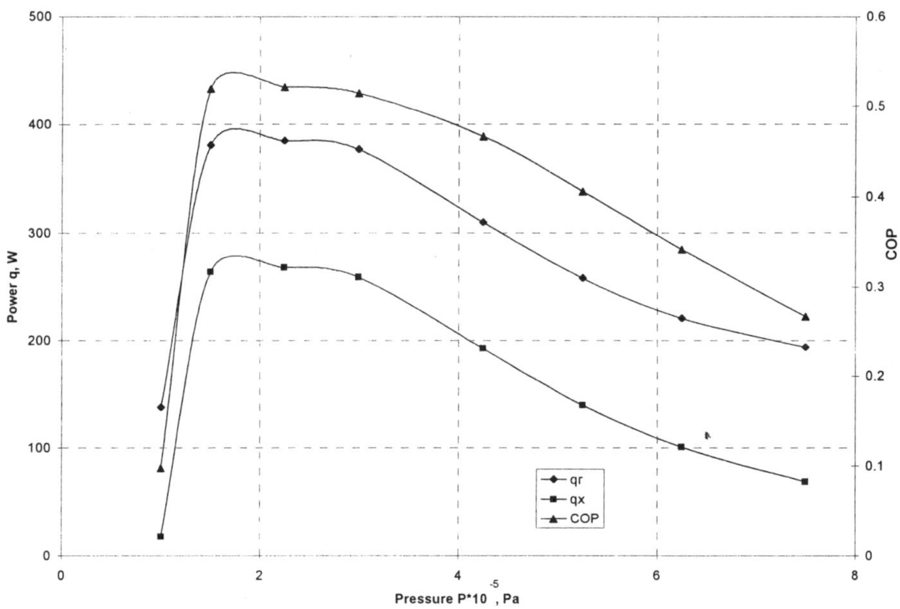

Figure 4. Influence of initial pressure of hydrogen on the power performance figures of the module: q_h - averaged power of heat supplied to HTH, q_c - averaged power of cold removed from LTH, COP - module efficiency (ratio of the energy removed from LTH Q_c to energy fed to HTH Q_h).

3.2.2. *Coolant temperature*

The increase of hot water temperature (at a level of temperature T_h) improves power of the pump. Simultaneously average integral powers of the module heating and cooling are

increased. Efficiency or COP of HHP slightly depends on a level of temperature, when the latter changes in a range of 60 to 105°C and has a flat maximum in area 70-90°C.
The increase of coolant temperature at the middle level (T_m) worsens power parameters, because of increasing of the first stage of a cycle time - hydrogen transfer from HTH- to LTH sorber due to slower cooling of the latter. At this time the minimum temperature of LTH-sorber will slightly increase. When this temperature is increased from 10 to 20°C, LTH-sorber changes temperature from minus 10°C up to minus 3°C.
Coolant temperature decrease at the low level (T_l) improves HHP parameters. If the initial temperature of input water in a cooling circuit is less than 5°C, the output water temperature in the circuit of LTH is near to 0°C.

3.2.3. *Water flow*
Water flow influences on coefficients of heat transfer in paths and on change of water temperature along sorbers length. In the investigated design the longitudinal water flow in channels is organized. Its heating during flowing results in change of temperature in various parts of sorbers causing axial skews of temperature and performance figures of sorbers. As it expected (and alternative computations have shown) the most essential is the mode of water flow - laminar or turbulent. If the hydrodynamic of laminar flow is stabilized, the heat transfer coefficient appears to be relatively small and practically does not depend on the flow. The mode of liquid flow essentially influences on instantaneous values of energy and power of heating/cooling. Influence of flow on power averaged on time for the realized module did not exceed 10 % and depend on in what heat exchangers flow was changed. The greatest influence on q_h and q_c has change of flow at the middle level of coolant temperature (performance figures increase by 5-10 % after change of flow from 0.01 kg/s to 0.04 kg/s). The identified characteristics have less dependence on changes of flow of hot water and water used for cooling. Similar result for averaged characteristics on the flow was found also in the experiments (in experiments the flow was changed from 0.013 to 0.026 kg/s per one module).

3.2.4. *Cycle time*
Calculations have shown, if module operations is controlled by time and times of half-cycles are the same, energy consumed by high-temperature hydride and energy released by low-temperature hydride will increase exponentially with increase of cycle time, Fig. 5. In addition, the growth practically stops if half-cycle time exceeds 8 min. At the same time average power of heating q_h and cooling q_c falls. The ratio of energy values of LTH/HTH (COP=Q_c/Q_h) increases a little with increase of time from 4 to 8 min being practically constant during further increase of half-cycle time. For a chosen modules design and excess of amount of high-temperature hydride over low-temperature hydride different durations of stages of process can be recommended for different modules: 8 minutes for heating of HTH and 6 minutes for cold generation in LTH. Experience has shown that approximately the same half-cycle times are optimal.

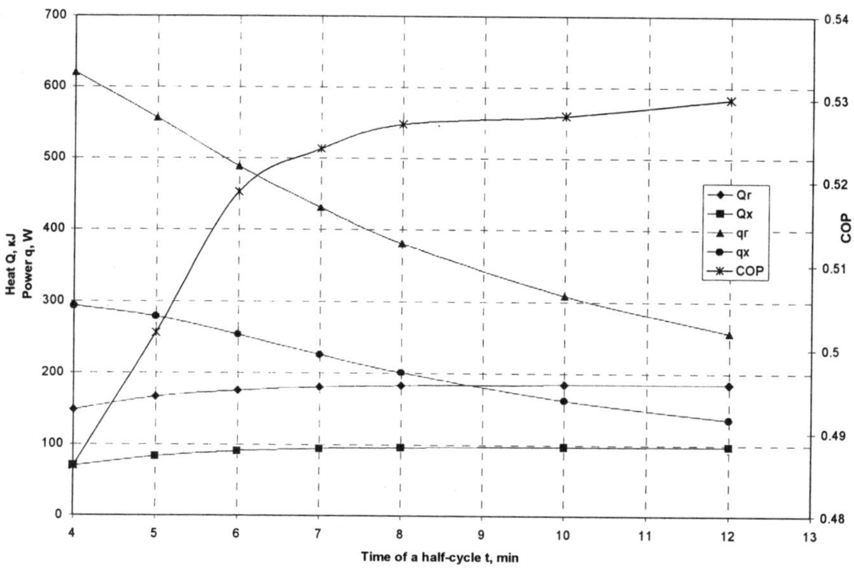

Figure 5. Influence of half-cycle time on the power characteristics of the module (see designations in Fig. 4).

3.3. COMPARISON WITH EXPERIMENTAL RESULTS

Comparison of results of calculation and experiments [3] was carried out for cold generation. In this comparison we measured experimentally cold which was transferred by a coolant (i.e. only useful cold). But in calculations all cold generated by the module was taken into account, including a cold used for cooling of construction (15-20 % of total value). The difference in methods of calculation of the generated cold in experiment and in calculations is especially clear comparing COP. The real, experimental efficiency as the ratio of the generated useful cold to the spent energy did not exceed 0.3. The calculated efficiency (as the ratio of the generated cold to the spent energy) came nearer to 0.5.

Good qualitative agreement of results was received for all performance figures of the module presented in Fig. 4,5. Quantitative excess of calculated results above experimental data did not exceed 1.3-1.8 times (or experimental data were by 30-50 of % lower in comparison to calculated values). Such quantitative discrepancies can be explained by several reasons:

1.Discrepancy of the idealized diagram of the module with the real design. We have sorbers with 7 modules incorporated in a bunch and made in a common shell of heat exchanger. The conditions of heat exchange differed from the scheme "pipe in pipe" to the worst side. In addition, there was an essential coolant temperature gradient along the module length. It especially concerns to initial stages of heating and cooling of the module, when the non-

stationary processes are most intensive. Though the one-dimensional model is more accurate than "dot" model [2], it is also still rather far from accounting of all three-dimensional, non-stationary processes going in real modules.
2.Difference of calculated and real properties and performance figures of hydrides and hydride layers. The maximum discrepancy could be caused by that in the majority of calculations the hysteresis of hydride sorption characteristics was not taken into account (that overestimates the performance figure by 30-50 %). Beside, the pressure concentration relationships obtained for hydrides in equilibrium conditions were used for calculations. It is known [5] that these characteristics deteriorate in dynamic conditions. The effective thermal conductivity of hydride layers was estimated by calculations, and was obviously overestimated in comparison with experimental data. Besides, in the real tube module there is always a contact thermal resistance on a contact of insert and pipe wall resulting in decrease of hydride layer heat exchange characteristics. This deterioration was not taken into account in calculations.
Comparison of information based on experiments and calculations, taking into consideration the above said, has shown their satisfactory fit.

4. Conclusions

Comparison of calculated and experimental results has shown that the developed one-dimensional model still is far enough from accounting of all real HHP design features. It can be used for qualitative study of influence of various parameters and approximate quantitative estimations. Nevertheless, developed mathematical modeling makes it possible to carry out operatively computations of various HHP designs. With use of modeling for short time and at minimum expenses (without manufacturing of apparatuses in "iron") it is possible to reveal the best approaches to development of HHP and their practically possible parameters.

5. References

1. S. Suda, S. and Sandrock, G. (1994) Three decades of intermetallic hydrides.- What happened to the applications?, *Zeitschrift fur Physikalische Chemie. Neue Folge* **183**, 149-156.
2. Fedorov, E.M., Shanin, Y.I. and Izhvanov, L.A. (1999) Simulation of hydride heat pump operation, *Inter. J. of Hydrogen Energy* **24**, 1027-1032.
3. Astakhov, B.A., Afanasyev, V.A., Bokalo, S. Yu., Izhvanov, L.A., Solovey, A.I., Frolov, V.P. and Shanin, Yu.I. (1999) Creation of small-sized refrigerating machineries with a metal hydride heat pump, *In proc. (Abstracts) of the 6-th Int. Conf. "Hydrogen Material Science and Chemistry of Metal Hydrides of ". Ukraine, Yalta, (02-08).09. 1999*, 306-307.
4. Astakhov, B.A., Izhvanov, L.A., Lysenko, A.A., Solovey, A.I. and Shanin, Yu.I. Creation heat and cold generating installation based on a hydride heat pump, *In proc. (Abstracts) of the 6-th Int. Conf. "Hydrogen Material Science and Chemistry of Metal Hydrides of ". Ukraine, Yalta, (02-08) .09.1999*, 306-307.
5. Nagel, M., Komazaki, Y., Matsubara, Y. and Suda, S. (1986) Dynamic behavior of paired metal hydrides.2: Analytic survey of the experimental results, *J. Less-Common Metals* **123**, 47-58.

HYDROGEN INTERACTION WITH ALLOYS OF Ti(Zr-)-Mn-V SYSTEMS

T.N.SMIRNOVA, S.V.MITROKHIN, V.N.VERBETSKY
Chemistry Department, Lomonosov Moscow State University,
Vorobjovy gory, Moscow 119899, Russian Federation

Abstract

The use of intermetallic hydrides in hydrogen energy applications is stipulated by the possibility of developing of hydrogen absorbing alloys with a wide range of sorption properties. Of the most perspective materials are the alloys of light transition elements (Ti, V, Cr, Mn) having Laves phase structure. Still, the metallic Ti-Mn-V and Ti-Zr-Mn-V systems are practically not described in literature, though they present great possibilities for practical applications as highly efficient hydrogen storage media. In present work the detailed investigation of the phase boundaries of Laves phase was performed by means of X-ray and electron-probe micro analysis methods. A rather wide homogeneity range of Laves phase was found for both systems. The dependence of hydrogen absorption properties on alloy composition was studied by PCT-isotherm method. The initial hydriding reaction takes place in rather mild PT-conditions without preliminary activation and resulting hydrides possess high hydrogen capacity (1.8-2.0 mass.% H_2). Studied alloys exhibit a strong dependencies of dehydriding reaction enthalpy and equilibrium desorption pressure on elementary cell volume. Found dependencies could form a basis for prediction of hydrogen absorption properties of Laves phase alloys for developing new highly efficient materials for metal-hydride technology.

1. Introduction

Intermetallic hydrides of Laves phase structure type are under intensive study now, since they are considered as rather promising highly effective chemical accumulators and sources of hydrogen. This is stipulated by the high rates of hydrogen interaction with alloys at mild PT-conditions and by relatively high mass content of hydrogen. Intermetallic compounds of Ti-Mn-V and Ti,Zr-Mn-V systems are formed of the lightest 3d-transition metals. This makes them rather attractive from the practical point of view.. Still they were not studied in detail.

This work was aimed on the systematic investigation of hydrogen absorption properties of C14 Laves phase alloys of Ti(Zr)-Mn-V systems. Recently these metallic systems were reported to have rather large homogeneity range of Laves phase (338-63 at.% Mn and 36-65 at.% Mn for ternary and quaternary systems respectively) [1, 2]. It should be noted that vanadium is dissolved in $TiMn_2$ C14 phase without any structural

N. Veziroglu et al. (eds.), Hydrogen Materials Science and Chemistry of Metal Hydrides, 107–112.

change up to 26 at.%. For the classical case of Laves phase formation its stability is determined considerably by the dimensional factor, that is R_A/R_B = 1.225.Since in the overwhelming number of cases intermetallic compounds of this type do not show any visible homogeneity range, the nonstoichiometry problem (B/A=2±x) was not considered in detail. In our case the presence of large homogeneity region leads to considerable deviations from stoichiometric composition AB_2 due to the location of atoms of one component in the crystallographic positions of the other. The hydrogen absorption properties of nonstoichiometric Laves phases were not studied in detail up to now. We systematically studied the hydrogen absorption properties of these alloys and applied a mathematical model to the description and prediction of these properties for Ti(Zr)-Mn-V alloys.

2. Experimental part

Alloys were melted from starting metals of >99.9% purity in an arc furnace and annealed at 850°C for 240 h. The PCT-isotherms were measured in a conventional Sieverts type apparatus.

The main features for prediction of hydrogen absorption properties of alloys are their cell parameters and thermodynamic functions of their reaction with hydrogen. These parameters usually change monotonously with concentration of components in the single phase solid solution region. Therefore we can apply mathematical methods of approximation of these parameters. In the simplest case the function describing the monotonous dependence "parameter-property" can be expressed by a full polynom of second power:

$$F(x)=Y_0+\Sigma A_i \cdot x_i + \Sigma B_{ij} \cdot x_i \cdot x_j \qquad (j \geq I), \tag{1}$$

where: Y, A, B – polynom coefficients, x – atomic concentrations of components.
Having a database of experimental values of parameters one can solve the system of equations for Y, A and B and then determine any parameter for not studied alloy.

3. Results and discussion

In our case we decided to use as a database entries own experimental results. This choice is explained by the fact that the literature results of different authors even for the same alloy can differ considerably because of the different purity of starting metals, methods of alloy preparation etc.

Because of the interpolating character of the method we chose for database entries firstly the alloys with maximum concentration of one of the components. The minimum number of entries is expressed by following relation:

$$N = 2i + (i^2 - 1) / 2, \tag{2}$$

where: i – number of independent components (for ternary system i = 5, for quaternary system i = 9).

We included in the database the experimental values for 19 alloys (Table 1). The calculation of the polynom coefficients gives the following equations for cell parameters and thermodynamic functions:

$$a(C) = 6.50 - 1.74C_{Ti} - 3.28C_{Mn} - 0.19C_V + 3.21C_{Ti}^2 - 1.73C_{Ti}C_{Mn} + 0.14C_{Ti}C_V + 2.59C_{Mn}^2 - 0.83C_{Mn}C_V - 1.21C_V^2 ; \tag{3}$$

$$c(C) = 9.56 - 1.40C_{Ti} - 1.78C_{Mn} - 1.57C_V + 0.53C_{Ti}^2 - 0.05C_{Ti}C_{Mn} - 0.76C_{Ti}C_V - 0.17C_{Mn}^2 + 0.57C_{Mn}C_V + 0.33C_V^2 ; \quad (4)$$

$$\Delta H(C) = 208.01 - 227.79C_{Ti} - 375.88C_{Mn} + 45.86C_V + 376.57C_{Ti}^2 - 101.15C_{Ti}C_{Mn} - 34.13C_{Ti}C_V + 292.58C_{Mn}^2 - 184.84C_{Mn}C_V - 73.88C_V^2 ; \quad (5)$$

$$\Delta S(C) = 191.99 + 22.09C_{Ti} - 361.43C_{Mn} + 94.15C_V + 316.57C_{Ti}^2 - 519.99C_{Ti}C_{Mn} + 359.43C_{Ti}C_V + 558.73C_{Mn}^2 - 275.25C_{Mn}C_V - 61.30C_V^2 ; \quad (6)$$

Table 1. Characteristics of alloys, included in the database

No.	Alloy composition (at.%)				ΔH (kJ/moleH$_2$)	ΔS (J/K·moleH$_2$)	a (Å)	c (Å)
	Ti	Mn	V	Zr				
1	30	50	20	0	28	116.8	4.860	7.956
2	42	55	3	0	35	109.9	4.888	7.977
3	38	50	12	0	35	121.6	4.881	7.971
4	35	42	23	0	41	134.7	4.908	7.998
5	30	51	19	0	28	111.6	4.850	7.945
6	41	44	15	0	43	127.8	4.999	8.010
7	31	46	23	0	34	126.2	4.870	7.990
8	40	49	11	0	33	112.06	4.880	7.98
9	36	44	20	0	41	119.9	4.915	8.011
10	27	57	13	3	30	121.7	4.858	7.978
11	32	59	5	4	29	112.9	4.860	7.959
12	32	50	14	4	33	116.36	4.888	7.990
13	36	55	5	4	34	114.58	4.909	8.011
14	28	44	24	4	36	117.8	4.918	8.039
15	28	50	18	4	31	116.4	4.870	7.991
16	26	51	20	3	32	123.01	4.859	8.001
17	34	50	11	5	36	117.34	4.910	8.061
18	36	50	10	4	38	115.5	4.928	8.069
19	39	50	5	6	39	109.3	4.940	8.070

The applicability of the mathematical model to the nonstoichiometric alloys was tested by comparison of calculated and experimental values of two ternary and two quaternary alloys, which were not included in the database (Table 2). The values for equilibrium desorption pressure were calculated using vant Hoff equation.

Good correlation shows that the model can be applied to nonstoichiometric alloys. Using equations (3-6) the hydrogen absorption parameters were calculated for alloys which were then studied experimentally. Table 3 shows the result for those alloys which did not absorb hydrogen in our experimental conditions. The calculated desorption pressures are rather high. Meanwhile, in order to initiate the reaction of hydrogen absorption the pressure should be somewhat higher than the equilibrium absorption pressure, which is due to the hysteresis effect is always higher than the desorption one. This clearly shows the reasons why the alloys did not absorb hydrogen in our experiments. The dependence of calculated equilibrium pressure values for whole Laves phase region of Ti-Mn-V system is shown in Fig.1.

Table 2. Comparison of experimental and calculated values for model test alloys

Alloy composition	a_{calc} Å	a_{exp} Å	c_{calc} Å	$c_{exp.}$ Å	ΔH_{calc} kJ/mol H_2	ΔH_{exp} kJ/mol H_2	ΔS_{calc} J/K·mol H_2	ΔS_{exp} J/Kmol H_2	P_{calc} atm at 295 K	P_{exp} atm at 295 K
$TiMn_{1.25}V_{0.5}$	4.909	4.887	7.993	7.994	36.4	37.0	124.0	117.9	1.1	1.3
$TiMn_{1.4}V_{0.7}$	4.88	4.889	7.984	7.986	34.2	34.6	123.5	113.0	2.5	3.1
$Ti_{0.9}Zr_{0.1}Mn_{1.3}V_{0.35}$	4.955	4.913	8.07	8.14	40.1	41.2	120.0	121.4	0.1	0.1
$Ti_{0.9}Zr_{0.1}Mn_{1.6}V_{0.4}$	4.865	4.880	7.98	8.015	30.0	32.7	116.5	125.9	6.1	4.5

The results obtained in this work allow to predict rather correctly the hydrogen absorption properties of stoichiometric and nonstoichiometric Laves phase alloys of Ti(Zr)-Mn-V systems. The main correlations which were found in this work show that the alloys of both metallic systems are very much alike. Therefore it is possible to look for some common dependences for both systems. As an examples of such dependences Fig.3,4 show the dependences of desorption reaction enthalpy and equilibrium desorption pressure vs. elementary cell volume of alloys.

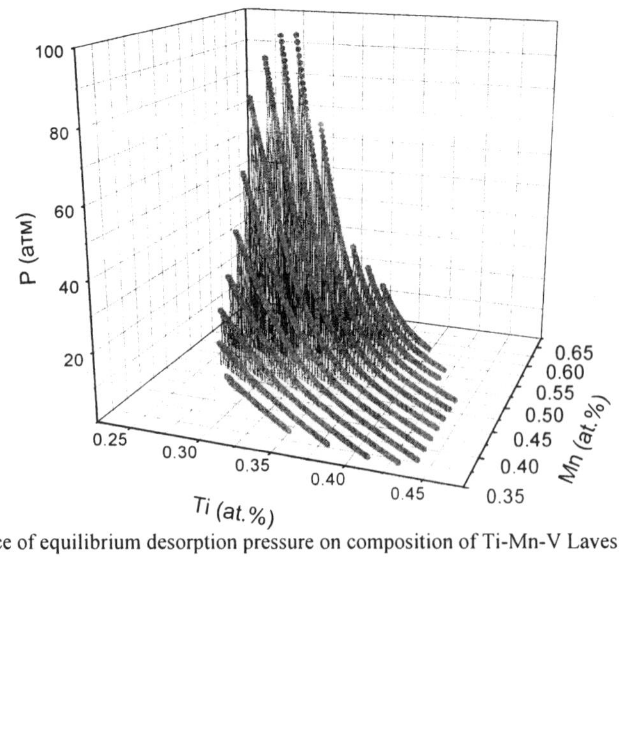

Fig.1. Dependence of equilibrium desorption pressure on composition of Ti-Mn-V Laves phase region

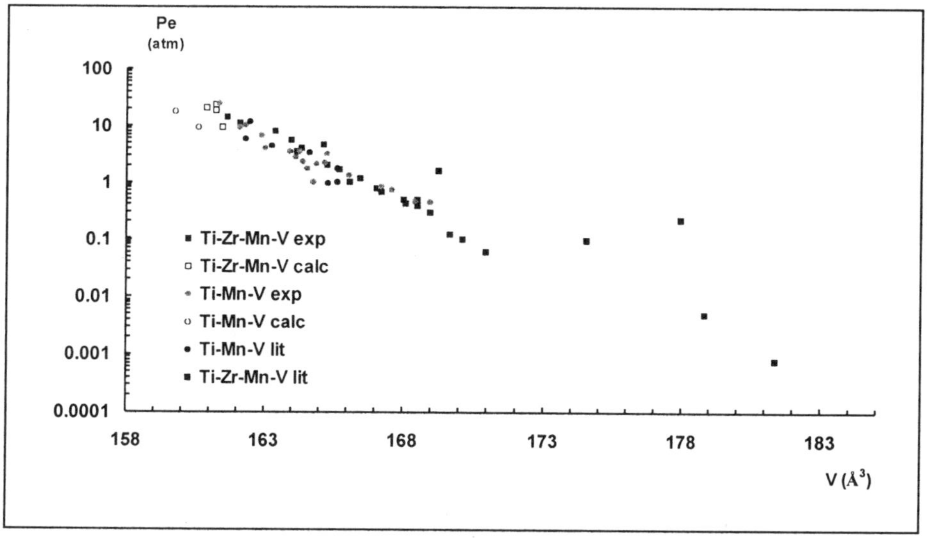

Fig.2. Dependence of equilibrium desorption pressure vs. elementary cell volume

Besides the data, obtained in this work (both experimental and calculated values) we plotted the literature values for alloys of Ti(Zr)-Mn-V system. It is clearly seen that both dependences are almost linear. The absolute values of reaction enthalpy increase with increasing cell volume, while the equilibrium pressure values decrease. The calculated values for non hydrided alloys fit well in these dependences. As follows from Fig.2., the addition of zirconium and vanadium to Ti-Mn alloys leads to decrease of equilibrium pressure and alloys with cell volume more than 160 Å^3 do not react with hydrogen at room temperature up to 100 atm.

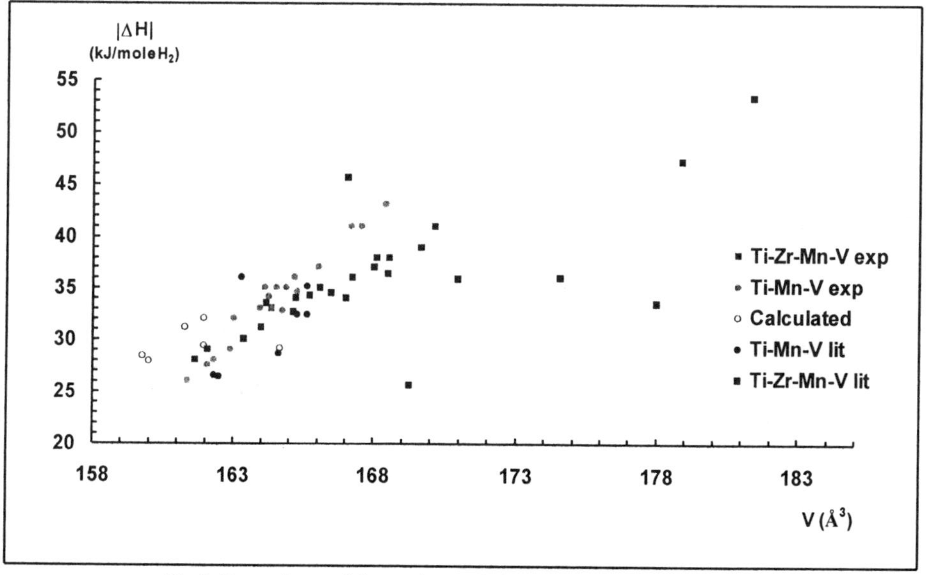

Fig.3. Dependence of desorption enthalpy vs. elementary cell volume.

4. Conclusions

In this work it was shown that the simple mathematical approach can be very useful for prediction of hydrogen absorption properties of Laves phase alloys. The experimental dependences give valuable information for search of new effective hydrogen storage materials.

Table 3. Calculated values for not hydrided alloys

Alloy composition	a_{calc} Å	a_{exp} Å	c_{calc} Å	$c_{exp.}$ Å	ΔH_{calc} kJ/mol H_2	ΔH_{exp} kJ/mol H_2	ΔS_{calc} J/K·mol H_2	ΔS_{exp} J/Kmol H_2	P_{calc} atm at 295 K	P_{exp} atm at 295 K
$TiMn_{1.8}V_{0.2}$	4.838	4.828	7.915	7.916	27.9		118.1		16.9	
$TiMn_{1.8}V_{0.1}$	4.841	4.839	7.911	7.92	29.4		117.8		8.9	
$TiMn_{1.85}V_{0.15}$	4.842	4.795	7.893	7.875	28.5		125.0		30.8	
$Ti_{0.9}Zr_{0.1}Mn_{2.2}V_{0.1}$	4.877	4.840	7.913	7.951	31.1		131.4		22.4	
$Ti_{0.9}Zr_{0.1}Mn_{2.1}V_{0.3}$	4.888	4.851	7.925	7.952	32.1		132.7		17.7	
$Ti_{0.9}Zr_{0.1}Mn_{2.0}V_{0.45}$	4.864	4.900	7.947	7.920	29.2		123.6		19.5	
$Ti_{0.9}Zr_{0.1}Mn_{2.1}V_{0.8}$	4.861	4.842	7.975	7.961	30.0		120.2		9.2	

Acknowledgements

This work was supported in part by RFBR Grants 99-02-16383, 00-15-97457 and 01-02-06024.

6. References

1. S.V.Mitrokhin, T.N.Bezuglaya and V.N.Verbetsky. Proc. 13th World Hydrogen Energy Conf., Beijing, 11-15 June 2000, p.534.
2. S.V.Mitrokhin, T.N.Bezuglaya and V.N.Verbetsky. J. Alloys and Compounds, to be published.

INFLUENCE OF ACCELERATED ELECTRON BEAM ON FORMATION OF IV GROUP METALS' HYDRIDES (Hf-H_2 SYSTEM)

S.K. DOLUKHANYAN, V.SH. SHEKHTMAN, N.N. AGHAJANYAN, KH.S.HARUTYUNYAN[1], A.G. ALEKSANYAN, H.G. HAKOBYAN, K.A. ABRAHAMYAN and
O.P. TER-GALSTYAN
Institute of Chemical Physics NAS of Armenia,
P. Sevak St., 5/2, Yerevan, 375044, Armenia
(1)Yerevan Physics Institute
Alikhanyan Br. St., 2, Yerevan, 375036, Armenia

Abstract

It is shown that at irradiation of hafnium in hydrogen environment by the accelerated electron beam of dose rate 0.3-1 Mrad/sec, the thermal-radiation synthesis of hafnium hydride occurs. Established is that at increase of irradiation dose, the temperature of sample raises up to 200-300°C after which a sharp jump of temperature is observed. Actually, the reaction Hf+H_2 proceeds in a thermal explosion regime. The mechanism of hafnium hydride formation in accelerated electron beam is obtained. It is shown that the characteristics of TRS do not depend essentially on irradiation dose rate. The phenomenon of cold synthesis of hafnium hydride is revealed. A superstoichiometric hydride of hafnium $HfH_{2.42}$ is synthesized for the first time.

1. Introduction

Recently, the linear accelerators of electrons (LAE) are widely applied to investigation of radiation effects in metals and alloys. The systematic study of radiation-chemical processes in solid state body have shown that, the ionization and excitation of electron and nuclear sublattices render considerable influence on physical and chemical properties of a starting material [1].

It is known also that depending on a beam dose, the energy of primary radiation is transformed appreciably to the thermal energy, causing warming of a material (thermal-radiation processes). One can say that the radiation-chemical processes taking place at interaction of metals and alloys with hydrogen practically are not yet investigated. In the work [2] the thermal-radiation synthesis (TRS) of hydrides on the basis of ZrMe (Me = Ni, Co, Fe) is described when an acute irradiation of intermetalides is performed.

Earlier we studied the combustion processes in system Hf-H in detail and found the basic regularities and mechanism of hafnium hydrides formation in SHS (self

N. Veziroglu et al. (eds.), Hydrogen Materials Science and Chemistry of Metal Hydrides, 113–121.

propagating high temperature synthesis) regime [3]. In this work the results of investigation of hafnium interaction with hydrogen in accelerated electron beam are presented.

2. Experimental

The influence of ionizing radiation on the hydrides formation was investigated on linear accelerator of electrons LAE-5 (beam energy and current intensity, respectively, up to 5 MeV and 500 μA). In Fig.1 the scheme of experiment is presented. A special camera for the electron-beam processing of materials in vacuum and in hydrogen atmosphere was designed and prepared.

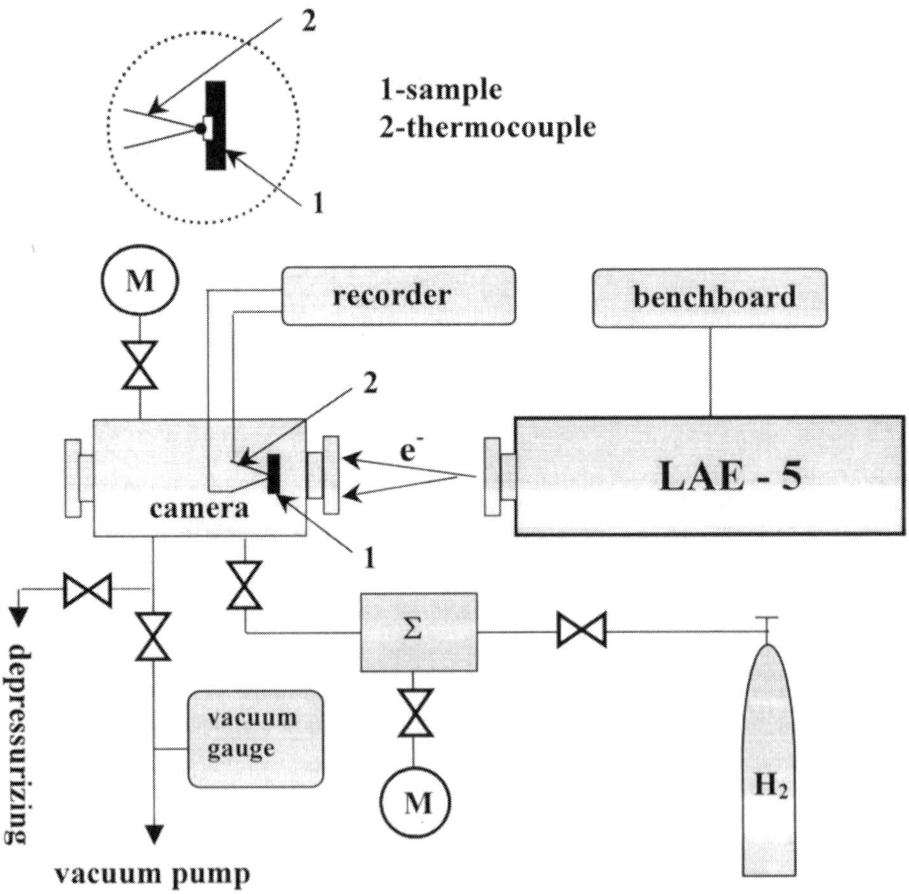

Figure 1. General process scheme for thermal-radiation investigations.

The camera was placed in the front of the high-current accelerator of electrons in the focused electron beam. The collimated beam provided a uniform irradiation over all surface of sample. The used working parameters of the accelerator were: beam energy 4 MeV, mean current 150 μA.

The systems for measuring the temperature and heat effects during irradiation were elaborated. The "S" type Pt/Pt-Rh thermocouple was used.

The tablet (d=20 mm, h=3 mm), pressed from hafnium powder (99.7%), disperse <100 μk), was placed in the camera on a special holder. Thickness of a sample was taken to ensure a uniform irradiation with electrons of all the sample volume. The total absorbed dose (Mrad) was determined as a product of dose rate (Mrad/sec) and of radiation time.

Certification of samples was carried out using chemical, X-Ray (Diffractometer DRON-2, Cu K_α irradiation) and differential thermal (Derivatograph Q-1500) analyses.

2.1. INVESTIGATION OF THERMAL-RADIATION PROCESSES IN THE Hf-H SYSTEM.

At irradiation of Hf in H_2 with accelerated electron beam of dose rate between 0.3 and 1 Mrad/sec, the TRS of hafnium hydride occurs. Below of 0.3 Mrad/sec, TRS is not accomplished. In Fig. 2, the thermogram of TRS process is shown at dose rate 0.7 Mrad/sec. TRS of hafnium hydride is similar to TRS of Ti and Zr hydrides [4,5]. In the beginning stage, the temperature of sample increases smoothly up to 200-300°C with the increase of the irradiation dose, i.e. of the irradiation time (irradiation dose rate is 20-30 Mrad/sec). This temperature has appeared sufficient for initiation of exothermal reaction $Hf+H_2$. Therefore at achieving this temperature interval, a sharp jump of temperature up to 653°C is observed (Fig. 2, curve 1). This temperature jump testifies the proceeding of an exothermal reaction, initiated by electron beam in all of the sample volume.

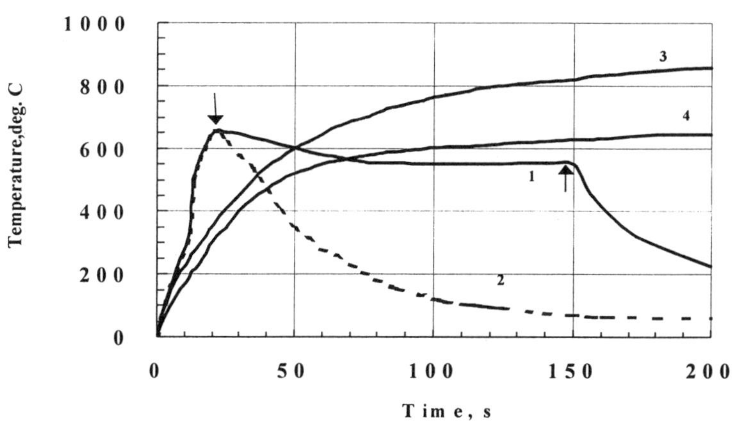

Figure 2. The TRS-thermogram in $Hf-H_2$ system at power dose 0.7 Mrad/s

Actually, the reaction proceeds in a «thermal explosion» regime. Then the temperature decreases smoothly. If the irradiation was stopped immediately in the achievement of the maximal temperature (curve 2), the more intensive cooling of the sample is observed. For comparison in Fig. 2, the dependencies of the temperature developed in TRS process (curve 2), and of the heating temperatures of Hf in vacuum (curve 3) and of hafnium hydride in H_2 (curve 4) against radiation time are presented at the irradiation dose rate of 0.7 Mrad/sec.

In Fig. 3, the thermogram of SHS process in system Hf-H is presented.

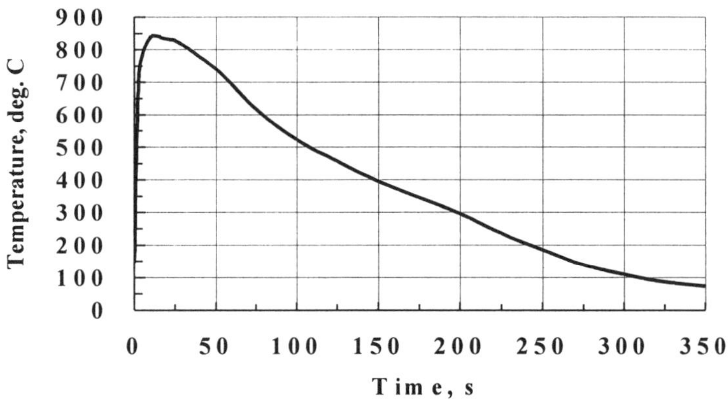

Figure 3. The SHS-thermogram in Hf-H_2 system.

The data of chemical and X-Ray analyses evidence that in TRS regime, the hafnium hydride of body-centered tetragonal (sct) structure and of high hydrogen content is formed. Let's note that the super-stoichiometric hafnium hydride of composition $HfH_{2.17}$ (i.e. H/Me > 2) has been obtained for the first time as a result of TRS in system Hf-H. It seems necessary to note also that the initial hafnium used in these experiments contained nearly 50 % HfH_X of FCC phase.

At combustion of such initial material in hydrogen (P_H = 10 atm) in SHS regime, the bi-phase hydride was formed: HfH_2 (bct) + 10-15 % HfH_X (fcc). Actually, the portion of fcc phase in the starting material decreased, but did not disappear, though one could expect that this phase should afterhydrogenate in SHS process to HfH_2. The hydride formation in SHS regime occurred in two stages: in the combustion front, the solid solution of hydrogen in metal was formed. In the second stage, in the process of afterhydrogenation, saturation by hydrogen took place along with cooling, forming hafnium hydride of fcc phase, then - of bct phase.

It is not absolutely clear, why the fcc phase in the initial hafnium was not saturated with hydrogen during afterhydrogenation up to HfH_2, when all requirements for it were available. Apparently, the high rate of combustion process (about 1cm/sec) did not provide completely the equilibrium requirements for single-phase hafnium dihydride formation. The same pattern was observed for TRS at irradiation dose rates 0.3-0.8 Mrad/sec. Probably in starting Hf, the fcc phase HfH_X is posed in the center of a grain and is "blocked". Therefore at hydrogenation in rapid SHS and TRS regimes, it can be saved. It was obtained that if the dose rate was increased up to 0.9-1 Mrad/sec, this phase could be excluded. Probably in a crystal lattice of hafnium at such high-power irradiation, the strains or diffusion barriers were removed, allowing the deep hydrogenation and hafnium hydride receiving of superstoichiometric composition.

In Table 1, the characteristics of TRS processes in systems Ti-H, Zr-H and Hf-H and hydrides obtained in TRS and SHS regimes are shown. One can note, that all the temperatures: as of the exothermal reaction beginning and the maximal, developed in TRS process, for system Hf-H are lower, than for Ti-H and Zr-H. We can say that TRS in these systems proceeds not only due to exothermal reaction, characteristic them. The contribution of the process of irradiation by accelerated electron beam is essential for interaction of metals with hydrogen. The results presented below evidence the special influence of irradiation on the process of synthesis.

The ability to TRS of hydrides is identical for all the IV group metals (Table 1). Here is worth to note that the temperatures of TRS process are always lower than of SHS approximately by 100-200°C. In Fig. 4, the dependencies of temperature

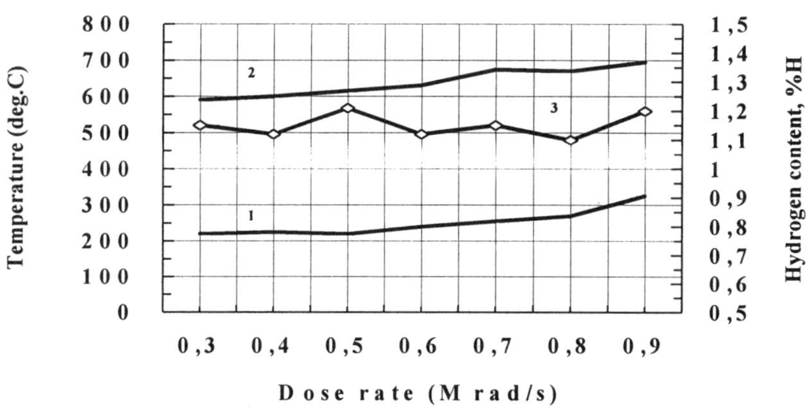

Figure 4. The dependence of reaction temperatures and hydrogen content in products on dose rate.

developed at TRS and of hydrogen content in final product versus dose rate are presented for system Hf-H. Curve 1 characterizes the temperature of exothermal reaction beginning, curve 2 – the temperature developed at exothermal reaction and curve 3 – the hydrogen content. As it is seen, the characteristics of TRS process weakly depend on irradiation dose rate.

TABLE 1. The characteristics of TRS processes in Ti-H, Zr-H and Hf-H systems

Mode	Dose rate, *Mrad/s, //* Dose, *Mrad*	*Thermal-radiation synthesis*					*Cold synthesis*					
		T_{begin} TRS, ^0C	T_{reac} ^0C	H_2, *wt. %*	Formula	Phase and parameters	Dose *Mrad*	$T_{heat.}$ in vac., ^0C	T_{begin} ^0C	$T_{reac.}$ ^0C	$H_{2,}$ *wt.%*	Formula
Titanium												
SHS	-	300-500	700-900	4.01	TiH_2	fcc a=4.44						
TRS	0.1	-	-	-	-	-	-	-	-	-	-	-
	0.2	-	-	-	-	-	40	420	70	725	3.95	TiH_2
	0.3	-	-	-	-	-	40	480	70	750	4.01	TiH_2
	0.4//40	295	640	4.08	TiH_2	a=4.447	40	540	60	760	4.08	TiH_2
	0.5//45	305	660	3.97	TiH_2	a=4.447	40	540	40	750	4.00	TiH_2
	0.6//30	330	680	3.98	TiH_2	a=4.445						
	0.7//30	357	695	4.03	TiH_2	a=4.458	50	640	80	730	3.99	TiH_2
	0.8//30	360	710	3.97	TiH_2	a=4.459	60	545	80	550	4.17	TiH_2
	0.9//30	375	735	4.01	TiH_2	a=4.458	60	570	80	545	4.16	TiH_2
	1//25	420	820	4.06	TiH_2	a=4.461	60	555	80	Didn't realized		
Zirconium												
SHS	-	300 - 500	900	2.07	ZrH_2	sct a=3.527; c=4.476						
TRS*	0.1		Didn't realized						Didn't realized			
	0.2	250	700	2.15	ZrH_2	sct	70	400	50	680	2.15	ZrH_2
	0.3	250	700	1.96	ZrH_2	-"-	47	650	25	720	2.08	ZrH_2
	0.4	250	730	2.06	ZrH_2	-"-	65	680	25	700	2.09	ZrH_2
	0.5	250	730	2.07	ZrH_2	-"-	60	720	25	680	2.10	ZrH_2
	0.6	280	750	2.06	ZrH_2	-"-	40	800	25	620	2.10	ZrH_2
	0.7	255	760	2.04	ZrH_2	-"-	70	840	25	625	1.97	ZrH_2
	0.8	280	800	1.99	ZrH_2	-"-	60	910	30	650	2.05	ZrH_2
	0.9	280	790	2.04	ZrH_2	-"-	70	895	25	620	1.92	ZrH_2
	1.0	280	790	2.11	ZrH_2	-"-						
Hafnium												
SHS	-	-	840	1.05	HfH_2	sct a=4.911 c=4.405						
TRS	0.05	Didn't realized					35	150	25	428	1.30	$HfH_{2.35}$
	0.1	-			-	-	35	220	25	428	1.23	$HfH_{2.22}$
	0.2	-			-	-	50	355	25	500	1.22	$HfH_{2.20}$
	3	220	590	1.15	$HfH_{2.08}$	sct	70	423	25	558	1.19	$HfH_{2.15}$
	0.4	225	600	1.12	$HfH_{2.02}$	-"-	70	490	25	513	1.18	$HfH_{2.13}$
	0.5	220	615	1.21	$HfH_{2.18}$	-"-	68	505	25	485	1.24	$HfH_{2.24}$
	0.6	240	630	1.21	$HfH_{2.18}$	-"-	84	707	25	603	1.22	$HfH_{2.20}$
	07	255	675	1.15	$HfH_{2.08}$	-"-	60	615	25	495	1.18	$HfH_{2.13}$
	0.8	270	670	1.10	$HfH_{2.06}$	-"-	100	915	25	625	1.22	$HfH_{2.20}$
	0.9***	325	695	1.20	$HfH_{2.17}$	-"-	100	840	***	620	1.23	$HfH_{2.22}$

* the TRS for Zr was carried out at dose 100 Mrad.

** the TRS for Hf was carried out at doses 45-55 Mrad.

*** the CS for HfH_2 was implemented at room temperature in 3 hours after irradiation under dose rate 0,9 Mrad/sec.

2.2. «COLD SYNTHESIS» OF HAFNIUM HYDRIDES.

During investigations, the new phenomenon of cold synthesis (CS) was revealed. Earlier we observed this phenomenon in systems Ti-H, Zr-H. The essence of CS is the next: after irradiation of sample in vacuum up to dose 100 Mrad (at dose rates between 0.05 – 0.9 Mrad/sec), it was cooled down to 25-60°C and Hydrogen was filled to camera (Fig. 5, curve 1). In 20-30 second the thermocouple recorded sharp jump of temperature up to 420-620°C (curve 2). The further course of the reaction proceeded similar to TRS for this system. It is very interesting, that CS reaction can proceed also at ambient temperature. Besides, if CS in Ti-H and Zr-H began at irradiation with dose rate 0.2 Mrad/sec, for Hf-H it begin at 0.05 Mrad/sec (Table 1). The nature of this phenomenon is still difficult to establish unequivocally. Probably, at irradiation, an «accumulation of energy» takes place, «particular radiation damages» (defects, strains in crystal lattice, etc.) happen and the new structural defects are formed resulting in strong activation of starting metals. Then, in hydrogen environment, the reaction of metals with it appeared possible at low temperature, close to the ambient, without any exterior intervention. The temperature developed at irradiation in vacuum and at CS increases with the dose rate of preliminary irradiation. And more power of beam, more brightly the CS proceeds and longer the arisen radiation damages are saved. Actually, there is an original activation of metal resulting in its ability to react with hydrogen at ambient temperature.

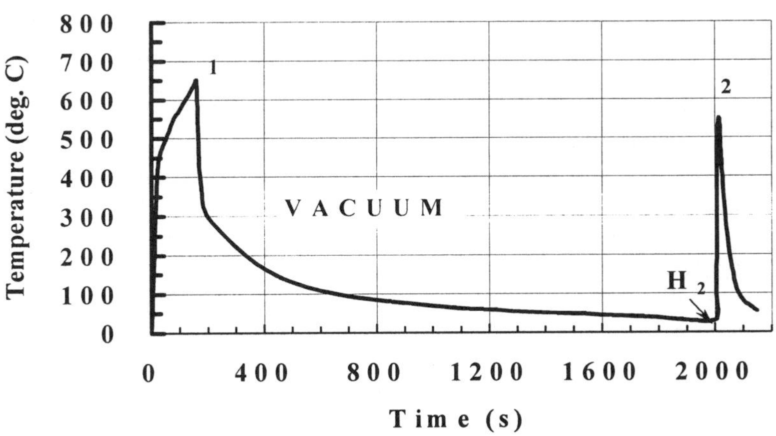

Figure 5. The CS-thermogram in Hf-H_2 system.

At hydrogenation by traditional methods, vacuum activation of metals is carried out at 1000°C during 10-12 hours. Then, the hydrogenation proceeds after long-therm keeping up in hydrogen at high temperature followed by a step-vise cooling with the rate 100°C/hour. Activation in accelerated electron beam occurs instantaneously. It is strong enough to allow the metals' reaction with hydrogen at ambient temperature and even in several hours after switch off the beam.

As it is seen from the data of X-ray analysis (Tables 1 and 2), in the system Hf-H, CS takes place at all irradiation dose rates bringing to the formation of single-phase bct hafnium hydride of superstoichiometric composition, $HfH_{2.15-2.42}$ (H/Me > 2). But at TRS conditions, a single-phase hydride $HfH_{2.22}$ is formed only at 0.9 Mrad/sec (see Table 1).

TABLE 2. The characteristics of cold synthesis in Hf-H system

Dose, *Mrad*	Temperature in vacuum, ^oC	Time of hydrogen feeling	Temperature of exothermic reaction, ^oC	Hydrogen content, *mas. %*	Formula, phase
68	505	15min(100°C)	485	1.24	$HfH_{2.24}$ sct
75	650	30 min	620	1.13	$HfH_{2.04}$ sct
75	680	1 hour	550	1.13	$HfH_{2.04}$ sct
75	590	2 hours	590	1.24	$HfH_{2.24}$ sct
100	630	3 hours	600	1.28	$HfH_{2.31}$ sct
75	655	4 hours	Didn't realized	-	-
100	685	4 hours	620	1.10	$HfH_{1.99}$ sct

The irradiation of starting bi-phase hafnium (fcc hydride containing) in vacuum has shown, that the developed temperature is lower than the temperature of decomposition of fcc hydride (850°C). Nevertheless, at irradiation by dose rates since 0.5 Mrad/sec, the portion of this phase decreases. It means that the decomposing of the fcc phase occurs mainly at the expense of radiation effect of electron beam in vacuum, but not at the expense of temperature increase (warming). At TRS of hafnium hydride, the fcc phase disappears only at high dose rate (0.9 Mrad/sec, $T_{react.}$ = 695°C). In this conditions afterhydrogenation occurs in high power electron beam. A series of experiments was carried out also for obtaining the life-times of these «radiation damages», i.e. for definition of the time of the activity saving in metal. It was found that in hafnium it is saved during 4 hours.

In Table 2, the data are presented on CS of hafnium hydride at hafnium irradiation with dose rate 0.5 Mrad/sec.

It was also obtained, that the ability to save the arisen radiation damages (life-time) mainly is determined by the dose rate of preliminary irradiation (either by total dose, or by the temperature developed at irradiation in vacuum). For example, at irradiation of Hf at dose rate 0.2 Mrad/sec (in vacuum, up to total dose 70 Mrad, $T_{warm.}$ = 355°C), CS takes place at 60°C. Thus, in consequence of the exothermal reaction, the temperature up to 500°C is developed. However, at these conditions the reaction does not proceed in three hours. At irradiation with dose rate 0.3 Mrad/sec (total dose 42 Mrad, $T_{warm.}$ = 380°C) CS proceeds even in 3 hours at ambient temperature.

As a result of CS, a hafnium hydride of superstoichiometric composition was obtained more hydrogen-rich than at above mentioned TRS. Its settlement formula is $HfH_{2.2-2.42}$. According to the differential thermal analysis data, the kinetics and the decomposition points of hafnium hydrides obtained by different methods - TRS, CS and SHS do not differ (Table 3).

TABLE 3. The differential thermal analysis data for of hafnium hydrides obtained by different methods

Method of synthesis	Hydrogen content, *mas.%*	Settlement Formula	Temperature of first endo-effect,	Temperature of second endo-effect
SHS	1.1	HfH_2	440	770
TRS at 0.3 Mrad/s	1.13	HfH_2	440	765
TRS at 0.8 Mrad/s	1.28	$HfH_{2.31}$	430	780
CS	1.34	$HfH_{2.42}$	430	780

3. Conclusions

1. For the first time, the thermal-radiation processes in system hafnium-hydrogen are investigated and the hafnium hydride is synthesized.
2. The mechanism and the main regularities of the process of hafnium hydride formation in the accelerated electron beam are established.
3. The new phenomenon of CS is revealed. As a result of it, the bct superstoichiometric hafnium hydride ($HfH_{2.42}$) is obtained.

Acknowledgements The work has been supported by ISTC. Project A-192.

4. References

1. Pikaev, A.K. (1986) *Modern Radiation Chemistry. Main State, Experimental Technique and Methods.* Nauka, Moscow.
2. Solovetsky, Yu.I., Lunin, V.V., Panteleeva, I.M. Formation and decomposition of Zr-Me-H (Me -Ni,,Co,Fe) hydrides under the beam of accelerated electron. International Symposium on Metal-Hydrogen System. Les Diablerets, 1996, Switzerland, Aug. 25-30, F5:02.
3. Dolukhanyan, S.K., Alexanian, A.G. and Hakobian, H.G.(1995) Interaction of hafnium with hydrogen and nitrogen in the combustion regime. *Int. J. Hydrogen Energy* **20** (5) 391-395.
4. Dolukhanyan, S.K., Shekhtman, V.Sh., Aghajanyan, N.N., Abrahamyan, K.A., Harutyunyan, Kh.S., Aleksanyan, A.G., Hakobyan H.G. and Ter-Galstyan, O.P. *(2000)* Radiation-thermal synthesis of titanium hydride. *Khimicheskaya Fizika, **19**(12) 21-25.*
5. Dolukhanyan, S.K., Shekhtman, V.Sh., Aghajanyan, N.N., Abrahamyan, K.A., Harutyunyan, Kh.S., Aleksanyan, A.G., Hakobyan, H.G., and Ter-Galstyan, O.P. Thermal-radiation processes in the Me^{IV}-H system. Synthesis of hydrides of metals. Abstracts of Internationale Symposium on Metal Hydrogen Systems, MH2000,1-6 Oct. 2000, Noosa, Queensland, Australia, p.96.

KINETICS OF THE HYDROGEN-ISOTOPES SHORT-RANGE ORDER IN INTERSTITIAL SOLID SOLUTIONS H.C.P.-*Ln*–H(D,T)

V.A. TATARENKO, T.M. RADCHENKO
G.V. Kurdyumov Institute for Metal Physics, N.A.S.U.,
36 Academician Vernadsky Blvd., UA-03680 Kyyiv-142, Ukraine

Abstract

The short-range order kinetics is studied by the using available data of measurements of residual electrical resistivity for interstitial solid solutions h.c.p.-Lu–H(D,T) during the thermal annealing. Within the framework of the first-order and (more realistic) second-order kinetics models, the relaxation times of this process in some α-Lu–H solid solutions at temperatures 160–180 K are evaluated on a base of the experimental results of P. Jung and R. Lässer (1992). The annealing-temperature dependent equilibrium residual resistivity, the pre-exponential factor in the Arrhenius law for the relaxation time, and energy of migration activation for these solid solutions are obtained.

1. Introduction

The rare earths absorb hydrogen readily and form solid solutions and/or hydrides exothermally at temperatures of several hundred Kelvin degrees. Their phase diagrams consist, in general, of three basic parts [1]: (a) the metallic solid solution, or α-phase, with the H atoms inserted in the tetrahedral interstices of the host-metal lattice and behave as impurities; (b) the equally metallic dihydride β-phase, where the two H atoms occupy ideally the two available tetrahedral sites; this phase crystallizes in the f.c.c. fluorite system; (c) the insulating trihydride, or γ-phase, which possesses an h.c.p. unit cell with both tetrahedral sites and the one octahedral site filled up. Exceptions are the divalent lanthanides Eu and Yb, whose dihydrides are already insulators and exhibit an orthorhombic structure, and Sc whose very small unit cell does not normally accept more than two H atoms. For the heavy lanthanides from Ho to Lu, as well as for Sc and Y, there exist two varieties in the metallic phase, α and α*, the latter being metastable at low temperatures and representing, in fact, an ordered configuration of the H sublattice [1, 2]. But quenching across the ordering temperature in the α*-phase creates disorder in the ordered configurations, *i.e.* hydrogen 'defects' which, during isothermal annealing, migrate and recombine in a recovery stage [1, 2].

To investigate a short-range order (SRO) kinetics, a direct approach consists of determining the SRO structure by diffuse-scattering experiments at a given

N. Veziroglu et al. (eds.), Hydrogen Materials Science and Chemistry of Metal Hydrides, 123–132.

temperature as a function of time. This suppose a large number of measurements, which are not easily performed by such a method. Another solution is to follow a physical property affected by the SRO evolution. The most used is electrical resistivity, which allows one to obtain the results with more simplicity and accuracy. Nevertheless such measurements could only be considered as macroscopic information compared with the determination of each Warren–Cowley parameter. Indeed the different effects contributing to the variations of the electrical resistivity are difficult to separate. This led to the development of several kinetic models for predicting the evolution of SRO.

In the present paper, we report an investigation of the SRO kinetics of the H-isotopes at tetrahedral interstices in h.c.p. lanthanoids (*Ln*), in particular, lutetium (Lu), using experimental data of the residual electrical-resistivity measurements during isothermal annealing at low temperatures (from 150 K to 200 K). This temperature interval is very important—an electrical resistivity anomaly takes place in it (see Refs. [1, 3, 4] and references therein).

Because of its macroscopic character, it is clear that method of 'electrical-resistance change' cannot bring an information as detailed as that deduced from diffuse-scattering intensity measurements [5–13]. However, the high accuracy of the resistivity measurement and its great sensitivity to any change in the atomic distribution, combined to its relative experimental simplicity, makes this technique a useful tool in studying both long-range order and short-range order kinetics [14–16].

2. Results and Analysis

Before presenting and analysing our results, let us consider the main features of the phenomenological models to interpret SRO-kinetics data for hydrogen isotopes in interstitial solid solutions h.c.p.-*Ln*–H(D,T) based on residual-resistivity measurements during the isothermal annealing.

2.1. FORMAL SRO-KINETICS MODELS

Because of the macroscopic character of the resistivity relaxation, all these models are phenomenological and macroscopic. They are based on the hypothesis that, for any interstitial mechanism of hydrogen diffusion in h.c.p.-LnH_c system ($0 \leq c \leq 0.5$), the reciprocal of relaxation time for the *i*-type channel of relaxation process, τ_i^{-1}, is directly proportional to the hydrogen jump frequency, ν_{Hi}, with an efficiency factor χ_i:

$$\tau_i^{-1} = \chi_i (1 - c) \nu_{Hi}. \quad (1)$$

As assumed, the relaxation time corresponds to the time required for the *i*-type channel migration of hydrogen to change the SRO-parameter value in *e* times. The temperature dependence of ν_{Hi} has its usual definition given by a Boltzmann distribution,

$$\nu_{Hi} = \nu_{0i} \exp\{- E_{mi}/(k_B T)\}. \quad (2)$$

Here ν_{0i} is a pre-exponential factor, k_B—Boltzmann constant, T—annealing temperature, E_{mi}—an energy of the *i*-type channel migration activation of H atoms.

In a case of spatial redistribution of H atoms between (tetrahedral) interstices, E_{mi} corresponds to H-atom total activation energy, E_{ai}: $E_{ai} \approx E_{mi}$. So, the temperature dependence of τ_i follows the so-called Arrhenius law [1]:

$$\tau_i = \tau_{0i}\exp\{E_{ai}/(k_B T)\} \text{ with } \tau_{0i} = \{\chi_i(1-c)\nu_{0i}\}^{-1}. \tag{3}$$

In this case, experimental data have to be described with only two parameters τ_{0i} and E_{ai} for each (ith) contribution to relaxation mechanism.

In residual-resistivity studies, the rate of isothermal variations of the degree of SRO is generally considered to be proportional to a 'thermodynamical force' as a function F of the distance to equilibrium.

So, the kinetic equation could be written in the form [16]

$$\frac{\partial\rho(t,T)}{\partial t} = -F\left(\{\tau_i^{-1}\}; \rho(t,T) - \rho_\infty(T)\right), \tag{4}$$

where $\rho(t,T)$ is the instantaneous residual resistivity and $\rho_\infty(T)$—the equilibrium residual resistivity at annealing temperature T.

Under stationary conditions, *i.e.* constant concentration of H atoms as interstitial defects, several different kinetic treatments are commonly used.

The simplest one is the first-order kinetics model with

$$F\left(\tau^{-1}; \rho(t,T) - \rho_\infty(T)\right) = \tau^{-1}\left(\rho(t,T) - \rho_\infty(T)\right), \tag{5}$$

where the SRO relaxation is described by a single-exponential behaviour of the normalized resistivity change [5, 17],

$$\frac{\rho(t,T) - \rho_\infty(T)}{\rho_0(T) - \rho_\infty(T)} = \exp\left[-\frac{t}{\tau(T)}\right] \tag{6}$$

($\rho_0(T)$—the initial residual resistivity at annealing temperature T) and characterised by a unique relaxation time $\tau(T)$, which is defined in terms of the hydrogen jump frequency (mobility), ν_H, with an efficiency factor χ, and the hydrogen concentration, c, in Ln (see Eq. (1)).

Higher-order (n) kinetics model resulting in a sum of several exponential relaxations is postulated as follows [14, 18, 19, 22]:

$$\frac{\rho(T) - \rho_\infty(T)}{\rho_0(T) - \rho_\infty(T)} = \sum_{i=1}^{n} A_i \exp\left[-\frac{t}{\tau_i(T)}\right], \tag{7}$$

with

$$\sum_{i=1}^{n} A_i = 1, \tag{8}$$

where A_i—the weight of the ith channel of relaxation process. Such a more complex relaxation behaviour, in the manner of a Khachaturyan approach [20] to the description of SRO-relaxation kinetics, might be also explained in the model of disperse order [18, 19]. As the number of independent fit parameters increases as $3n-1$, in order to keep the analysis trustworthy, the number of exponentials (n) necessary to fit the data, as a rule, should not exceed 2 or 3 [15].

When the SRO relaxation is not really an first-order (single-exponential) kinetics (6), and when it cannot be described by a sum of 2 or 3 exponentials (7), a model lying on a log-normal Gaussian distribution of the relaxation times around an average value $\bar{\tau}(T)$ given by an Arrhenius law, with a half-width, β, of the distribution, is used:

$$f(\ln\tau,T) = \frac{1}{\beta\sqrt{\pi}}\exp\left[-\left(\frac{\ln\{\tau/\bar{\tau}(T)\}}{\beta}\right)^2\right]. \tag{9}$$

The resistivity data are simulated by

$$\frac{\rho(t,T)-\rho_\infty(T)}{\rho_0(T)-\rho_\infty(T)} = \int f(\ln\tau,T)[1-\exp(-t/\tau)]d\ln\tau. \tag{10}$$

In this model developed by Nowick and Berry [23] and first applied to SRO-induced resistivity relaxation by Balanzat and Hillairet [24], three parameters are necessary to reproduce these data: τ_0, E_a (as parameters of an Arrhenius law for $\bar{\tau}(T)$), and β.

At last, a power law, corresponding to the SRO-relaxation parameter γ, with no definite physical meaning, in which derivative $\partial\rho(t,T)/\partial t$ is proportional to $\tau^{-1}\{\rho(t,T)-\rho_\infty(T)\}^\gamma$ as follows

$$\frac{\partial\rho(t,T)/\partial t}{\rho_0(T)-\rho_\infty(T)} = -\tau^{-1}\left|\frac{\rho_0(T)-\rho_\infty(T)}{B}\right|^{\gamma-1}\left[\frac{\rho(t,T)-\rho_\infty(T)}{\rho_0(T)-\rho_\infty(T)}\right]^\gamma. \tag{11}$$

Equation (11) deduced from the theory of chemical-reaction rates has been proposed by Schulze and Lücke [25] assuming that the rate of change of the parameter $\tilde{\alpha}$ that characterises the degree of SRO is described by a power law, and that resistivity is a linear function of $\tilde{\alpha}$. Here B is a proportionality factor, with the dimension of a resistivity, between the quantity $\tilde{\alpha}$ and the SRO-induced residual electrical resistivity ρ_{SRO} in the form

$$\rho_{SRO} = B\tilde{\alpha} \text{ with } \rho_{SRO} = \rho - \rho_{dis}, \tag{12}$$

where ρ_{dis} is residual electrical resistivity of the disordered h.c.p.-*Ln*–H alloy. It is easy to see that if $\gamma = 1$, then power-law kinetics Eq. (11) turn into first-order kinetics Eqs. (4), (5) with solution (6).

For a binary interstitial alloy, in terms of the degree of SRO, *i.e.* considering the other contributions (magnetic effects, changes in electronic structure) to be negligible, the residual-resistivity formula comes to the expression, following to Rossiter and Wells [26],

$$\rho = \rho_{dis}\left(1+\sum_{l=1}^{\infty} Z_l\alpha_l Y_l\right), \tag{13}$$

where α_l are usual Cowley parameters of the Z_l atoms in the lth shell at distance r_l and Y_l is an integral function which may either be positive or negative depending on the radius of the co-ordination shell and the number of conduction electrons per atom.

The sign of $\partial\rho/\partial T$ calculated from the last equation will thus depend on the signs of the parameters α_l and also of the integrals Y_l [26].

2.2. APPLICATIONS TO EXPERIMENTAL RESULTS

The kinetics of short-range order of the H-isotopes (H, D, T) in h.c.p. Lu was studied in Ref. [27] by electrical resistivity measurements during isothermal (T= =const) anneals performed with two specimens. Their hydrogen concentrations, given as the ratio c of H to Lu atoms, were determined from pressure drop in the

TABLE I. Values of relaxation time, equilibrium resistivity, and difference in initial and equilibrium resistivities, $\rho_0-\rho_\infty$, within the framework of the first-order kinetics model (see Eq. (6)) for α-Lu–H solid solutions at different temperatures.

Temperature T, K	$LuH_{0.180}$ Relaxation time τ, 10^4 s	Equilibrium resistivity ρ_∞, μΩ·cm	$\rho_0-\rho_\infty$, μΩ·cm	Standard deviation δ_ρ, μΩ·cm	$LuH_{0.254}$ Relaxation time τ, 10^4 s	Equilibrium resistivity ρ_∞, μΩ·cm	$\rho_0-\rho_\infty$, μΩ·cm	Standard deviation δ_ρ, μΩ·cm
161.8	2.07 ± 0.23	51.87 ± 0.04	1.29 ± 0.05	$5\cdot10^{-5}$	1.71 ± 0.17	66.39 ± 0.04	1.76 ± 0.04	$5\cdot10^{-5}$
167.3	1.01 ± 0.07	52.95 ± 0.01	1.11 ± 0.03	$7\cdot10^{-6}$	0.72 ± 0.08	67.63 ± 0.04	1.69 ± 0.05	$8\cdot10^{-5}$
180.2	0.12 ± 0.02	56.33 ± 0.01	0.47 ± 0.01	$1\cdot10^{-5}$	0.05 ± 0.01	71.41 ± 0.01	1.15 ± 0.01	$3\cdot10^{-5}$

reaction volume during loading (charging) and during degassing after the measurements as follows: $c=0.180$ and $c=0.254$; another estimates of corresponding H-to-Lu ratios for these two specimens were derived from measurement of the quenched-in resistivities ρ at 4.2 K and 296 K and are $c_\rho=0.259$ and $c_\rho=0.280$, respectively (*i.e.* above the T-independent solubility limit, $c^{max}\approx0.20$, in the metastable low-T short-range-ordered α*-phase [1]). Considerable changes in resistivity $\rho(t,T)$ of LuH_c ($c=0.180$ and $c=0.254$) during isothermal annealing within reasonable times, t, were only observed at temperatures between about 150 K and 200 K [27]. As evaluated, an activation enthalpy of 0.45 eV corresponds to the migration of H-atoms at temperatures from about 160 K to about 190 K [27].

The relaxation times, $\{\tau_i\}$, for alloys $LuH_{0.180}$ and $LuH_{0.254}$ were estimated by using experimental results of Ref. [27] within the framework of the first-order kinetics model and the (more realistic) second-order kinetics model (see Tables I and II, respectively). The normalized resistivity changes in these solid solutions during isothermal annealing at the given temperatures were obtained. Typical results are given in Figs. 1, 2 showing a decrease of normalized resistivity change at a decreasing rate, eventually attaining an asymptotic value, when residual resistivity attains the value $\rho_\infty(T,c)$ ('for $t\to\infty$'; see Fig. 3). The resistivity decrease is caused by the migration of hydrogen to induce the equilibrium short-range order.

The Arrhenius-type dependences (with parameters $\{\tau_{0i}\}$ and $\{E_{ai}\}$; see Table III) of the relaxation times τ (within the framework of the first-order kinetics

TABLE II. Values of relaxation time, equilibrium resistivity, and difference in initial and equilibrium resistivities, $\rho_0-\rho_\infty$, within the framework of the second-order kinetics model ($n=2$ in Eq. (7)) for α-Lu–H solid solutions at different temperatures.

Temperature T, K	$LuH_{0.180}$ First relaxation time τ_1, 10^4 s	Second relaxation time τ_2, 10^4 s	Equilibrium resistivity ρ_∞, μΩ·cm	$\rho_0-\rho_\infty$, μΩ·cm	Standard deviation δ_ρ, μΩ·cm	$LuH_{0.254}$ First relaxation time τ_1, 10^4 s	Second relaxation time τ_2, 10^4 s	Equilibrium resistivity ρ_∞, μΩ·cm	$\rho_0-\rho_\infty$, μΩ·cm	Standard deviation δ_ρ, μΩ·cm
161.8	3.09 ± ± 0.46	0.14 ± ± 0.08	51.76 ± ± 0.05	1.12 ± ± 0.10	$1\cdot10^{-5}$	4.65 ± ± 1.09	0.57 ± ± 0.08	66.04 ± ± 0.12	1.67 ± ± 0.12	$4\cdot10^{-6}$
167.3	2.52 ± ± 0.54	0.08 ± ± 0.03	52.67 ± ± 0.06	1.05 ± ± 0.16	$4\cdot10^{-6}$	1.90 ± ± 0.37	0.26 ± ± 0.04	67.47 ± ± 0.04	1.49 ± ± 0.14	$1\cdot10^{-5}$
180.2	0.54 ± ± 0.06	0.05 ± ± 0.01	56.29 ± ± 0.01	0.40 ± ± 0.02	$1\cdot10^{-7}$	0.36 ± ± 0.07	0.03 ± ± 0.01	71.37 ± ± 0.01	0.60 ± ± 0.09	$1\cdot10^{-6}$

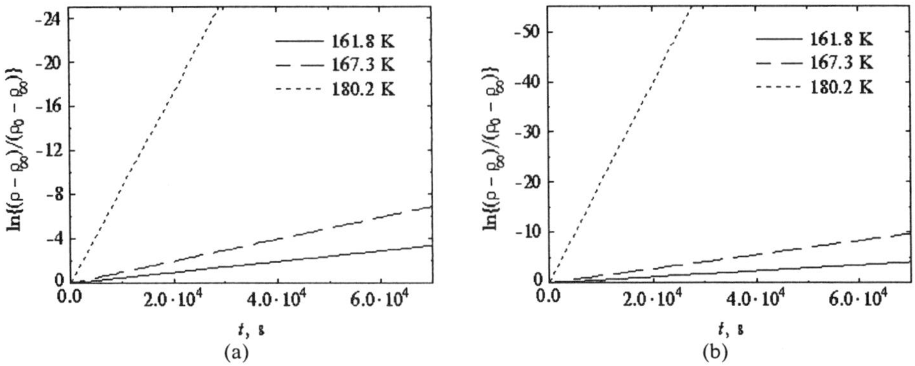

Figure 1. Logarithm of the normalized resistivity changes versus isothermal-annealing time within the framework of the first-order kinetics model (see Eq. (6)) for the interstitial solid solutions h.c.p.-$LuH_{0.180}$ (a) and h.c.p.-$LuH_{0.254}$ (b) at the given annealing temperatures.

model) and τ_1, τ_2 (within the framework of the more realistic second-order kinetics model) on a temperature, T, for alloys $LuH_{0.180}$ and $LuH_{0.254}$ are presented in Fig. 4. In any case, the obtained values of activation energies of migration $\{E_{ai}\}$ (0.13–0.49 eV at $T \in$ (160 K, 180 K); Table III) differ from comparatively high-temperature ($T \in$ (380 K, 540 K)) data [28] about diffusion of H atoms in the enough weak solution $LuH_{0.05}$: $E_a = 0.574 \pm 0.015$ eV (see also Refs. in [1]). As in Ref. [29], this result suggests that the H atoms in Lu migrate more easily at lower temperatures due to tunnelling-mechanism diffusion contribution.

The diffusivity of H atoms in Lu was studied by the residual electrical-resistivity measurements at temperatures from about 135 K to 170 K in Ref. [29].

Disordered H atoms, which are formed by the quenching from 273.2 K, migrate to order during annealing in the above temperature range [29]. Isothermal resistance curves at various annealing temperatures for $LuH_{0.06}$ and $LuH_{0.12}$ against annealing time were obtained in this work. Besides, the reciprocal annealing-temperature dependence of the reciprocal relaxation time of the process for these

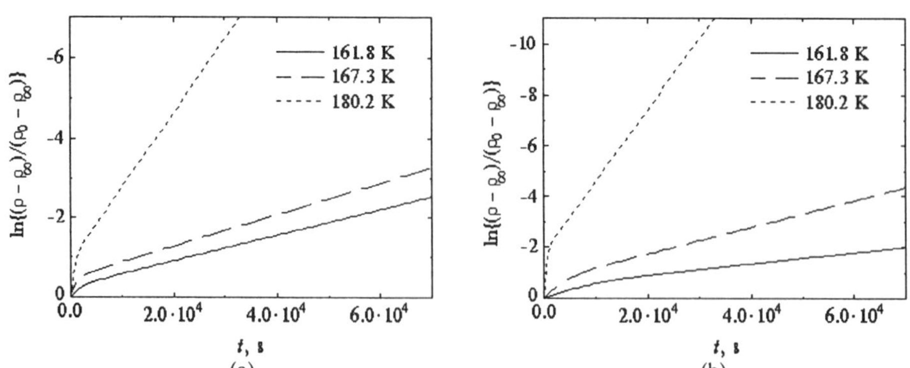

Figure 2. Logarithm of the normalized resistivity changes versus isothermal-annealing time within the framework of the second-order kinetics model ($n = 2$ in Eq. (7)) for the interstitial solid solutions h.c.p.-$LuH_{0.180}$ (a) and h.c.p.-$LuH_{0.254}$ (b) at the given annealing temperatures.

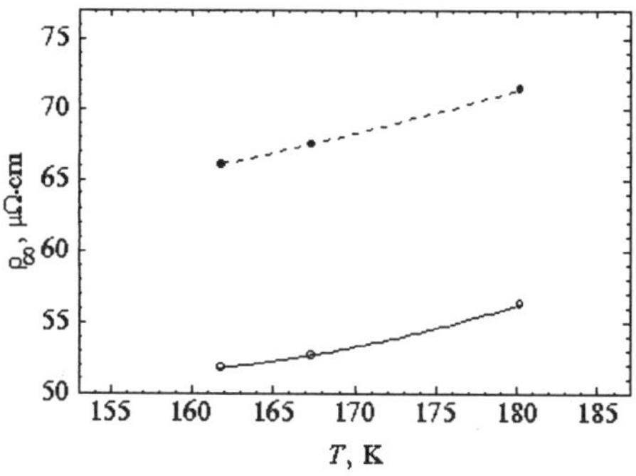

Figure 3. Equilibrium resistivity, ρ_∞, versus annealing temperature, T, for the h.c.p-$LuH_{0.180}$ (open circles on solid line) and h.c.p.-$LuH_{0.254}$ (solid circles on dotted line) solid solutions (see Table II).

interstitial solid solutions is also obtained. The rate of the resistivity decrease during the annealing depends on the H-ordering rate [29]. From the resistance decrease during isothermal annealings, the activation energies of H migration were determined as 0.428 ± 0.003 eV and 0.433 ± 0.003 eV for both solid solutions $LuH_{0.12}$ and $LuH_{0.06}$, respectively. So, such an energy did not depend on the H concentration in the range of the study in [29] within the experimental error.

As expected, the time dependence of the normalized resistivity change for $LuH_{0.06}$ and $LuH_{0.12}$ (see Figs. 2 and 3 in Ref. [29]) can be described within the framework of the first-order (or higher-order) kinetics model.

At these hydrogen concentrations (as in Ref. [29]), hydride is not formed, and the hydrogen order is caused by long-range diffusion such as it was revealed in the measurements of Gorsky effect [2, 28].

3. Summary

A short-range order kinetics in the interstitial solid solutions h.c.p.-Lu–H is studied numerically by the residual electrical-resistivity change during isothermal annealing.

Experimental results obtained in Ref. [27] for $\rho = \rho(t)$ in interstitial solid solutions h.c.p.-Lu–H(D,T) have been described by its authors within the framework of the *power*-law ($1 < \gamma < 2.5$) kinetics model (which is most probably justified at long times and/or high temperatures). In a given paper, we proposed to describe these experimental data (and maybe data of Ref. [29]) within the framework of the alternative

TABLE III. Parameters obtained by the least-squares method within the framework of the first-order and second-order kinetics models for two α-Lu–H solid solutions at issue.

Alloy	First-order kinetics model		Second-order kinetics model			
	Pre-exponential parameter τ_0, s	Activation energy of migration E_a, eV	First pre-exponential parameter τ_{01}, s	Second pre-exponential parameter τ_{02}, s	First activation energy of migration E_{a1}, eV	Second activation energy of migration E_{a2}, eV
$LuH_{0.180}$	$8.59 \cdot 10^{-12}$	0.41	$6.02 \cdot 10^{-4}$	$8.77 \cdot 10^{-2}$	0.25	0.13
$LuH_{0.254}$	$4.05 \cdot 10^{-9}$	0.49	$6.68 \cdot 10^{-7}$	$1.15 \cdot 10^{-9}$	0.35	0.41

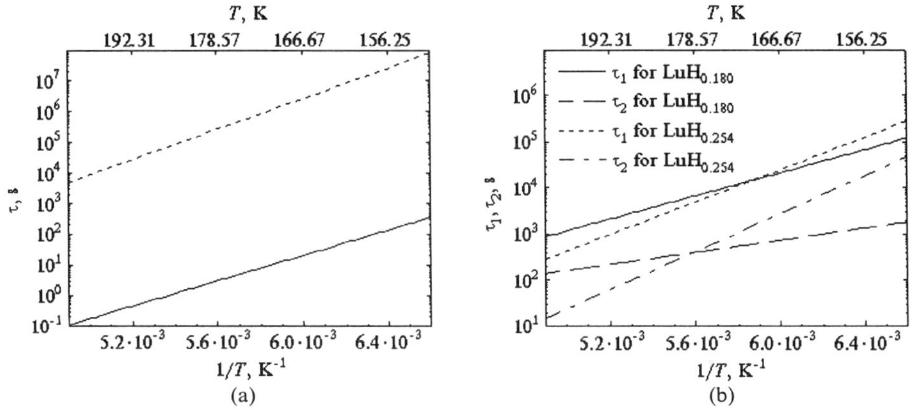

Figure 4. Relaxation times, τ, τ_1 and τ_2, against reciprocal annealing temperature, $1/T$, within the framework of the first-order (a) and second-order (b) kinetics models for h.c.p.-Lu–H solid solutions. In Fig. 4(a), solid line corresponds to $LuH_{0.180}$ and dotted line corresponds to $LuH_{0.254}$.

first-order ($\gamma = 1$) and second-order kinetics models, which are justified starting with small times (early stages) when SRO-induced relaxation of $\rho(t)$ has a single-*exponential* or two-*exponential* behaviour. (The rate of the resistivity decrease during isothermal annealing correlates on the H-ordering rate [29].) The time-evolution curves of normalized resistivity changes for α-$LuH_{0.180}$ and α-$LuH_{0.254}$ interstitial solid solutions (during isothermal annealing [27]) are plotted here in Figs. 1, 2 by estimating the annealing-temperature dependent resistivity-relaxation times. The parameters ($\{\tau_{0i}\}$ and $\{E_{ai}\}$) of the Arrhenius-type temperature dependence of relaxation times, $\{\tau_i\}$, corresponding to the times required for the migration of hydrogen to change the SRO value in *e* times, are evaluated for these solid solutions. The values of $\{E_{ai}\}$ revealed by us for h.c.p.-Lu–H solutions, containing more amount of H ($c = 0.180$, 0.254 [27]) as compared with such a solid solution ($c = 0.05$) studied in Ref. [28], are clearly smaller than a value from Ref. [28]. This decrease of $\{E_{ai}\}$ values may be caused by the switching on of the tunnelling-diffusion mechanism which takes place predominantly at lower temperatures [29].

On the other hand, in comparison with results of Ref. [29] for $c = 0.06$ and 0.12, we link a deceleration of a relaxation of a resistivity revealed here with a fact that in more concentrated interstitial solutions at issue ($c = 0.180$, 0.254) the short-range order (spatial fluctuation of a concentration) and its thermodynamical fluctuations can be more developed. In fact, as well known, in concentrated solutions h.c.p.-Lu–H the anomaly of a resistivity in dependence of $[\partial\rho_\infty(T,c)/\partial T]\big|_{c=\mathrm{const}}$ on T within the range 160–190 K is expressed more clearly than in weak solutions [1–4, 8–13, 21, 30, 31] (alongside with salient features of diffuse scattering [5–7, 20, 32–36]).

4. Acknowledgements

The authors wishes to thank Prof. V.E. Antonov from the Inst. of Solid State Physics, R.A.S. (Chernogolovka), Prof. T.N. Veziroglu from Univ. of Miami (Coral

Gables), Dr. I.G. Ratishvili from the Inst. of Physics, G.A.S. (Tbilisi), Drs. V.M. Nadutov and R.V. Chepul's'ky from the Inst. for Metal Physics, N.A.S.U. (Kyyiv) for valuable conversations regarding the short-range ordering of alloys. We are grateful to Dr. P. Vajda from the École Polytechnique (Palaiseau), Prof. H. Wipf from Inst. für Festkörperphysik der Technischen Hochschule Darmstadt (Darmstadt), and Prof. Y. Fukai from Chuo Univ. (Tokyo) for communicating applicable results and important references. A grant #2412 from the STC in the Ukraine is gratefully acknowledged for the support in part.

5. References

1. Vajda, P. (1995) Hydrogen in rare earth metals, including RH_{2+x}-phases, in K.A. Gschneidner, Jr. and L. Eyring (eds.), *Handbook on the Physics and Chemistry of Rare Earths* **20**, Elsevier Science, Amsterdam, ch. 137, pp. 207–292.
2. Fukai, Yuh. (1993) *The Metal–Hydrogen System: Basic Bulk Properties*, in *Springer Series in Materials Science* **21**, Springer-Verlag, Berlin–Heidelberg–New York.
3. Daou, J.N., Lucasson A., and Lucasson P. (1976) Electrical resistivity of the α-phase lutetium–hydrogen at low temperature, *Solid State Commun.* **19**, 895–898.
4. Burger, J.P., Daou, J.N., Lucasson, A., Lucasson, P., and Vajda, P. (1985) Observation of short-range ordering in rare earth-H(D) solid solutions, *Z. Phys. Chem. Neue Folge* **143**, 111–128.
5. Cook, H.E. (1969) The kinetics of clustering and short-range order in stable solid solutions, *J. Phys. Chem. Solids* **30**, 2427–2437.
6. Chen, H. and Cohen, J.B. (1977) *J. Physique* **C7**, 314.
7. Bley, F., Amilius, Z., and Lefebvre, S. (1988) Wave vector dependent kinetics of short-range ordering in $^{62}Ni_{0.765}Fe_{0.235}$, studied by neutron diffuse scattering, *Acta. metall.* **36**, 1643–1652.
8. Bonnet, J.E., Ross, D.K., Faux, D.A., and Anderson, I.S. (1987) Long-range and short-range ordering of hydrogen or deuterium atoms in solid solution in yttrium, *J. Less-Common Metals* **129**, 287–295.
9. Metzger, T.H., Vajda, P., and Daou, J.N. (1985) Debye temperature and static displacements in LuH_x single crystals from energy dispersive x-ray diffraction, *Z. Phys. Chem. Neue Folge* **143**, 129–138.
10. Blaschko, O., Krexner, G., Pleschiutschnig, J., Ernst, G., Daou, J.N., and Vajda, P. (1989) Hydrogen ordering in α-LuD_x investigated by diffuse neutron scattering, *Phys. Rev. B* **39**, 5605–5610.
11. Blaschko, O., Krexner, G., Daou, J.N., and Vajda, P. (1985) Experimental evidence of ordering of deuterium in α-LuD_x, *Phys. Rev. Letters* **55**, 2876–2878.
12. Pleschiutschnig, J., Schwarz, W., Blaschko, O., Vajda, P., and Daou, J.N. (1991) Hydrogen ordering and local mode in α-LuD_x at low concentrations, *Phys. Rev. B* **43**, 5139–5142.
13. Blaschko, O. (1991) Hydrogen ordering in metal–hydrogen systems, *J. Less-Common Metals* **172–174**, 237–245.
14. Dahmani, C.E., Cadeville, M.C., and Pierron-Bohnes, V. (1985) Temperature dependences of atomic order relaxations in Ni–Pt and Co–Pt alloys, *Acta metall.* **33**, 369–377.
15. Afyouni, M., Pierron-Bohnes, V., and Cadeville, M.C. (1989) SRO relaxation in dilute NiAl solid solution through 4 K resistometry, *Acta metall.* **37**, 2339–2347.
16. Sitaud, B. and Dimitrov, O. (1990) Equilibrium and kinetics of local ordering in Ni(Al) solid solutions, *J. Phys: Condens. Matter* **2**, 7061–7075.
17. Kidin, I.I. and Shtremel', M.A. (1961) Kinetics of changes of the short-range order in binary alloys, *Fiz. Met. Metalloved.* **11**, 641–649 (in Russian).
18. Shtremel', M.A. and Satdarova, F.F. (1969) The short-range order in alloys with a body centred cubic lattice (equilibrium and kinetics), *Fiz. Met. Metalloved.* **27**, 396–407 (in Russian).
19. Püschl, W. and Aubauer, H.P. (1980) The kinetics of stable heterogeneous binary systems, *Phys. Status Solidi b* **102**, 447–457.
20. Khachaturyan, A.G. (1983) *Theory of structural transformations in solids*, Wiley, New York.
21. Vajda, P., Daou, J.P., Burger, J.P., Kai, K., Gschneidner, K.A., Jr., and Beaudry, I.J. (1986) Investigation of the anisotropic deuterium ordering in α-LuD single crystals by resistivity and heat-capacity measurements, *Phys. Rev. B* **34**, 5154–5159.
22. Bessenay, G. (1986) *Doctoral dissertation*, Univ. of Paris VI.

23. Nowick, A.S. and Berry, B.S. (1972) *Anelastic Relaxation in Crystalline Solids*, Academic Press, New York.
24. Balanzat, E. and Hillairet, J. (1981) A quench and irradiation study of the vacancy parameters in α-Cu 30 at.% Zn, *J. Phys. F: Met. Phys.* **11**, 1977–1990.
25. Schulze, A. and Lücke, K. (1972) The influence of vacancies on short-range order formation in Au–Ag-alloys, *Acta metall.* **20**, 529–542.
26. Rossiter, P.L. and Wells, P. (1971) The dependence of the electrical resistivity on short-range order, *J. Phys. C: Solid State Phys.* **4**, 354–363.
27. Jung, P. and Lässer, R. (1992) Short-range ordering of hydrogen isotopes in lutetium, *J. Alloys & Compounds* **190**, 25–29.
28. Völkl, J., Wipf, H., Beaudry, B.J., and Gschneidner, K.A., Jr. (1987) Diffusion of H and D in lutetium, *Phys. Status Solidi b* **144**, 315–327.
29. Yamakawa, K. and Maeta, H. (1997) Diffusion of hydrogen in Lu at low temperatures, *Defect and Diffusion Forum* **143–147**, 933–938.
30. Daou, J.N. and Vajda, P. (1988) Solutions solides d'hydrogene dans des terres rares a basses temperatures, *Ann. Chim. (Fr)* **13**, 567–574 (in French).
31. Daou, J.N., Vajda, P., Lucasson, A., Lucasson, P., and Burger, J.P. (1986) Low-temperature defect study of α-LuH(D)$_x$ solid solutions, *Philos. Mag. A* **53**, 611–625.
32. Tatarenko, V.A. and Tsinman, C.L. (1999) Strain-induced effects on interactions of non-metal atoms introduced in metals having h.c.p. structure. 2. Interstitial atoms on tetrahedral interstices, *Met. Phys. Adv. Tech.* **18**, 273–311.
33. Tatarenko, V.A. (2001) Interaction of interstitial hydrogen atoms within h.c.p. metals and distribution of diffuse radiation-scattering intensity in reciprocal space, *Int. J. Hydrogen Energy* **26**, 469–475.
34. Tatarenko, V.A. and Tsynman, C.L. (1997) Strain-induced and blocking effects in thermodynamics of the ordering and precipitation reactions within the off-stoichiometric close-packed-metal hydrides, *Solid State Ionics* **101–103**, 1061–1067.
35. Krivoglaz, M.A. (1996) *Diffuse Scattering of X-Rays and Neutrons. Fluctuations in Solids*, Springer, Berlin.
36. Tatarenko, V.A. and Radchenko, T.M. (2002) Direct and indirect methods of the analysis of interatomic interaction and kinetics of a short-range order relaxation in substitutional (interstitial) solid solutions, *Uspehi Fiziki Metallov* **3**, 1–126 (in Ukrainian).

INTERACTION OF TiH_X (x>2) WITH SOLUTIONS OF SOME ACIDS AND ALKALIES

I.M. ASTRELIN, G.N. PROKOFYEVA, V.I. SUPRUNCHUK,
YU.V. KNYAZEV, V.M. PANASHENKO [1], I.A. MOROZOV[1],
R.A. MOROZOVA[1]
National Technical University of Ukraine «Kiev Polytechnic Institute»,
37 Peremohy av., Kiev, 03056 Ukraine
[1] *Institute for Problems of Material Science of NAS Ukraine,*
3 Krzhyzhanovsky st., Kiev, 03142 Ukraine

Abstract

Comparative studies were made of the resistance of $TiH_{2.4}$ and $TiH_{1.8}$ titanium hydride in H_2SO_4 , HCl, NaOH and KOH water solutions at different temperatures and reagent concentrations. It was stated that the chemical processes of $TiH_{X>2}$ and $TiH_{X<2}$ dissolution were similar only differing in the rate and activation energy.

1. Introduction

The recent decades an increasing interest is being paid to investigations of hydrogen compounds of metals and intermetallides. Researchers are mainly interested in state diagrams, thermodynamic parameters and hydrogen sorption-desorption kinetics. This is explained by specific properties of hydrides, in particular those which shall allow their application as hydrogen accumulators, and also as a material to be used in chemical power sources. However, there is a few knowledge of chemical properties of hydrides as a whole, and in connection to acids and alkalis in particular. Only separate data can be found in publications concerning chemical properties of hydrides of metals and intermetallic compounds [1-5]. On the other hand, these data are of great interest useful for hydride application in various industries including reduction agents and catalysts.

The Institute for Problems of Materials Science has developed a proper technology of hydration that provided production of titanium hydrides described by the following formula: TiH_X, where x>2 [6-8]. This work studies samples of such hydrides using the dissolution kinetics. The interaction of TiH_X (x=2.40) with water solutions of H_2SO_4, HCl, NaOH or KOH was investigated as function of temperature and solution concentrations.

N. Veziroglu et al. (eds.), Hydrogen Materials Science and Chemistry of Metal Hydrides, 133–140.

There is a rare opportunity for direct comparison of results provided by various works because of details and conditions of such researches. Therefore, new $TiH_{2.4}$ hydride was compared with conventional commercial $TiH_{1.8}$ hydride under similar conditions.

2. Experimental details

To prepare $TiH_{2.4}$ titanium hydride, a titanium of «iodide» purity and hydrogen gas were used. Titanium was in the form of a thin chip of 0.048±0.007 mm as cut from the rods of this metal. The resulted hydride was of scale-like shape. To minimize the effect of the surface change when dissolution, a fraction of non-damaged chip was selected for the experiments, and the samples of $TiH_{2.4}$ and $TiH_{1.8}$ have been dissolved to the degree of not above than 40 % and 10 %, respectively. High-purity hydrogen was produced by thermal decomposition of $TiH_{1.8}$ titanium hydride. A unit of Sieverts-type was used for hydration.

The dissolution kinetics of samples in acids and alkalis was studied using the volumetric method of hydrogen evolution. The hydrogen volume was referred to the normal conditions and was converted to the actual concentration of Ti^{3+} ions in the solution and variation rate. The titanium ion condition in the solution was determined by spectrophotometric method (spectrophotometer SF-20) in a pressurized cell.

The samples were dissolved in a reaction vessel equipped with backflow condenser. The reaction vessel had a water thermostat for temperature control of ±0.5 K accuracy. The weighed portion of titanium hydride was 1.000 g, the volumes of solution were 10.0 and 20.0 ml. The concentrations of acid and alkali solutions, and also dissolution temperatures were rationalized due to preliminary studies.

3. Results and discussion

3.1. TITANIUM IONS

The resultant relationships of solution light absorption of the wave length of light flux $A=f(\lambda)$ revealed the presence of a single band of light absorption having peak $\lambda=500$-510 nm, witch corresponds to Ti^{3+} ions. The identification of these spectra shall be supported by a full similarity with the relation $A=f(\lambda)$ for the $Ti_2(SO_4)_3$ titanium sulfate solution of high purity. These light absorption spectra are similar in nature but only differing in intensity.

The oxidation degree of titanium ions in acid solutions was also determined through the volume of hydrogen evolved during dissolving of sample metallic titanium in solutions of sulfuric or hydrochloric acid. As was stated, the process fully corresponded to the stoichiometry equations.

3.2. FORMULAE

The titanium hydride formulae were calculated in this study using the stoichiometry of the dissolution process in the sulfuric or hydrochloric acid:

$$2TiH_X + 6H^+ = 2Ti^{3+} + H_2 + x\, H_2$$

The content of hydrogen in TiH_X was (in H_2SO_4/in HCl): x=1.77±0.04/1.81±0.03 for the conventional hydride, x=2.40±0.042/2.38±0.03 for the hydride prepared by advanced technology. A comparison of formulae provided from different acids showed their identity.

3.3. INTERACTION WITH ACIDS

The TiH_X relation to the acids was determined at 353 and 363 K levels for H_2SO_4 solution, and 347, 353 and 363 K for HCl. The acid concentrations during the interaction with hydrides decreased beginning from $C(H_2SO_4)$=3.76 mole/l, and C(HCl)=7.423 mole/l. Fig. 1 and 2 show titanium ions accumulation curves in H_2SO_4 and HCl solutions.

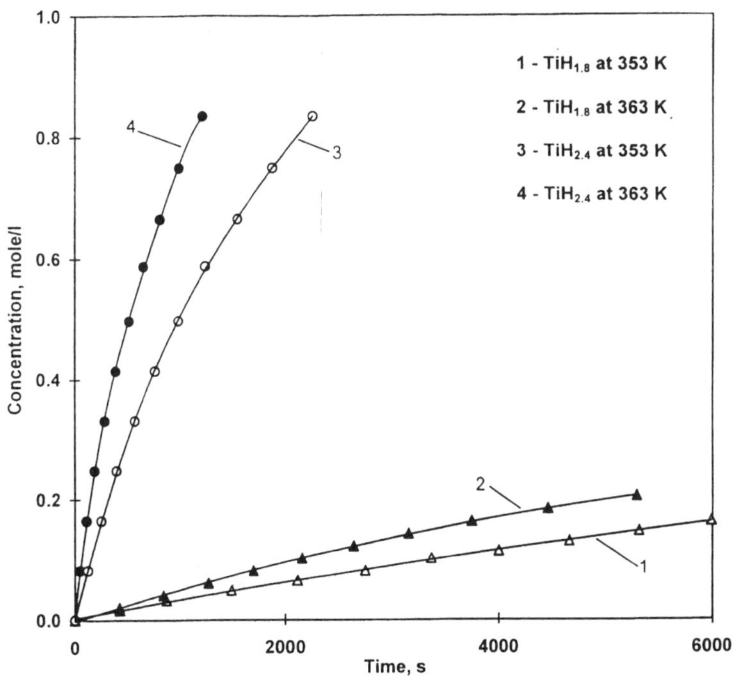

Figure 1. Ti^{3+}-ions accumulation in H_2SO_4 solution

The dissolution rates are referred in «dissolution rate-concentration» coordinates. This affords a more correct comparison of these as the interaction rates of samples are dependent of the reagent concentration, which can be different for each sample in the same instant. Thus, the dissolution rate of $TiH_{2.4}$ in H_2SO_4 shall vary within $5*10^{-4}$ and $2.5*10^{-3}$ mole/(l*s), as it seen from Fig. 3. The temperature rise from 353 to 363 K shall result then in 2-2.5 time dissolution rate gain. A concentration decrease shall yield a dissolution rate slowing. Thus, at 353 K, the H_2SO_4 concentration decrease from

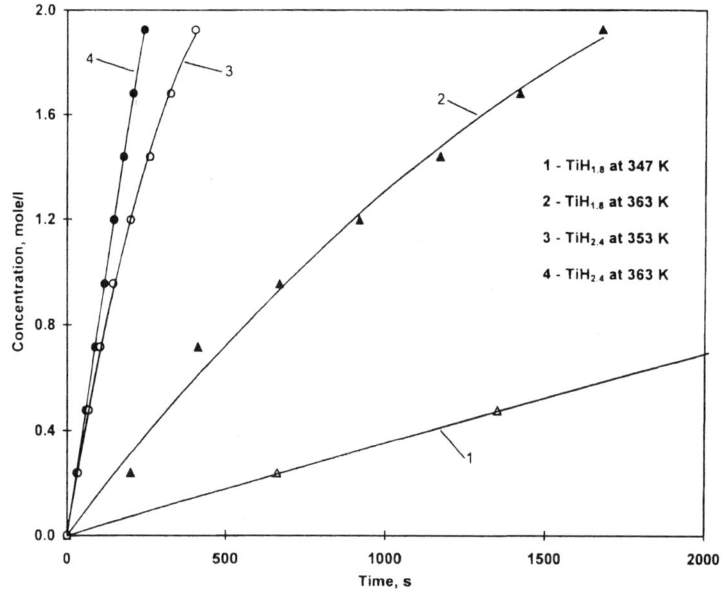

Figure 2. Ti^{3+}-ions accumulation in HCl solution

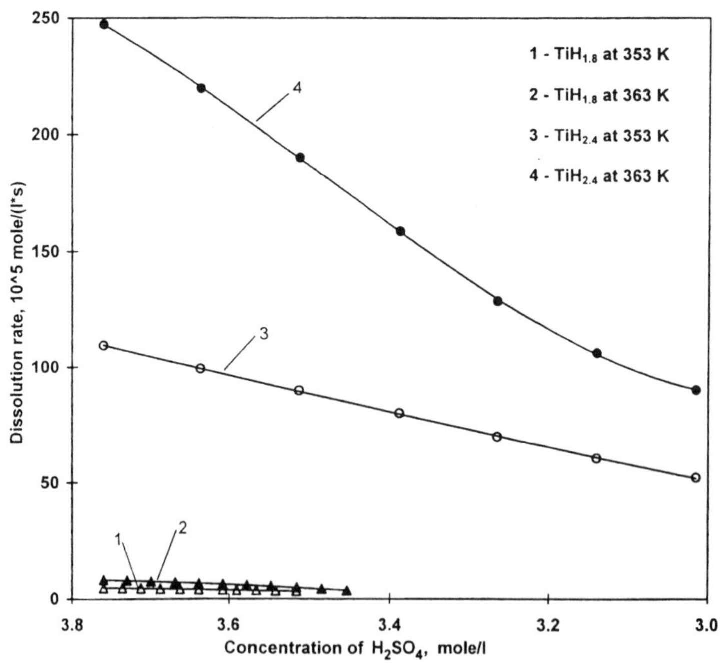

Figure 3. Curves of titanium hydrides dissolution rate in H_2SO_4

$C(H_2SO_4)$=3.76 mole/l to 2.51 mole/l shall give a nearly 3 time decrease of this rate, whereas at 363 K, similar concentration variations yield a 8 time decrease of dissolution rate. This is an evidence of a considerable influence of the acid concentration, as well as temperature on the interaction of $TiH_{2.4}$ and H_2SO_4. The $TiH_{1.8}$ behavior in H_2SO_4 is similar, however their dissolution rates are 1-2 order lower under the same conditions, which can be only suggested by an elevated content of hydrogen in $TiH_{2.4}$.

Like in H_2SO_4 solutions, the $TiH_{2.4}$ sample exhibits a higher activity in HCl solutions than $TiH_{1.8}$ sample (Fig. 4). Thus, at 363 K, the initial rate of dissolution of $TiH_{2.4}$ is $8.1*10^{-3}$ mole/(l*s) versus $9.2*10^{-4}$ for $TiH_{1.8}$. The dissolution rate of $TiH_{2.4}$ at 363 K with lowering of HCl concentration shall then vary slightly, whereas at 353 K, it shall drop strongly showing a 3 time decrease. The effect of HCl on the both samples is more efficient as that of H_2SO_4 as a whole.

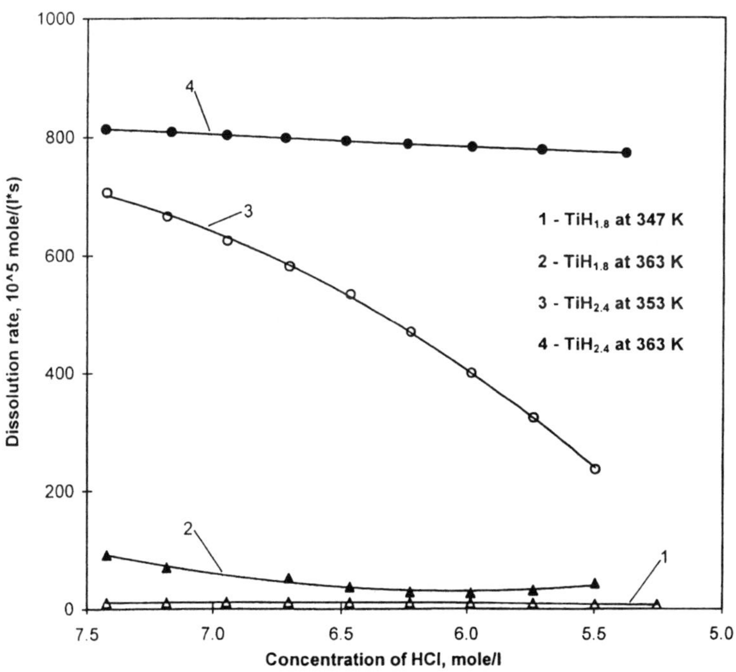

Figure 4. Curves of titanium hydrides dissolution rate in HCl

The fact that $TiH_{2.4}$ has a higher solubility in acids than $TiH_{1.8}$ at other similar conditions, can be explained by the structure of crystalline lattice of titanium hydride. An increase of hydrogen content causes the lattice to expand [8], the Ti-Ti bonds to elongate and these atom interactions to weaken.

It was found that the experimental activation energies of the dissolution processes for these hydrides reveal a great difference. In both acids these values amount to 17-23 and 100-106 kJ/mole. However, if in H_2SO_4, the activation energy of $TiH_{2.4}$ dissolution was higher as for $TiH_{1.8}$, the same was lower in HCl.

3.4. INTERACTION WITH ALKALIS

Similar investigations of interaction process and chemical resistance of TiH_X in acids were also made in alkalis. For experiments, solutions of NaOH and KOH having initial concentrations of 7.4 and 12.8 mole/l, respectively, were used. Figures 5-7 shows that the initial TiH_X dissolution rates in acids and alkalis are comparable. A passivation of solid phase can be observed after the dissolution of about 2-3 % of both sample hydrides leading to a slowing of the reaction rate and hydrogen evolution during the following 4-5 hours. The hydride surfaces have evidently low soluble titanium hydroxides formed on them. The temperature rise and hydrogen content in the hydride have then the same influence on the TiH_X decomposition as in the case of acids: these factor growth yields an increase of hydride dissolution rates.

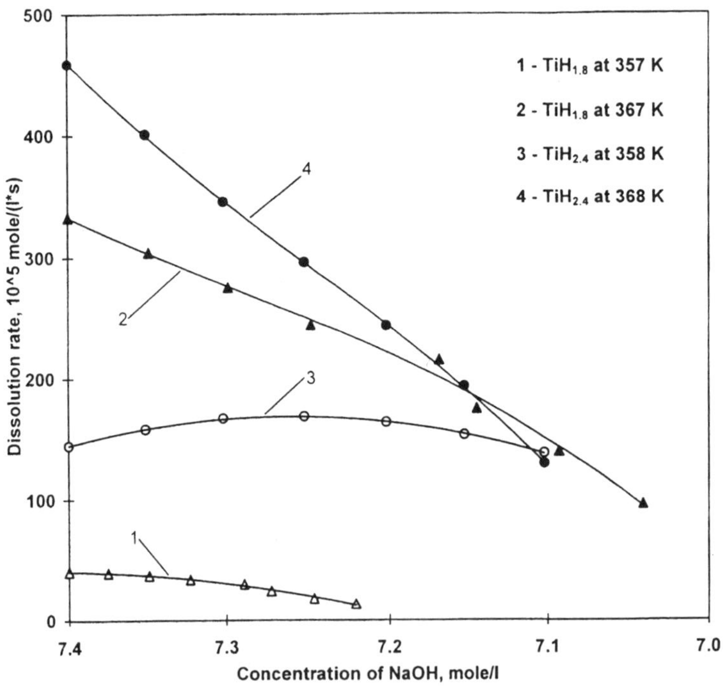

Figure 5. Curves of titanium hydrides dissolution rate in NaOH

4. Conclusion

- It was stated that the chemical process of $TiH_{X>2}$ dissolution in H_2SO_4 and HCl solutions was identical to the process of Ti and $TiH_{X<2}$ dissolution, and occurs through formation of titanium compounds (III).
- It was stated that the $TiH_{X>2}$ dissolution rate in acids and alkalis was 3-60 times higher than of $TiH_{X<2}$ hydride.

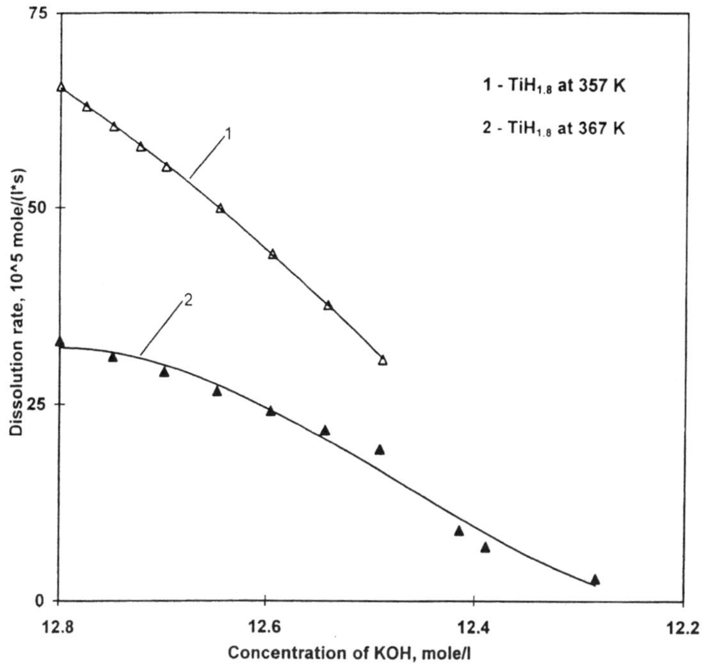

Figure 6. Curves of $TiH_{1.8}$ dissolution rate in KOH

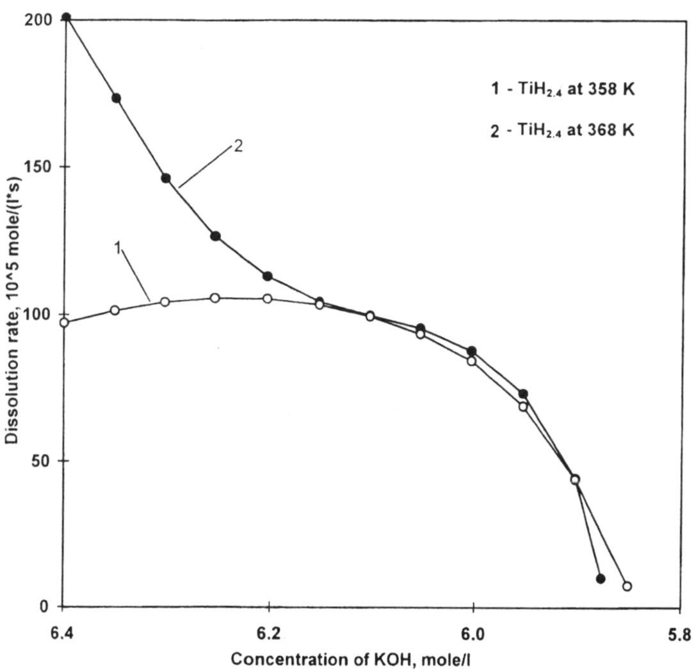

Figure 7. Curves of $TiH_{2.4}$ dissolution rate in KOH

- It was shown that the investigated samples of titanium hydride had higher resistance in alkali media than in acids. The chemical resistance of hydride in acids and alkalis was then decreased with temperature. Higher concentrations of both H^+ and OH^- ions resulted in a stronger interaction between the solution and TiH_X.

5. References

1. Kost, M.E. (1957) On chemical properties of cerium hydrides, *Zhurnal neorganicheskoj khimii* **2**, 2689.

2. Kost, M.E. (1978) Chemical properties of hydrides of some rare-earth metals, *Izv AN SSSR Ser Neorgan Materialy* **14**(9), 1697.

3. Antonova, M.M. (1985) *Properties of hydrides: Handbook*, Naukova Dumka Publishing, Kiev.

4. Brinza, A.P., Antonova, M.M., Patrusheva, A.G. and Golovnyova, L.B. (1972) On interaction of different phases of Ti-H-system with hydrochloric acid, in *Physics and Chemistry of Hydrides*, Naukova Dumka Publishing, Kiev, pp. 200-210.

5. Antonova, M.M. (1969) Synthesis methods and chemical resistance of transition metals of IV and V groups, in *Chemichal Properties and Analysis Methods of Refractory Compounds,*: Naukova Dumka Publishing, Kiev, pp. 45-52.

6. Trefilov, V.I., Lavrenko, V.A., Morozova, R.A., Momot, G.G. and Shemet, V.Zh. (1987) Influence of hydrogen content in titanium hydride and helium saturation on peculiarities of high-temperature oxidation, *Dokl AN SSSR,* **295**, 431.

7. Trefilov, V.I., Morozov, I.A., Morozova, R.A., Dobrovolsky, V.D., Zaulichny, Ya.A., Kopylova, E.I. and Khyzhun, O.Yu. (1999) Peculiarities of interatomic interaction in titanium hydrides with different content of hydrogen, *Int J Hydrogen Energy,* **24**(2-3), 157-161.

8. Trefilov, V.I., Timofeeva, I.I., Klochkov, L.I., Morozov, I.A. and Morozova, R.A. (1996) Effects of temperature change and hydrogen content on titanium hydride crystal lattice volume, *Int J Hydrogen Energy***21**(11-12), 1101-1103.

FULLERENE FORMATION IN ARC DISCHARGE

N.I. ALEKSEYEV, G.A. DYUZHEV
Ioffe Physico-Technical Institute of RAS
26 Politechnicheskaya str., Saint Petersburg, 194021, Russia

1. Introduction

Unique properties of fullerene has stimulated intense researches of fullerene-based materials in a wide range of applications, hydrogen power industry among them [1]. But too high price of fullerene prevents from their using as a hydrogen storage. The only industrial method of the fullerene production remains so far the arc discharge with graphite electrodes [2]. But its optimal efficiency [3–5] is either far from necessary.

Surprising transformation of chaotic carbon vapor to fullerene is not quite clear. As was proven in [5] with using isotope ^{13}C, carbon in the arc column can only exist in a form of atoms and atomic ions. On the contrary, the answer is known to the question of where in arc fullerene arise [6]. Between these stages a number a clusters is observed with different mobility [7]. The authors of [7] proposed the following evolution scheme of these clusters: chains – rings – milti-ring clusters – closed fullerenes. In their experiments clusters interfered neither one with another nor with carbon atoms and molecules. This is not right as applied to the arc but no detailed experiments in arcs had been carried out.

Attempts to build up a theory of the fullerene formation may be divided into two groups: the modeling of the transformation of multi-ring clusters to fullerene [8–10] and the modeling of the whole route from atoms to fullerene in a turbulent jet, formed by the fullerene arc [11–13].

This paper is a review of theoretical and experimental researches on the fullerene formation mechanism in arc discharge carried out in Ioffe Institute of RAS in 1998–2001.

2. Fullerene Formation Mechanism in Arc Discharge. Quantitative Review

Given in Fig. 1 are the fullerene formation stages as they seem to the authors of this paper.

In the very arc carbon is a plasma forming substance. Beyond the arc gap it is moved off with a gas-plasma jet. The existence of the jet is proven in experiments with probes collecting carbon particles from the arc [14–15] and with admixing $^{13}CO_2$ to helium [16]. Characteristics of the origin of the jet (carbon concentration N_0, gas temperature T_0 and initial velocity of the jet V_0) had been obtained from calculations of the fullerene arc discharge, unusual for discharge physics [17].

In the jet the carbon atoms condense to clusters as the gas cools. The topology of the clusters is reliably known from the experiments in drift tubes up to the stage of rings, these rings being not the immediate precursors of fullerene. In [9] we proved two-ring clusters to be such precursors basing on quantum chemistry CC methods. The papers

N. Veziroglu et al. (eds.), Hydrogen Materials Science and Chemistry of Metal Hydrides, 141–150.

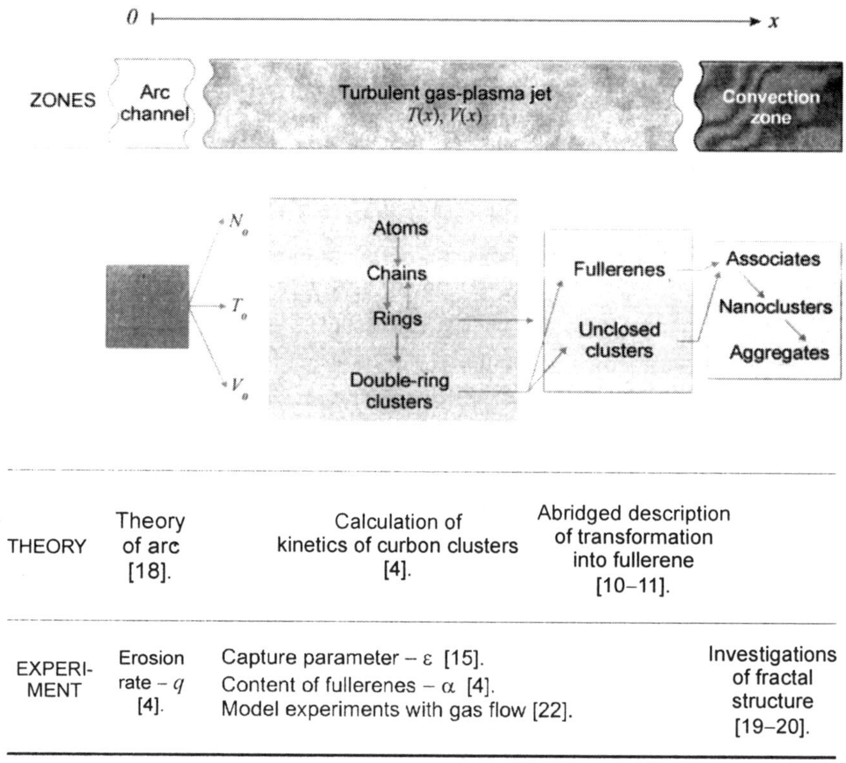

Fig. 1. Fullerene formation scheme in arc discharge

[9–10] are devoted to calculations of transformation of two-ring clusters to fullerene (assembling), the whole space of assembling routes is taken into account.

On the base of these results a semi-analytical calculation had been performed in [14] for the condensation process as a whole. The calculation made it possible to predict the fullerene yield depending on the parameters at the origin of the jet.

From here on some results of the declared papers are outlined.

3. The Spatial Change of the Soot Structure in the Jet

Information of the structure of the fullerene-containing soot is important for understanding of the fullerene formation. In [16] the soot was researched by electron microscopy and electron diffraction combined with ideas of fractal geometry. Micro-images had shown that "elementary particles" of the soot are either fullerenes or clusters of the same size, failed to close to fullerene because of "architectural" mistake. The "elementary" particles are united to associates, accosiates – to nanoclusters, nanoclusters form aggregates, which microscopic particles of the soot consist of [18].

The same method was applied to the soot collected on the probes at different distances x from the arc axis [19]. At $x < 4\ cm$ the density of nanoclusters wss uniform, i. e. associates do not form inside the clusters. At $x \approx 4\ cm$ the associates are visible and reveal some mesh-like structure. At $x \geq 5\ cm$ the density of nanoclusters is hardly dependent on x. Hence, if the structure remains fractal as deeply as to the elementary structural unit of the soot [19], one can conclude that: fullerene form in a region of

$4\ cm \geq x \geq 3\ cm$; the formation of associates is finished by $x \leq 5\ cm$.

These conclusions correlate well with the results obtained with using different methods.

4. Fullerene Arc

The following preposition were made to simplify the theoretical consideration of fullerene arc.

The arc parameters do not depend on the distance from the arc axis. Buffer gas is immovable and not ionized. Concentration of carbon atoms is much lower than that of buffer gas (his preposition does not cover all the regimes but sufficient to formulate qualitative tendencies). The carbon transfer has the anode-to-cathode direction and is described in terms of diffusion. The plasma in the gap is in a local thermodynamic equilibrium (LTE) except narrow electrode layers.

The system of equations accounted the transfer of energy, of electrons, of the heavy component in a plasma bulk and margin conditions that imitate the solution of layer problem at the electrodes.

The preset external parameters were cathode emission and the buffer gas pressure.

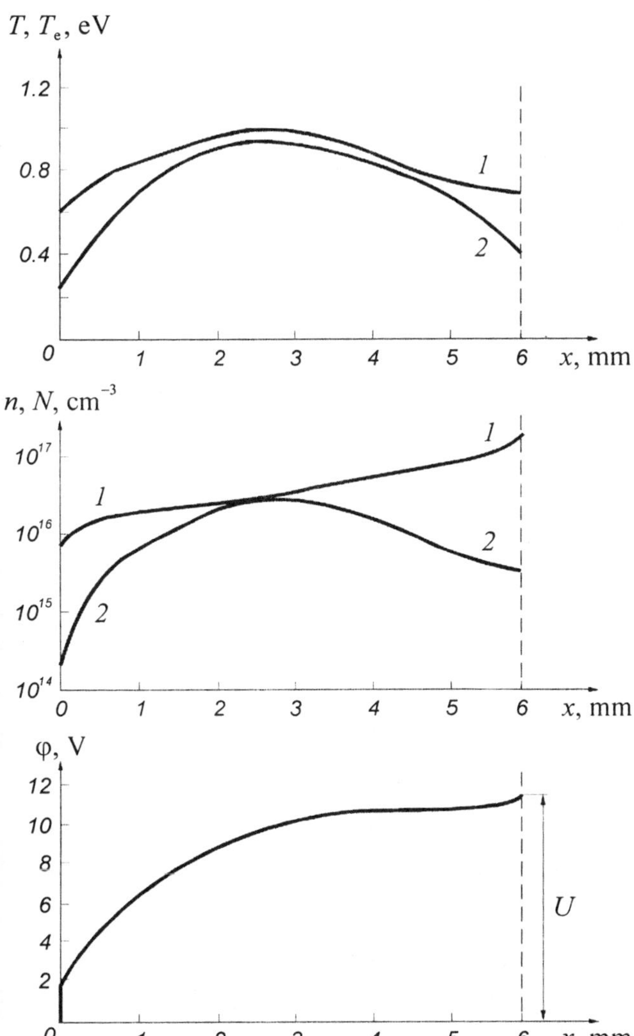

Fig. 2. Distributions along the discharge gap of electron temperature T_e (*1*), heavy component temperature T (*2*), concentration of carbon atoms N (*1*), ions n (2), and of plasma potential φ. Helium $P_{He} = 100$ *Torr*, $I = 80$ A, $3d = 6$ *mm*.

Given in Fig. 2 are the profiles of the temperatures of electrons T_e and of the gas T, of carbon atomic N and ionic n concentrations and of the plasma potential φ through the gap at a helium pressure of $100\ Torr$ and the current I of $80\ A$ (current density $j = 284\ A/cm^2$).

The electrode diameter $2r_0 = 0.6\ cm$ and the gap $2d = 0.6\ cm$. The point $\varphi = 0$ is the cathode, U is the whole voltage drop.

It is seen that the difference $T_e - T$ is essential near the electrodes only. In the bulk of the gap the plasma is really close to total LTE.

The carbon atom concentration decreases monotonously from the anode to the cathode. At the same time the concentration of charged particles has high gradients, especially near electrodes. The cathode drop φ_{Cath} is not high ($\varphi_{Cath} \sim 3\ V$), the anode drop $\varphi_{Cath} \sim T_e$ is positive.

Characteristics of some calculated regimes for different gases are given in Table. The table gives the values of I, φ_{Cath}, U, the maximal temperature $T_{\max}$, the anode erosion rate q and experimental date on the values of the fullerene content in the soot α and the ratio of the cathode deposit growth to the anode erosion m_c/q.

TABLE

Gas		I, A	φ_{Cath}, eV	U, V	$T_{\max}$, eV	$q, mg/s$	$\alpha, \%$	m_c/q
He	Calc. (LTE)	80	1.8	11.1	0.99	2.32		
	Calc.*	80	1.6	20	1.08	2.4		
	Exper.	80		21.5		1.3	9.8	0.41
Ne	Calc.	85	3.1	13.4	1.2	0.95		
	Exper.	85		14		0.14	4.3	0.1
Ar		102	3.6	13.7				
		105		14	1.26	0.70		

Quite a good agreement is seen between the calculations and the experiments for helium. Some discrepancies were obtained in the arc voltage. They are connected with the need for consideration of the anode non-equilibrium ionization region in helium first of all with a drop of $\varphi_A \sim (5-10)\ V$, while not more than the tenth of V in argon, according to the estimations in [17]. The calculated arc voltage, added by the value of this estimation, corresponded the experimental date quite well (Line "Calc." in Table).

In heavy buffer gases the same erosion is reached at higher current, which fact is qualitatively confirmed either. The reason is that carbon mobility in heavy gases is lower and its escape to the electrodes is more difficult. This fact is either seen from the experiments: the ratio $m_c/q = 0.4$ in helium and 0.1 in argon.

The calculation enabled to make clearer the role of buffer gas. Buffer gas was usually considered to be necessary in order to accept the surplus of vibration energy of growing carbon cluster. But this explanation it is difficult to make more detailed for different sorts of the gas. Our calculations show that its role is crucial at a level of the jet parameters first of all. So, in heavy gases the same level of erosion is reached at much higher current. But an increase in current brings about an increase in the initial jet velocity. As it is shown in [21] and is discussed below such increase lowers the fullerene formation efficiency drastically.

5. The Gas Jet Impact on Cluster Kinetics and Fullerene Formation. Experimental Date

In [21] an additional helium flow with consumption δg was blown into the arc through an opening of $2.6\ mm$ diameter in the cathode. The dependence of the fullerene content in the soot α on δg is given in Fig. 3 $(\alpha_0 = \alpha\ (\delta g = 0))$. It is seen that the fullerene content drops to almost nil in a narrow interval of δg and the flow velocity V, correlating with asymptotic velocity of the plasma transfer in free-burning high current arc from the source of small diameter $2r_0$ [20]:

$$V_0 = 1.5 \times 10^3 \frac{I}{r_0}\left(\frac{T}{P_{He}}\right)^{1/2} \tag{1}$$

As soon V gets close to V_0, the flying time through the fullerene formation region decreases for those clusters which are the immediate fullerene precursors and the fullerene assembling process breaks. The same value of V is pointed at by the experiments on the mass increase rate on cylindrical probes at different distances from the arc axis [14], performed with a purpose of independent determination of the distances where principal stages of carbon transformation are passed by, including the formation of carbon associates.

The process of carbon particle deposition (heavy impurity) is usually characterized by the ratio of the flow to the probe to the flow determined by rectilinear trajectories of the impurity particles (capture parameter) ε. At small x $\varepsilon(x) > 1$. Then ε decreases with x; at $x > 5\ cm$ $\varepsilon < 1$.

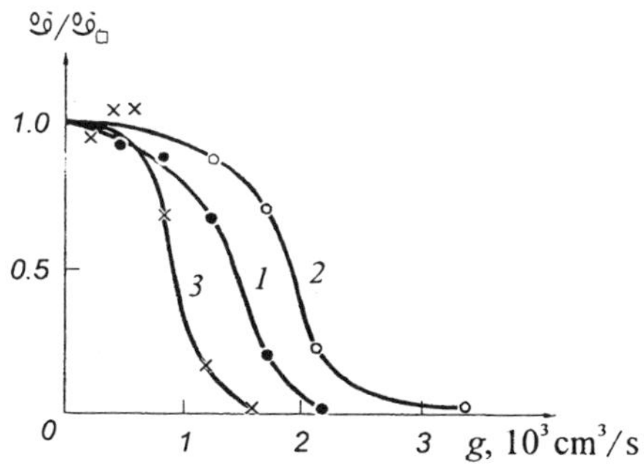

Fig. 3. Fullerene content α in the soot deposited on chamber walls versus helium flow rate.
1 – P_{He} = 100 *Torr*, I = 100 A;
2 – P_{He} = 100 *Torr*, I = 200 A;
3 – P_{He} = 600 *Torr*, I = 100 A;

Estimations showed that at a small x atoms can only satisfy the experimental values of ε. On contrary, at a great x the condition $\varepsilon < 1$ can only be satisfied by associates.

The analysis of experimental date [14] had been performed in [15] with three alternative prepositions of the nature of carbon mass carriers.

1. All carbon atoms are contained in chains and rings, whose distribution function has a form

$$f(n)/N_0 = \alpha_c \cdot \left(1 - (n/n_1)^{\zeta}\right) \tag{2}$$

where $n_1 = n_1(x)$ is the maximal number of carbon atoms in cluster, $\zeta < 1$, N_0 is the carbon concentration at $x = x_0 = \vartheta r_0$, ϑ is a model turbulent constant, depending on the geometry (in our geometry $\vartheta \sim 3.5$).

2. All carbon atoms are concentrated in rings whose distribution is given by (2) with fixed margin size $n_1 = n_{10} = 38$, and in fullerenes with $n_F = 60$ atoms. Normalization coefficients of chains α_c and of fullerenes α_F were supposed unknown:

$$f(n)/N_0 = \alpha_c \cdot \left(1 - (n/n_1)^{\zeta}\right) + \alpha_F \delta_{60,n} \tag{3}$$

3. All clusters reaching the probe are associates of fullerenes and of fullerene-like "elementary" particles.

The total concentration of carbon impurity in all clusters, the temperature and the velocity were given from the models of developed turbulence

$$(V, N, T - T_\infty) = (V_0, N_0, T_0 - T_\infty) \cdot \vartheta \cdot r_0 / x \tag{4}$$

Herewith, basing on the supposed character of change in distribution function it turned out possible to determine concrete distances from the arc axis where types of clusters transfer one to another.

Intersection of the curves for the calculated values of $\alpha_C = \alpha_C^{(1)}$ for chains-rings and $\alpha_C = \alpha_C^{(2)}$ for the mixture of rings and fullerenes indicates the existence of a place where fullerene appear in spectrum alongside with chains and rings. At $V_0 < 2.5 \times 10^3\ cm/s$ the curves $\alpha_C^{(1)}$ and $\alpha_C^{(2)}$ do not intersect. At greater velocities the value n_1 in an intersection point approaches n_{10}. The difference $n_{10} - n_1$ was smallest at $V_0 = 4.6 \times 10^{-3}\ cm/s$ and corresponded $x_1 = 3.2\ cm$ from the arc axis. This value of V_0 corresponded well with the estimation (1). If the mass of carbon is supposed to be contained in associates, they start to form at $x_2 \approx 4\ cm$ (the number of particles in associates about $S_a > 1$). So the length of the free fullerene lifetime zone until they merge to associates is about $x_2 - x_1 \approx (0.5 - 1)\ cm$.

These results are in a good correspondence with fractal analysis of micro-images in [18–19].

6. Immediate Precursors of Fullerene. Fullerene Assembling from Two-Ring Clusters

The probe measurements give no information of immediate precursors of fullerene, i.e. the principal link necessary for modeling the process as a whole. The solution to this problem was based on the CC methods. As it is said above, two-ring clusters with one ring-to-ring bond were reasoned as such precursors [9]. Then a statistics on speculative routes of transformation of such clusters to fullerene (we call this stage by term "assembling" from now) was collected and analyzed.

As a great number of routes and intermediate clusters exist in the process of assembling, an abridged description scheme [9] had been proposed. A small number of parameters is used for intermediate clusters. Possible variants of configurations of such clusters, that require more detailed parameters, are supposed to be in equilibrium.

So, in [9] the intermediate clusters were characterized by three parameters: the number of pentagons, of hexagons and of two-bond atoms at the verge of already assembled frame of the surface (radicals). All possible variants of bonding the ring fragments to the frame were supposed to be in equilibrium. In [10] the choice of parameters of abridged description was varied depending on the purposes of a calculation. The simplest variant used the full number of polygons as the only parameters.

The main results of this consideration are the following:

1. The assembling rate decreases with the number of atoms in cluster n at least up to $n \sim 60$ and has an order $\tau_F \sim 10^{-4}\ s$ at $T = 0.25\ eV$.
2. τ_F decreases with temperature but at $T \geq 0.26\ eV$ the share of fullerene is very low.
3. Owing to the features of final reactions of the assembling (discarding spare ring fragments [14]) the number of fullerenes with $n < 60$ is negligible.

7. The Calculation of the Carbon Cluster Kinetics

As already said, the evolution of carbon clusters occurs in the turbulent jet, flowing from the region of arc source.

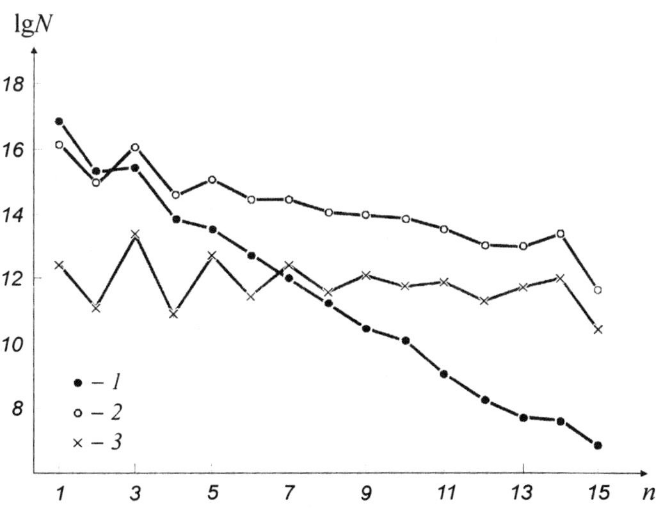

Fig. 4. Equilibrium spectra of chains at different distances x from the arc axis. *1* — $x = 1.3\ cm$; *2* — $x = 1.7\ cm$; *3* — $x = 2.9\ cm$.

Fig. 4 shows a typical equilibrium spectrum of chains on their size at different distances from the source, calculated in [13] on the base of CC methods (AM1). The total concentration of atoms, of molecules C_2, C_3, of rings and two-ring clusters was determined by (4), as well as the temperature. At a small distance from the source the spectrum decreases within all the size range (curve 1), starting with atoms. At lower temperatures molecules C_3 are dominant as compared to C_2 and atoms; at $n > 3$ the spectrum decreases monotonously.

When calculating the evolution of carbon clusters it is important to know the length scales for equilibrium between different clusters to be established. The equilibrium is reached through the reactions of coagulation (when clusters collide) and the decay of clusters.

The analysis performed in [13], showed that with concentrations, corresponding the origin of the jet, large clusters come to an equilibrium among them very quickly. As to the small clusters (molecules C_2, C_3) and atoms, the time of their interference is determined by the buffer gas, playing the role of the third body, and by the lifetime of a complex C_2^* (or C_3^*). So, this time can be in a great excess of flying time of pair interference. A preposition was made in [17] that by the moment of the formation of rather complicated clusters the small ones and atoms already came to equilibrium.

The calculation of equilibrium concentrations of different types of clusters performed in [17] and the estimations of scales necessary for the equilibrium among different groups of clusters were a base for calculation of spatial evolution of clusters. The following prepositions were used:

1. The formation of two ring clusters and their decay to separate rings can hardly effects the ring kinetics; in turn, the fullerene decay can hardly effects the two-ring cluster kinetics.
2. The growth of chains and rings by the capture of atoms and molecules C_2, C_3 can be described in terms of diffusion in the size space. Clusters formed as a result of the capture has time to be annealed to the topology of chains or rings by the next collision.
3. Before the moment of the equilibrium disturbance in the large cluster spectrum (because of decrease in the molecule concentration) the atoms, the molecules C_2, C_3 have time to come to equilibrium between themselves.
4. The difference between the properties of chains and rings is insufficient. This fact enables to operate with the total chain-ring spectrum.

Two-ring cluster can not be formed by the capture reactions only. So, a coagulation of two rings is necessary. Herewith diffusion part, principal for chain-ring balance equation, is omitted.

Basing on the results of [9], the process of the fullerene assembling from two-ring clusters was characterized by the only parameter – the rate of assembling which takes place in a narrow temperature interval δT near the value $T = 0.25\ eV$.

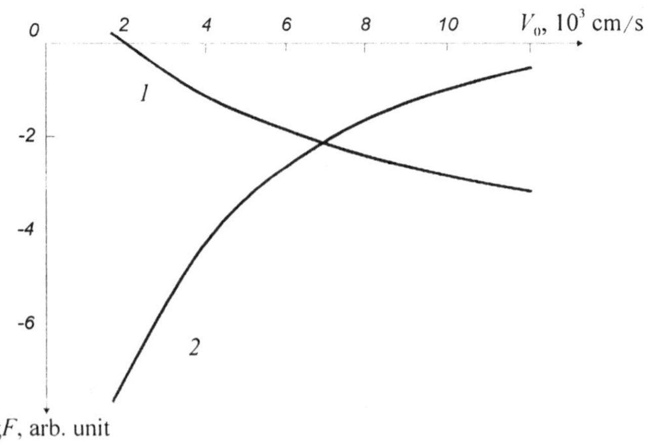

Fig. 5. Dependence of the fullerene yield (arb. units) on parameters at the origin of the jet. 1 – initial velocity ($N_D = 10^{17}$ cm^{-3}, $T_0 = 0.7\ eV$). 2 – carbon concentration ($T_0 = 0.7\ eV$, $V_0 = 4\times10^3$ *cm/s*.

On solving the balance equations for chains-rings and two-ring clusters the fullerene yield

$$F = \int_0^\infty \frac{dx}{V}\frac{dF}{dt}$$

in arbitrary units was calculated. The results of the integrating are given in Fig. 5.

It is seen, that both factors V_0 and N_0, which determined the fullerene yield (as well as α), increase with the current, their effect on α being opposite [13]. Therefore the dependence α on I has a maximum, observed experimentally.

The Fig. 5 and (1) explain the reason for decrease in α when heavy buffer gases are used instead of helium [13]. At a same current (and the same velocity V_0 subsequently) the lower erosion in heavy gas brings about smaller value of N_0 and the fullerene yield. Some increase in current enables to reach the same level of erosion as it is in helium but leads to simultaneous increase in the velocity and decrease in N_0. Both tendencies evolve an enormous decrease in the fullerene yield.

8. Conclusions

The researches made possible further advance in understanding the fullerene formation and the improvement of a plasma reactor for obtaining the fullerene- containing soot. The characteristics of our reactor, assembled in Saint Petersburg (Russia) are the following: total efficiency is ~ 100 g/h, fullerene content is more than 12% and the self-cost is 0.2 $/g for the soot with taking all expenses.

Certainly these parameters are far from ideal ones, but the reactor admits further improvements. Besides this, very prospective are hybrid schemes [2], "laser + gas discharge" among them.

The study was carried out in the framework of the Russian program «Fullerenes and Atomic Clusters» and supported by the Russian Foundation for Basic Research (Project No. 00-02-16928).

9. References

1. Tarasov, B.P., GoldShleger, N.F., Moravskij, A.P. (2000) Hydrogen-countering carbon nanoclusters. Synthesis and properties, *Progress in Chemistry (Rus.)*, **70**, 149-166.
2. Bogdanov, A.A., Deininger, D., Dyuzhev, G.A. (2000) Development Prospects of the Commercial Production of Fullerenes, *Techn. Physics (Rus.)* **45**, 521-527.
3. Parker, D.H., Wurz, P. *et al.* (1991) High-Yield Synthesis, Separation and Mass-Spectrometric Characterization of fullerene C_{60} to C_{266}, *J. Am. Chem. Soc.*, **113**, 499-7503.
4. Afanas'ev, D.V., Blinov, I.O., Bogdanov, A.A. *et al.* (1994) Formation of Fullerene in an Arc Discharge. Part I., *Techn. Physics (Rus.)* **39**, 76-90; Part II. *ibid.*, **42**, 234-237.
5. Belz, T., Find, J. *et al.* (1997) On the Production of Different Carbon Form by Electric Arc Graphite Evaporation, *Ber. Bunsenges. Phys. Chem.* **101**, 712-725.
6. Dyuzhev, G.A., Karataev, V.I. (1994) Where in Arc Fullerene Form, *Solid State (Rus)* **34**, 2795.

7. Hunter, J.M., Fye, J.L., Jarrold, M.F. (1993) Annealing and Dissociation of Carbon Rings, *J. Chem. Phys.* **3**, 1785-1795.
8. Strout, D.L., Scuseria, G.E (1996) A Cycloaddition Model for Fullerene Formation, *Journ. Phys. Chem.* **100**, 6492-6498.
9. Alekseev, N.I., Dyuzhev, G.A. (2001) A Statisyical Model of Fullerene Formation Based on Quantum-Chemistry Calculations. The Most Probable Precursors of Fullerenes, Verification of the Model and Kinetics of Transformation into Fullerene, *Techn. Phys. (Rus.)* **46**, 573-583.
10. Alekseev, N.I., Dyuzhev, G.A. (2002), *Techn. Phys. (Rus.)* **47**. *To be published.*
11. Nerushev, O.A., Sukhinin, G.I. (1997) Kinetics of Fullerene Formation by Electric-Arc Vaporization of Graphite, *Techn. Phys. (Phys.)* **47**, 41-49.
12. Krestinin, A.V., Moravskij, A.P., Tesner, P.A. (1998) *Journ. Chem. Physics (Rus.)* **17**, 70-86.
13. Alekseev, N.I., Dyuzhev, G.A. (2002) *Techn. Phys .(Rus.)* **47**. *To be published.*
14. Afanas'ev D.V., Dyuzhev G.A., Kruglikov A.A. (2001) Carbon Flows from the Arc Discharge in Regimes Optimal for Production of Fullerene, *Techn.Phys. (Rus.)* **71**, 134-135.
15. Alekseev, N.I., Chibante, F., Dyuzhev, G.A. (2001) On the Transformation of Carbon Vapor in the Gas-Plasma Jet of an Arc Discharge, *Techn. Phys. (Rus.)* **46**, 761-766.
16. Afanas'ev, D.V., Baranov, V.I., Beljaev, A.A. (1999) ^{13}C Enrichment of Fullerenes. *Techn. Phys. Lett. (Rus.)*, **25**, 12-18.
17. Alekseev, N.I., Dyuzhev, G.A. (2001) Arc with a Vaporizable Anode: Why is the Fullerene Formation Affected by the Kind of Buffer Gas?, *Techn. Phys. (Rus.)* **46**, 1247-1255.
18. Gorelik, O.P., Dyuzhev, G.A., Novikov, D.V. *et al.* (2000) Cluster Structure of Fullerene-Containing Soot and C_{60} Fullerene Powder, *Techn. Phys. (Rus.)* **45**, 1489-1495.
19. Gorelik, O.P., Dyuzhev, G.A., Novikov, D.V. (2002) *Techn. Phys. (Rus.)* **47**. *To be published.*
20. Ramakrishnan, S., Stokes, A., Lowke, J. (1978) Model for High-Current Free Burning Arc, *J. Phys. D.* **11**, 2267-2280.
21. Afanas'ev, D.V., Dyuzhev, G.A., Kruglikov, A.A. (2001) Effect of Gas Flows on Fullerene Formation Process, *Techn. Phys. (Rus.)* **47,** 137-138.

THERMAL-RADIATION SYNTHESIS OF HYDRIDONITRIDES ON THE BASIS OF TITANIUM AND ZIRCONIUM

A.G. ALEKSANYAN*, S.K. DOLUKHANYAN,
V.S. HAIRAPETYAN[(1)], O.P. TER-GALSTYAN
AND N.L. MNATSAKANYAN
Institute of Chemical Physics of Armenian NAS,
P. Sevak St., 5/2, Yerevan, 375044, Armenia
[(1)]Yerevan Physics Institute,
Alikhanyan Br. St., 2, Yerevan, 375036, Armenia

Abstract

The results of systematic investigations of thermal-radiation processes in the system $Ti_kZr_{1-k}N_x$-H at irradiation by the accelerated electron beam are presented. The possibilities of thermal-radiation hydrogenation of nitrogen solid solutions in the bimetallic matrix titanium - zirconium were investigated. The parameters of the reaction of thermal-radiation synthesis in these systems were determined. The influence of a second metal on the post-radiation-stimulated interaction of the matrix with hydrogen was demonstrated. It was shown, that in the studied systems, the presence of a non-metallic sub-lattice, as well as of a second metal did not hinder the proceeding of radiation processes and the synthesis of HCP hydridonitrides of high hydrogen content (2.07-2.42 wt %).

1. Introduction

In the complex process of thermal-radiation interaction of metal matrix with hydrogen as a result of irradiation by the accelerated electron beam, the following main macro-processes might be highlighted: the transformation of electron beam energy to the thermal energy; the generation of defects and of relaxation processes in the crystal lattice; the creation of conditions, when stored in the system energy is sufficient for initiation of chemical interaction. In the case of strong regularity of atoms replacement in the crystal lattice, the irradiation can result in the typical lattice effects: in the mass and energy transfer inside of the lattice by definite directions.

Interaction of nitrogen-containing bimetallic alloys with hydrogen in the electron beam was studied for the first time. The systems under investigation had an original crystal-chemical structure and a composite character of chemical bonds.

The combustion of these alloys in hydrogen had been investigated earlier [1]. It was shown that the hexagonal closed packed α-alloy and the nitrogen content of up to 6 wt.% (the solid solution of nitrogen in metals of IV group and in their alloys) perfectly interacted with hydrogen in the combustion regime resulting in HCP hydridonitride synthesis.

N. Veziroglu et al. (eds.), Hydrogen Materials Science and Chemistry of Metal Hydrides, 151–160.

The possibility of chemical interaction in the system Me-H immediately in the accelerated electron beam (thermal-radiation synthesis, TRS) was stated in work [2]. It was founded also that after irradiation in pure metals, the defects were formed providing cold synthesis (CS) of hydrides, when the exothermic interaction of metal and hydrogen proceeds at room temperature without any external initiation.

The results of investigation of TRS and of post-radiation stimulated interactions in triple systems Ti-N-H, Zr-N-H were presented in works [3,4]. The mechanism of interaction of hydrogen and solid solution of nitrogen in titanium and zirconium at TRS and CS reactions was reviled. The single-phase hydridonitrides of compositions $TiN_{0.18}H_{1.4-1.54}$ and $ZrN_{0.17}H_{1.37-1.65}$ were synthesized. It was shown, that the interaction of nitrogen solid solutions in titanium and zirconium with hydrogen in electron beam was promoted not only by thermodynamic parameters as in SHS, but also by processes of defect-formation in the crystal lattice of metal matrix with the nitrogen sub-lattice.

The goal of the present study was: to study the thermal-radiation processes in the system $Ti_kZr_{1-k}N_x$–H under the accelerated electron beam; to state the possibility of TRS realization and of post radiation-stimulated interaction of metal matrix and hydrogen resulting in formation of HCP hydridonitrides of bimetallic α-alloys. We would like also to detect the features of thermal-radiation processes concerning the hethero-atomic and hetero-desmic nature of chemical bonds in the compounds under investigation. In hetero-desmic crystals, the displacement of shifted atoms after irradiation depends on the inter-atomic bonds.

2. Experimental Part and Discussion

To perform the present investigations, we synthesized the following bimetallic solid solutions by SHS technique: $Ti_{0.88}Zr_{0.12}N_{0.22}$ (with traces of cubic phase of titanium nitride); $Ti_{0.81}Zr_{0.19}N_{0.22}$ + TiN (30%); $Ti_{0.74}Zr_{0.26}N_{0.22}$ + TiN (30%).

The investigation on thermal-radiation processes in the system $Ti_kZr_{1-k}N_x$–H was carried out in the special camera providing experiments in vacuum and pressure conditions. It was equipped with a device for measuring the temperature of sample during irradiation and any possible post thermal-radiation processes. The temperature profiles of all processes were registered with the "S" type Pt/Pt-Rh thermocouple. The system $Ti_kZr_{1-k}N_x$–H under accelerated electron beam was studied using a standard technique designed in our laboratory for investigation of binary Me-H mixtures [2]. The experiments were carried out on a linear accelerator of electrons LAE-5 at the electron beam energy 4 MeV and the current 150 μA.

For each composition, the following experiments were carried out:

a) the irradiation of an alloy by the accelerated electron beam in H_2;

b) the interaction of an alloy with H_2 at low temperature after preliminary irradiation in vacuum.

2.1. THERMAL-RADIATION SYNTHESIS OF COMPOSITE HYDRIDONITRIDES IN $Ti_{0.88}Zr_{0.12}N_{0.22}$ - H, $Ti_{0.81}Zr_{0.19}N_{0.22}$ – H and $Ti_{0.74}Zr_{0.26}N_{0.22}$ - H SYSTEMS

The alloys were irradiated in the presence of hydrogen at dose rates between 0.2-1.0 Mrad/s up to total dose 100 Mrad. It was obtained that in these systems like to the binary mixtures Me-H, the interaction of a metal matrix with hydrogen under the accelerated electron beam is possible. The thermogram of the process testified the exothermic character of the interaction (Fig. 1 a, b). The registered temperatures, the chemical and X-ray analyses data proved that as a result of this interaction, composite HCP lattice hydridonitrides were formed.

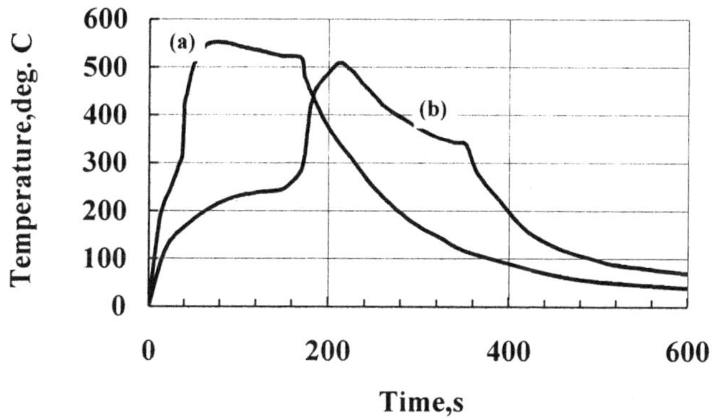

Figure 1. The thermograms of TRS at dose rate 0.6 Mrad/sec (a) and 0.3 Mrad/sec (b).

In Table 1 the characteristics of thermal-radiation processes in the studied systems are presented. As it is seen from the table at low dose rates of the beam, some temperature threshold exists for the TRS reaction beginning. Below this temperature, the alloys did not interact with hydrogen. The dose rate of beam plays an essential role for initiation of hydrogenation reaction in alloys.

For all alloys with the increase of dose rate, the dose necessary for initiation of hydrogenation reaction decreases. The lower limit of the radiation dose rate for TRS realization was equal to 0.3 Mrad/s. With dose rate increasing, the temperature of TRS beginning and the maximal temperature of TRS increased in all alloys (Table 1). Besides, it was obtained that with increase of the zirconium portion in an alloy, the TRS temperature also increased. By the same time, the H/Me ratio practically did not vary. In the system $Ti_{0.74}Zr_{0.26}N_{0.22}$ at dose rate 0.2 Mrad/s and total dose 100 Mrad, TRS was not realized in hydrogen, though the sample under beam heated up to 225°C during 8 minutes. With rise of beam dose rate, TRS proceeded. Interesting is, that the temperature of TRS initiation at dose rate 0.3 Mrad/s, is higher than that at 0.2 Mrad/s only by 20°C. The hydrogenation reaction started in 3.5 min after switching on the beam at temperature of a sample 245°C. The total dose was 62 Mrad. The thermogram of the process in Fig. 1b describes the classical TRS process. With the increase of the dose rate, the initial dose of TRS initiation in the indicated system decreased more significantly, and the temperature of TRS was higher than in the first two compositions (Table 1). The temperature range of TRS in the system $Ti_{0.74}Zr_{0.26}N_{0.22}$-H is in limits of 580-800°C. Probably, it is a result of zirconium influence.

The obtained composite hydridonitrides are single-phase and have HCP structure. The diffractograms of composite TRS hydridonitrides differ by the form, intensity and width of peaks from the identical SHS products (Fig. 2a, b,c).

Table 1. The characteristics of RTS in the Ti-Zr- N-H system

Dose power, Mrad/s	$T_{begin.}$, °C	$T_{react.}$, °C	Dose of begin., Mrad	Total dose. Mrad	N_2 content, wt. %	H_2 content, wt. %	Compound (accounted formula)	Phase content
				$Ti_{0.88}Zr_{0.12}N_{0.22}$ – H				
SHS		700	-	-	5.77	2.43	$Ti_{0.88}Zr_{0.12}N_{0.22}H_{1.3}$	HCP
0.2	205	-	-	100	-	0.32	$Ti_{0.88}Zr_{0.12}N_{0.22}$	Solid Sol., HCP
0.3	250	500	48	100	5.88	2.44	$Ti_{0.88}Zr_{0.12}N_{0.22}H_{1.3}$	HCP
0.5	360	540	31	100	5.96	2.31	$Ti_{0.88}Zr_{0.12}N_{0.22}H_{1.23}$	HCP
0.6	305	555	22.5	100	5.94	2.52	$Ti_{0.88}Zr_{0.12}N_{0.22}H_{1.34}$	HCP
0.8	335	620	18	100	5.89	2.54	$Ti_{0.88}Zr_{0.12}N_{0.22}H_{1.35}$	HCP
0.9	335	640	18	100	5.90	2.12	$Ti_{0.88}Zr_{0.12}N_{0.22}H_{1.35}$	HCP
				$Ti_{0.81}Zr_{0.19}N_{0.22}$ – H				
SHS	-	700	-	-	5.36	2.3	$Ti_{0.81}Zr_{0.19}N_{0.22}H_{13}$	HCP
0.2	225	-	-	100	5.63	-	$Ti_{0.81}Zr_{0.19}N_{0.22}$	Solid Sol., HCP + TiN, FCC
0.4	260	570	44	100	5.41	2.16	$Ti_{0.81}Zr_{0.19}N_{0.22}H_{1.02}$	HCP
0.5	270	630	35	100	5.4	2.36	$Ti_{0.81}Zr_{0.19}N_{0.22}H_{1.32}$	HCP
0.6	290	590	24	100	5.42	2.33	$Ti_{0.81}Zr_{0.19}N_{0.22}H_{1.25}$	HCP
0.9	365	785	13.5	100	5.39	2.18	$Ti_{0.81}Zr_{0.19}N_{0.22}H_{1.22}$	HCP
1.0	355	785	13	100	5.35	2.25	$Ti_{0.81}Zr_{0.19}N_{0.22}H_{1.25}$	HCP
				$Ti_{0.74}Zr_{0.26}N_{0.22}$-H				
SHS		700			5.11	2.2	$Ti_{0.74}Zr_{0.26}N_{0.22}H_{1.3}$	HCP+TiN, FCC
0.2	225	-	102	102	5.12	0.32	$Ti_{0.74}Zr_{0.26}N_{0.22}H_{0.18}$	Solid Sol., HCP + FCC
0.3	245	580	62	101	5.08	1.98	$Ti_{0.74}Zr_{0.26}N_{0.22}H_{1.17}$	HCP
0.5	280	590	22.5	100	5.14	2.11	$Ti_{0.74}Zr_{0.26}N_{0.22}H_{1.24}$	HCP
0.8	295	685	14	100	5.14	2.21	$Ti_{0.74}Zr_{0.26}N_{0.22}H_{1.31}$	HCP
1.0	335	800	12	100	5.13	2.18	$Ti_{0.74}Zr_{0.26}N_{0.22}H_{1.28}$	HCP + traces FCC

The special attention should be paid that the obtained composite hydridonitrides are mono-phase, though some of the initial samples contained up to 30 % of FCC phase. Similar to the SHS case, a complete homogenization of the system happens during hydrogenation, therefore, the FCC nitride phase disappears. The diffraction peaks are wide and their intensity is lower than of $TiN_{0,18}H_{1,4-1,.54}$ peaks (Fig. 2). The parameters of cell are larger: $\underline{a}$ = 3.038, $\underline{c}$ = 5.062 in $TiN_{0.18}H_{1.38}$, $\underline{a}$ = 3.099, $\underline{c}$ = 5.125 in $Ti_{0.74}Zr_{0.26}N_{0.22}H_{1.3}$. At the same time, the hydrogen concentrations are a little lower.

2.2. POST–RADIATION STIMULATED INTERACTION OF NITROGEN-CONTAINING ALLOYS WITH HYDROGEN

In this series of experiments, the possibility a post - radiation stimulated interaction of above-mentioned alloys with hydrogen at low temperature, i.e. in the "cold synthesis" mode had been shown. The alloys had been preliminary irradiated in vacuum in limits of dose rates 0.2 - 1 Mrad/s (Table 2). The radiation heating temperatures achieved 400-950°C. After switching off the beam and cooling the sample down to 50 - 75°C, the hydrogen was filled into the camera and a jump of temperature was observed. The thermogram of the process proves the existence of exothermal interaction in these systems (Fig. 3). Below the indicated temperatures CS is not realized. For example, at dose rate 0.2 Mrad/sec, the sample of alloy $Ti_{0.81}Zr_{0.19}N_{0.22}$ was heated up to 415°C

a

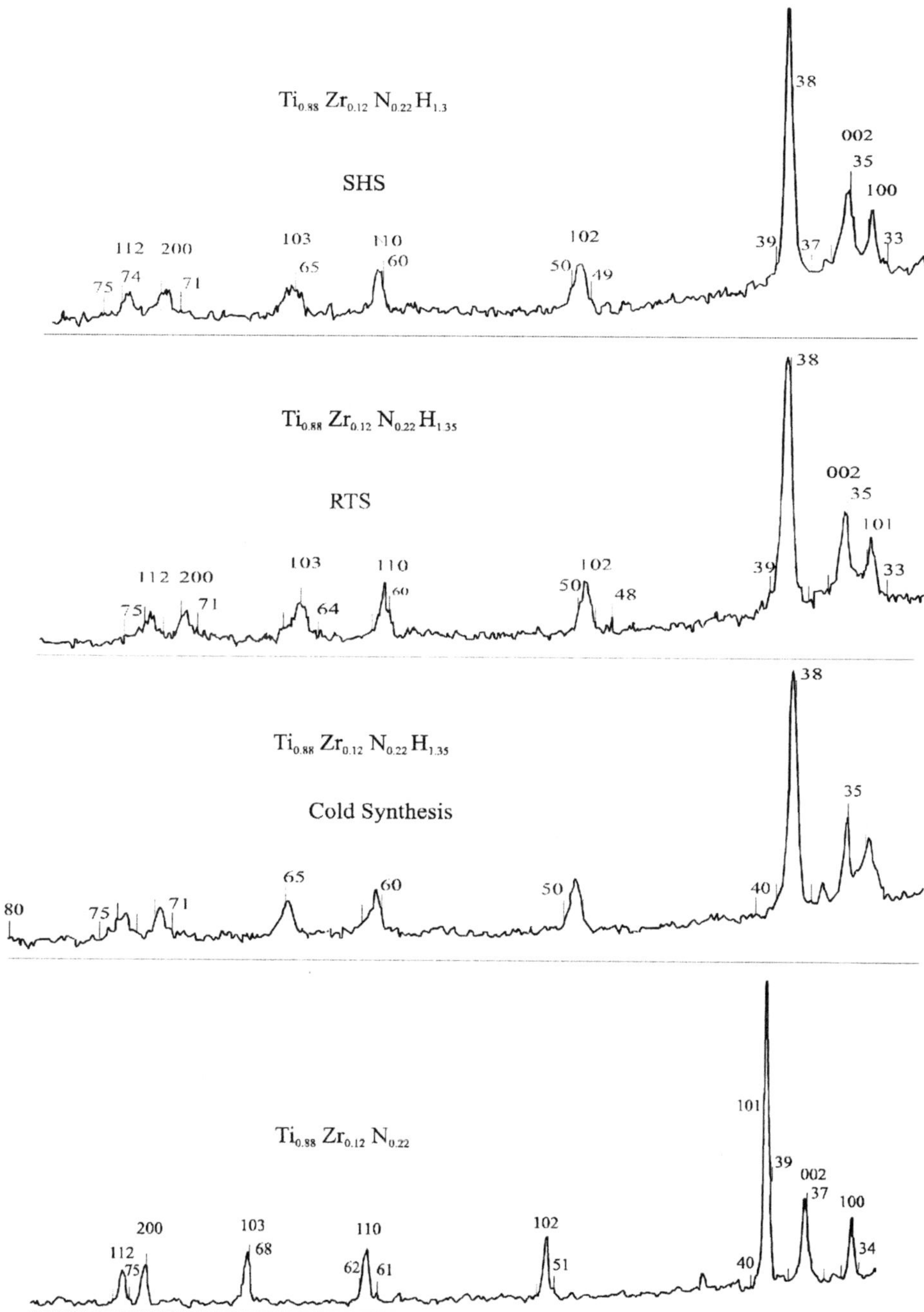

$Ti_{0.88}Zr_{0.12}N_{0.22}H_{1.3}$
SHS
38
002
35
100
39
37
33
102
50
49
110
60
103
65
112
200
74
71
75
$Ti_{0.88}Zr_{0.12}N_{0.22}H_{1.35}$
RTS
38
002
35
101
39
33
102
50
48
110
60
103
64
112
200
71
75
$Ti_{0.88}Zr_{0.12}N_{0.22}H_{1.35}$
Cold Synthesis
38
35
40
65
60
50
80
75
71
$Ti_{0.88}Zr_{0.12}N_{0.22}$
101
39
002
37
100
34
40
102
51
110
62
61
103
68
200
112
75

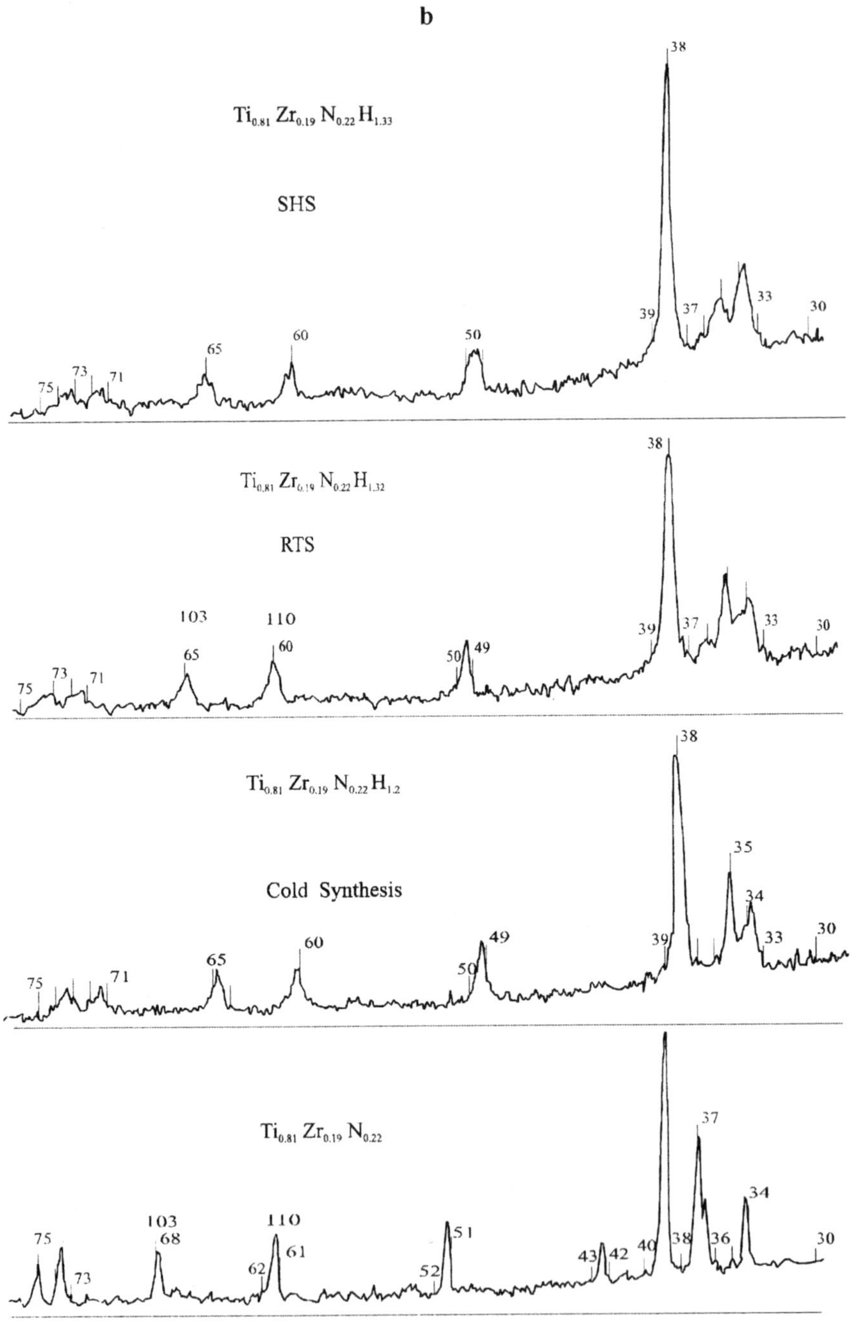

b
Ti0.81 Zr0.19 N0.22 H1.33
SHS
Ti0.81 Zr0.19 N0.22 H1.32
RTS
Ti0.81 Zr0.19 N0.22 H1.2
Cold Synthesis
Ti0.81 Zr0.19 N0.22

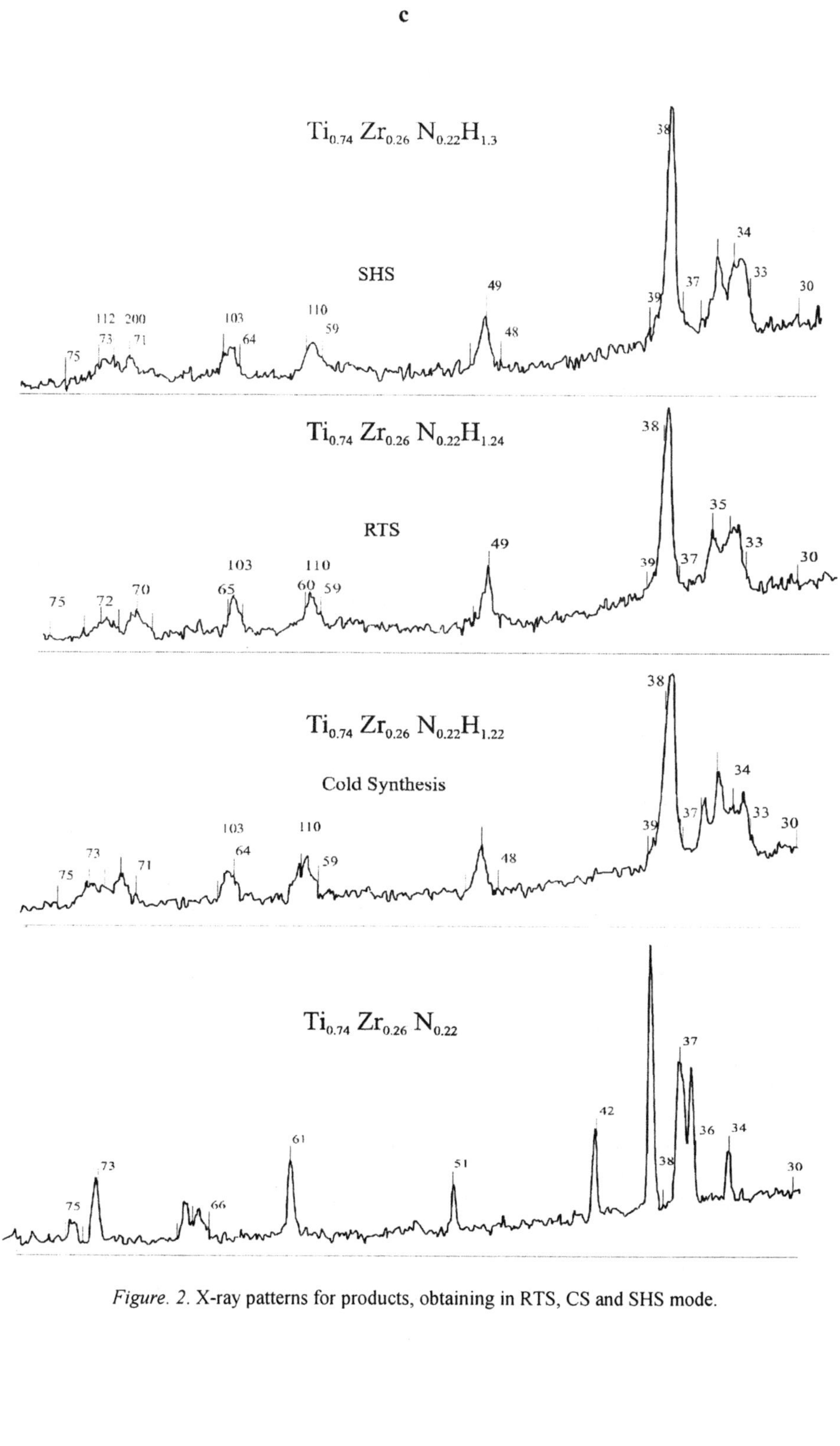

Figure. 2. X-ray patterns for products, obtaining in RTS, CS and SHS mode.

(total radiation dose 100 Mrad). After switching off the beam and cooling in vacuum down to 35°C, the hydrogen filling did not bring to CS. However if the sample was cooled down to 50°C, the hydrogenation reaction proceeded, the temperature of process achieved 430°C. The rise of dose rate did not influence the temperature of CS.

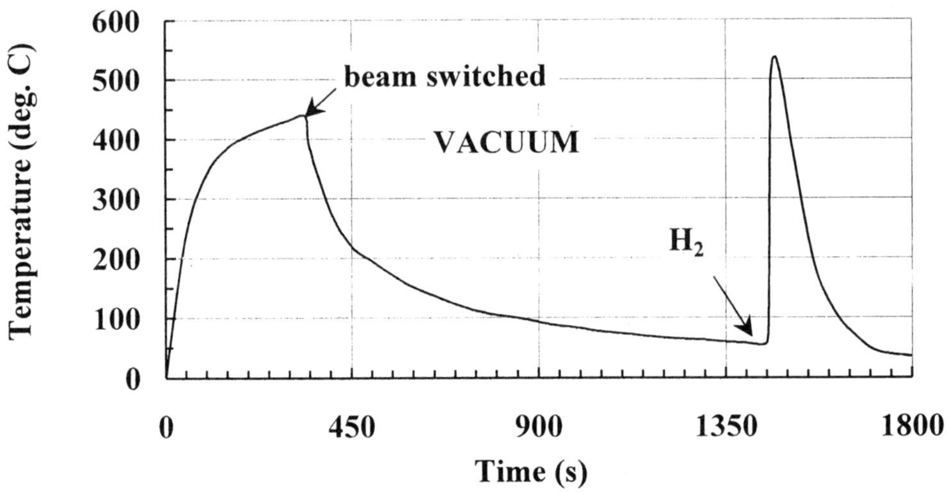

Figure. 3. The thermogram of CS at dose rate of preliminary irradiation 0.3 Mrad/sec.

As a result of this interaction, composite mono-phase hydridonitrides of alloys were obtained (Table 2).

Comparing the diffractograms, one can see that the diffraction peaks are wider and their intensities are lower than in the previous system (Fig. 2).

It was obtained that the presence of a second metal (in this case - of zirconium) strongly influenced on the radiation processes in nitrogen-containing bimetallic α-alloys. In the systems Ti-N-H and Zr-N-H the CS started at 30-35°C at very low dose rate of preliminary irradiation (0.025 Mrad/s). But in the case under consideration, the reaction started at higher dose rates and temperatures of hydrogen filling. It testifies that the defect formation in these alloys proceed under tighter conditions.

The Zr addition into the system raises the temperature of post-radiation interaction with hydrogen till 50-75°C. The obtained composite hydridonitrides are similar to the TRS hydridonitrides by the phase composition and hydrogen content. At the post-radiation interaction of a bimetallic nitrogen-containing matrix with hydrogen, the hydrogenation temperatures were close to that at TRS and depended weakly on dose rate, total dose and temperature of a sample heating at the preliminary irradiation. It testifies that the process of interaction has a chemical character and the further rise of temperature is a result of exothermal interaction of a metal matrix with hydrogen.

Thus, at preliminary irradiation of a metal matrix, a critical concentration of rather steady radiation defects was created, resulting in rise of its reactivity. The created in a metal matrix radiation defects are partially saved during some time. We can suggest that in an alloy the relaxation processes in its crystal lattice are more intensive at the expense of composite character of a chemical bond. Thereof the post-radiation interaction in this system starts at higher temperature, than in binary Me-H and triple Me-N-H systems. To determine the "lifetime" of the created defects, the radiation dose rate was raised. At 0.5 Mrad/sec and total radiation dose 100 Mrad, the temperature of preliminary heat in vacuum of $Ti_{0.74}Zr_{0.26}N_{0.22}$ achieved 590°C. The camera was filled by hydrogen after sample cooling to 60°C, but the CS reaction was not observed. The interaction of this sample with hydrogen at 60°C did not proceed also at higher radiation dose rate, 1.0 Mrad/s, when the sample was heated up to 950°C.

Table 2. The characteristics of the cold-synthesis in the Ti-Zr-N-H system

Dose power, MRad/s	T_{heat}. in vacuum, °C	T of H_2 filling, °C	T_{react}., °C	Total dose, MRad	N_2 content, wt. %	H_2 content, wt. %	Compound (accounted formula)	Phase content
				$Ti_{0.88}Zr_{0.12}N_{0.22}$ – H				
SHS	-	-	700	-	5.77	2.43	$Ti_{0.88}Zr_{0.12}N_{0.22}H_{1.3}$	HCP
0.2	452	60	-	100	6.08	-	$Ti_{0.88}Zr_{0.12}N_{0.22}$	Sol.Sod., HCP + traces FCC
0.2	385	75	510	100	5.82	2.34	$Ti_{0.88}Zr_{0.12}N_{0.22}H_{1.35}$	HCP
0.3	440	60	540	100	5.87	2.36	$Ti_{0.88}Zr_{0.12}N_{0.22}H_{1.25}$	HCP
0.3	410	60	400	50	5.86	2.3	$Ti_{0.88}Zr_{0.12}N_{0.22}H_{1.22}$	HCP
0.3	440	75	420	100	5.9	2.42	$Ti_{0.88}Zr_{0.12}N_{0.22}H_{1.29}$	HCP
0.4	515	60	510	100	5.83	2.4	$Ti_{0.88}Zr_{0.12}N_{0.22}H_{1.28}$	HCP
0.4	465	60	400	50	5.89	2.25	$Ti_{0.88}Zr_{0.12}N_{0.22}H_{1.19}$	HCP
				$Ti_{0.81}Zr_{0.19}N_{0.22}$ – H				
0.2	415	30	-	100	5.57	0.23	$Ti_{0.81}Zr_{0.19}N_{0.22}H_{0.13}$	Sol.Sod., HCP + TiN, FCC
0.2	400	50	430	100	5.45	2.19	$Ti_{0.81}Zr_{0.19}N_{0.22}H_{1.23}$	HCP
0.3	460	50	430	100	5.52	2.15	$Ti_{0.81}Zr_{0.19}N_{0.22}H_{1.2}$	HCP
0.3	400	50	460	50	5.45	2.15	$Ti_{0.81}Zr_{0.19}N_{0.22}H_{1.2}$	HCP
0.2	400	100	400	100	5.17	2.07	$Ti_{0.74}Zr_{0.26}N_{0.22}H_{1.22}$	HCP
0.5	590	60	-	100	5.28	0.47	$Ti_{0.74}Zr_{0.26}N_{0.22}H_{0.27}$	Sol.Sod., HCP + TiN, FCC
1.0	950	60	-	100	3.32	-	$Ti_{0.74}Zr_{0.26}N_{0.22}$	Sol.Sod., HCP + TiN, FCC
				$Ti_{0.74}Zr_{0.26}N_{0.22}$ – H				
0.2	400	100	400	100	5.17	2.07	$Ti_{0.74}Zr_{0.26}N_{0.22}H_{1.22}$	HCP
0.5	590	60	-	100	5.28	0.47	$Ti_{0.74}Zr_{0.26}N_{0.22}H_{0.27}$	Sol.Sod., HCP + TiN, FCC
1.0	950	60	-	100	3.32	-	$Ti_{0.74}Zr_{0.26}N_{0.22}$	Sol.Sod., HCP + TiN, FCC

Apparently, in the systems Me-H, Me-N-H, etc., the arisen radiation damages had a longer lifetimes: the ability to interact with hydrogen in TiNx was saved during 8 min after beam switching off, and in ZrNx - 1 hour (dose rate of beam was low, 0.05 Mrad/s).

From the results of these experiments we can conclude, that in bimetallic N-containing alloys the mechanisms of TRS and of post-radiation stimulated interaction are of character identical to that of Me-N-H systems. In the alloys, the Ti/Zr ratio influences only on absolute limiting values of parameters of thermal-radiation processes. Either the presence of Zr, or the mutual influence of these two metals makes the alloy less susceptible to irradiation. With increase of the zirconium amount in the used interval of dose rates, the post-radiation interaction with H_2 at temperatures lower of 50°C becomes impossible. Besides, TRS was not realized at dose rate lower of 0.3 Mrad/s, but at $P \geq 0.3$ Mrad/s, the presence of zirconium raised the temperature of TRS process.

In two-phase nitrogen-containing alloys, the process of homogenization during TRS and CS proceeded very actively, probably, due to the presence of zirconium. Thus, some observed peculiarities of thermal-radiation processes in the above-described systems can be explained by hetero-atomic nature of their crystalline structure and by hetero-desmisity of chemical bonds.

3. Conclusions

As a result of radiation effects, the thermal-radiation interaction immediately in beam and the post-radiation interaction at the expense of created during preliminary irradiation defects took place. In bimetallic alloys, the titanium/zirconium ratio played also an important role. The optimal parameters of TRS of composite hydridonitrides $Ti_kZr_{1-k}N_xH_y$ were stated. It was shown that the presence of a non-metallic sub-lattice as well as of a second metal did not hinder the proceeding of thermal-radiation processes and the formation of hydridonitrides with HPC lattice of high hydrogen content (2.19 - 2.55 wt. %).

Acknowledgements: The work is supported by ISTC. Project A - 192.

4. References

1. Dolukhanyan, S.K., Shekhtman, V.Sh., Aghajanyan, N.N., Abrahamyan, K.A., Harutyunyan, Kh.S., Aleksanyan, A.G., Hakobyan, H.G., and Ter-Galstyan, O.P. (2000) Radiation-thermal synthesis of titanium hydride, *Khimicheskaya Fizika*, **19** (12) 21-25

2. Aleksanian, A.G., Dolukhanyan, S.K., Hakobian, H.G., Beibutian, V.M. and Shekhtman, V.Sh.. Combustion processes in Ti-Zr-N_2-H_2 system. Hydrogen Energy Progress X1"Proceedings of 11-th World Hydrogen Energy Conference, 23-29 Jun. 1996, Stuttgart, Germany, **2**, 1341-1346.

3. Aleksanyan, A.G., Dolukhanyan, S.K., Shekhtman, V.Sh., Abrahamyan, K.A., Hayrapetyan, V.S., Mnatsakanyan N.L. (2001) Radiation-thermal synthesis of titanium hydridonitrides. *Khimicheskaya Fizika*, **20**, 1, 32-36.

4. Aleksanyan, A.G., Aghadjanyan, N.N.,. Dolukhanyan, S.K, Mnatsakanyan, N. L., Harutyunyan, Kh. S, Hayrapetyan, V.S.. (2002) Thermal-radiation synthesis of zirconium hydridonitrides and carbohydrides. Thermal-radiation synthesis of zirconium hydridonitrides and carbohydrides. *Int. J. Alloys and Compounds*, **330-332,** 559-563.

ON HYDROGEN FUGACITY IN METALS UNDER ELECTROLYTIC CHARGING

Yu.S. NECHAEV
Bardin Central Research Institute of Ferrous Metallurgy
9/23, 2-nd Baumanskaja St., Moscow 107005, Russia

Abstract

On the basis of using the experimental data on apparent solubility and diffusivity of hydrogen in electrolytically charged maraging steel and iron-'Armco', the fugacity (pressure) of molecular hydrogen in metallic materials are evaluated ($\sim 3\cdot 10^7 - 3\cdot 10^8$ Pa) for typical current densities (10-100 A/m^2). The apparent quantities are described within the approximation of the local equilibrium distribution of absorbed hydrogen between the traps (dislocations' and vacancies' regions) and the normal lattice. The hydrogen fugacity has been also estimated by using the deformation-mechanism maps with respect to the experimental data on blistering and cracking of low alloyed steel, under electrolytic hydrogen charging.

1. Introduction

The problem of high fugacity (pressure) of molecular hydrogen in metals and alloys under electrolytic charging was considered in a plenty of works, for instance, in [1, 2]. Nevertheless, the problem has still been as an open question, despite of its importance both for theory and for practice.

As it was shown in [3], large anomalies of hydrogen solubility and diffusivity in electrolytically charged maraging steel and iron-'Armco' were mainly caused by compound-hydride-like nanosegregation at dislocations, and also by complexes of hydrogen with vacancies (under the conditions of high hydrogen fugacity). Detailed study results on blistering and cracking in some low alloyed steel under electrolytic charging (and the corresponding hydrogen fugacity) were obtained in [4].

In the present contribution the methodological and conceptual aspects of evaluation of the hydrogen fugacity in electrolytically charged metallic materials are illustrated on the examples of using the data [3, 4]; particularly, the critics is made of the [1, 2] approach.

2. The case of one type of hydrogen traps (dislocations)

The martensitic structure of the maraging steel specimens [3] can be considered as the case when the main hydrogen traps are the near-dislocation segregation regions (NDSR). All the rest traps can be neglected, that is the case of one type of the traps.

N. Veziroglu et al. (eds.), Hydrogen Materials Science and Chemistry of Metal Hydrides, 161–164.

For such a case, the apparent solubility (C_Σ) of hydrogen in the electrolyticaly charged material can be described in the framework of the local equilibrium approximation [5], with using Sievert's law and Boltzman's type distribution of hydrogen between the NDSR and the normal lattice [5, 6], as follows:

$$C_\Sigma \approx C_j (1 + \eta_\perp K_\perp), \tag{1}$$

$$C_j \approx C^0 (P_j / P^0)^{1/2}, \tag{2}$$

$$P_j \approx A\, j, \tag{3}$$

where C_j is the hydrogen concentration (atomic fraction) in the metal normal lattice corresponding to hydrogen fugacity (pressure) of molecular hydrogen (P_j); A is the proportionality coefficient in equation (3), within the analogy [1] to the kinetic theory of gases; j is the electrical current density; C^0 is the standard solubility (atomic fraction) in the normal lattice corresponding to the standard pressure of molecular hydrogen ($P^0 = 1 \cdot 10^{5}$ Pa); $\eta_\perp$ is the volume (atomic) fraction of the NDSR; $K_\perp$ is the apparent equilibrium constant of the distribution of hydrogen between the NDSR and the normal lattice.

Also, for the case of materials with one type of the traps, within the same approximation [5, 6], the apparent diffusion coefficient (D_Σ) of hydrogen can be described as follows:

$$D_\Sigma \approx D / (1 + \eta_\perp K_\perp), \tag{4}$$

where D is the diffusion coefficient of hydrogen in the normal lattice.

It can be noted that the equation (4) is valid if $D >> (\eta_\perp K_\perp D_\perp)$, where $D_\perp$ is the diffusion coefficient of hydrogen in the NDSR, [5, 6].

From equations (1-4) the desirable expression for the fugacity evaluation is followed:

$$P_j \approx P^0 (C_\Sigma D_\Sigma / C^0 D)^2. \tag{5}$$

By using the values [3] of $j = 40$ A/m^2, $C_\Sigma \approx 2.5 \cdot 10^{-3}$, $D_\Sigma \approx 7.2 \cdot 10^{-13}$ m^2/s, $C^0 \approx 7 \cdot 10^{-7}$, and $D \approx 8 \cdot 10^{-11}$ m^2/s, the quantities for the charged maraging steel are obtained as $P_j \approx 1 \cdot 10^8$ Pa, $A \approx 3 \cdot 10^6$ Pa·m^2/A, $C_j \approx 2 \cdot 10^{-6}$.

3. The case of two types of traps (dislocations and vacancies)

The main hydrogen traps in the cold worked (before the charging) Fe-Armco specimens [3] are obviously the NDSR and the excess (deformation) vacancies, that is the case of two types of the traps.

For the case of materials with the two types of the traps, the apparent solubility of hydrogen can be described (within the approximation [5, 6]) as:

$$C_\Sigma \approx C_j (1 + \eta_\perp K_\perp) + C_{VH}, \tag{6}$$

where C_{VH} is the atomic fraction of hydrogen bound with vacancies.

The apparent diffusion coefficient of hydrogen in materials with the two types of the traps can be approximated [3, 5, 6] as:

$$D_\Sigma \approx D / \{(1 + \eta_\perp K_\perp)[1 + (C_V - C_{VH}) K_V]\}, \qquad (7)$$

$$K_V \approx C_{VH} / \{C_j (C_V - C_{VH})\}, \qquad (8)$$

where C_V – is the total concentration of the excess (deformation) vacancies; K_V is the equilibrium constant of the hydrogen distribution between the vacancies' regions and the normal lattice.

From (2, 6-8) the expression for the fugacity evaluation is followed:

$$P_j \approx P^0 C_{VH} C_\Sigma D_\Sigma / (C^0)^2 D. \qquad (9)$$

By using the values [3] of $j = 50$ A/m^2, $C_V \approx 1\cdot10^{-4}$, $C_\Sigma \approx 1.1\cdot10^{-3}$, $D_\Sigma \approx 7\cdot10^{-14}$ m^2/s, $C^0 \approx 3.6\cdot10^{-8}$, $D \approx 3.2\cdot10^{-9}$ m^2/s, and also the value of $C_{VH} \approx 7.4\cdot10^{-5}$, [7], the quantities for the charged Fe-specimens are obtained as $P_j \approx 1.5\cdot10^8$ Pa, $A \approx 3\cdot10^6$ Pa·m^2/A, and $C_j \approx 1.3\cdot10^{-6}$. The obtained values are in the accordance with the previous evaluations for the maraging steel specimens. By substituting the values of C_{VH}, C_j, and C_V in equation (8), the quantity is obtained of $K_V \approx 2.2\cdot10^6$ which is close to the experimental K_V-value [7] for iron.

It is relevant to note that the fugacity values [1, 2] are seems overstated of one-two orders due to using an ambiguous equation as $P_j = P^0 (C_\Sigma / C^0)^2$, instead of equation (2).

4. The fugacity estimates from the blister growth rate

The fugacity estimates can be also done on the basis of using the deformation-mechanism maps [8] for the data [4] on blistering and cracking of low alloyed steel under electrolytic hydrogen charging ($j = 100$ A/m^2), as an example.

By substituting the above obtained quantity of $A \approx 3\cdot10^6$ Pa·m^2/A and the j-value in equation (3), the value is obtained of $P_j \approx 3\cdot10^8$ Pa (for the low alloyed steel specimens, [4]).

In Figs. 4 and 3 in [4], the kinetic data are presented on the growth of the near surface blisters, obviously, under the hydrogen fugacity effect ($P_j \approx 3\cdot10^8$ Pa). From these data (Figs. 3 and 4, [4]), the rate of the local deformation of the material can be evaluated as 10^{-3}-10^{-4} s^{-1}.

According to the deformation-mechanism maps [8] for iron and low alloyed steels (with the suitable grain size), such deformation rates (at the room temperature) can have place in the case of the stresses of the P_j order.

It proves the acceptability of the used above concept and methodology for the fugacity evaluation in metallic materials under electrolytic charging.

5. Conclusion

The fugacity of molecular hydrogen in metallic materials under electrolytic charging (at the room temperature and typical current densities $j = 10$ -100 A/m^2) can be evaluated (with the uncertainty of the order of the magnitude) as P_j (Pa) $\approx 3 \cdot 10^6$ j; it results to $3 \cdot 10^7$-$3 \cdot 10^8$ Pa.

The blistering of metallic materials under electrolytic charging is mainly caused by the high fugacity of molecular hydrogen (~10^8 Pa, in microcracks, micropores, and so on).

6. References

1. Geld, P.B., Rjabov, R.A. and Kodes, E.S. (1979) *Hydrogen and imperfections of metal structure*, Metallurgija Publishers, Moscow.
2. Brass, A.M. and Chanfreau, A. (1996) Accelerated diffusion of hydrogen along grain boundaries in nickel, *Acta Mater.* **44** (9), 3823-3831.
3. Nechaev, Yu.S. and Filippov, G.A. (2001) Hydride-like segregation at dislocations in α-iron and steels, *Defect and Diffusion Forum* **194-199**, 1099-1104.
4. Isakov, M.G., Izotov, V.I., Karpeljev, V.A. and Filippov, G.A. (2000) Kinetics of damage formation in low alloyed steel under hydrogen charging. *Phys. Met. Metall.* **90** (3), 97-103.
5. Nechaev, Yu.S. and Egziabher, K.H.G. (1988) On gas permeability, diffusivity and characteristics of near-dislocation segregation regions in metals. *Phys. Stat. Sol.(a)* **106**, 399-406.
6. Nechaev, Yu.S. (2001) Characteristics of hydride-like segregation at dislocations in palladium. *Progresses Phys. Sciences* **171** (11), 1251-1261.
7. Kedzierzawski, P. (1985) Hydrogen trapping in iron and iron alloys. In: Oriani, R.A., Hirth, P. and Smialowski, M., editors. *Hydrogen degradation of ferrous alloys*, Noyes Publishers, New Jersey, pp. 271-288.
8. Frost, H.J. and Ashby, M.F. (1982) *Deformation-mechanism maps*, Pergamon Press.

HIGH-DISPERSED HYDRIDE PHASES IN RT_5 - NH_3 SYSTEMS, WHERE R = La, Sm; T = Ni, Co

V.N.FOKIN, B.P.TARASOV, YU.M.SHUL'GA, E.E.FOKINA,
I.I.KOROBOV, A.G.BURLAKOVA, S.P.SHILKIN
Institute of Problems of Chemical Physics of RAS
Chernogolovka 142432, Moscow Region, Russia

Abstract

The interaction of the intermetallic compounds RT_5 (R = La, Sm; T = Ni, Co) with ammonia was investigated at an initial pressure of ammonia of 0.6-0.7 MPa in the presence of 10 mass% (from the quantity of intermetallide entered in reaction) of NH_4Cl as a process activator. It was established, that depending on temperature of interaction in $RT_5 - NH_3$ systems there are both the processes of hydrogenation with a formation of hydride phases of initial intermetallic compounds, and the processes of disproportionation with an isolation the hydride phases of new intermetallides. The reaction products are the high-dispersed powders.

1. Introduction

The intermetallic compounds of RT_5-type (R = La, Sm; T = Ni, Co) are used as starting materials for a manufacture of the permanent magnets and as the catalysts of pyrolysis of hydrocarbon compounds. The intermetallides $LaNi_5$ and $SmCo_5$ ($CaCu_5$-type structure), crystallizing in a hexagonal lattice with parameters $a = 5.017$ Å, $c = 3.982$ Å and $a = 5.003$ Å, $c = 3.978$ Å, accordingly, absorb hydrogen with good kinetics with a formation of hydride phases of the compositions $LaNi_5H_{6.8}$ and $SmCo_5H_4$ [1-2]. At removal of pressure the hydride phase $LaNi_5H_{6.8}$ isolates hydrogen to α-solid solution of hydrogen [1], and $SmCo_5H_4$ loses hydrogen in the same requirements up to composition $SmCo_5H_{\sim 2}$, further decomposition of which proceeds at rise of a temperature on stages: at 70-80°C – up to $SmCo_5H_1$, at 190-200°C hydrogen is isolated completely [3]. The specific surface of the hydrogenation products is 0.2-0.3 m^2/g [3]. The size of the particles of powders prepared by a method of hydride dispergation is restricted to several microns.

On the basis of the results of interaction examinations in the $Ti - NH_3$, $ScFe(Ni)_2 - NH_3$ systems [4-5] it is followed to hope that the introduction of hydrogen and nitrogen simultaneously in a matrix of intermetallic compound RT_5 will allow to receive the hydride phases containing some of nitrogen, and also the products of a

N. Veziroglu et al. (eds.), Hydrogen Materials Science and Chemistry of Metal Hydrides, 165–170.

disproportionation of intermetallic compounds under an action of ammonia – transitive metal (Co or Ni) and hydrides or nitrides of metal R– in high-dispersed state in the form of homogeneous mixtures.

In this connection in the present work the experimental examination of a possibility of introduction of hydrogen and nitrogen, generated by NH_3, in a lattice of intermetallic compounds RT_5 in the presence of NH_4Cl was carried out at various temperatures.

For the resolution of the defined problem the process of interaction of intermetallic compounds $LaNi_5$ and $SmCo_5$ with ammonia was investigated at initial pressure of ammonia of 0.6-0.7 MPa in an interval of the temperatures of 150-350°C in the presence of ammonium chloride added in the quantity of 10 mass. % from the quantity of the intermetallide entered in the reaction.

2. Experimental

The initial samples of intermetallic compounds RT_5 were prepared by alloying a charge from the corresponding metals with purity of 99.8% (La, Sm) or 99.99% (Ni, Co) under ~0.2 MPa of pyre argon in an electroarc furnace with non-spent tungsten electrode. The homogenizing annealing of alloys was carried out in evacuated quartz ampoules at 600°C for 250 h with subsequent quenching in cold water. Directly before an examination the alloys were refined by a mechanical procedure from a surface oxide layer, washed out by ethanol, diethyl ether and grinded in an argon atmosphere up to particles by the size of 1-5 mm.

The ammonium chloride was dried by an evacuation for 9 h at 150°C. The ammonia, desiccated by metallic sodium, had a purity of 99.99%.

The X-ray graphical investigations were performed on an automatic assembly consisting of a diffractometer ADP-1 (Cu K_α- and Cr K_α-emissions for the compounds of La and Sm, accordingly) and a computer. The determination error of the crystal lattice constants did not exceed 0.005 Å.

A specific surface of the samples was determined on an amount of the low-temperature adsorption of krypton after the removal of volatile products from a solid phase under a vacuum of 1.3×10^{-3} Pa at a temperature of 300°C for 15 h and was calculated by the Brunauer-Emmet-Teller method. The determination error was ± 10%.

The thermoanalyses were carried out on a derivatograph of type 3434C with microprocessor control in a stream of argon.

The contents of hydrogen in the reaction products were defined by a standard method of burning the samples in a stream of oxygen. The contents of nitrogen were determinated by the Kjeldal method. The contents of chlorine were analyzed by a method of turbidimetry.

The interaction of RT_5 with ammonia was conducted in the laboratory plant of high pressure by a capacity of 60 cm^3. Before the investigation the mixture of a powder of RT_5 with 10 mass% of NH_4Cl was mixed by pounding with a pestle in argon atmosphere. The given quantity of prepared mixture (1-2 g), placed in a stainless-steel container, was loaded into a reactor-autoclave and evacuated until residual pressure of ~1 Pa at room temperature for 30 min; ammonia was brought up to a pressure of 0.6-

0.7 MPa and the reactor-autoclave was kept at the same temperature for 30 min. Further, the reactor was heated up to various given temperatures for 3 h, was cooled up to ~20°C and was again heated up. After the realization of a given quantity of heating-cooling cycles, the ammonia was discharged into a buffer vessel, and ammonium chloride was removed under a vacuum of ~1 Pa, if necessary, by heating the reactor up to 180-200°C for 2 h. The reaction products were unloaded in an inert atmosphere and analyzed.

3. Results and discussion

By the X-ray graphical analysis the homogeneity of the intermetallide $LaNi_5$ was established and determined the constants of crystal lattice $a = 5.013$ Å, $c = 3.986$ Å ($CaCu_5$-type structure). It was shown also that $SmCo_5$ ($CaCu_5$-type structure, $a = 5.001$ Å, $c = 3.971$ Å) contains the insignificant quantity of the second phase – Sm_2Co_7, which occupies the intergranule space and promotes cracking and refining the basic fields of $SmCo_5$ at hydrogenation of a material.

The conditions and results of the investigation of an interaction of the intermetallic compounds $LaNi_5$ and $SmCo_5$ with ammonia in the presence of NH_4Cl are given in Table 1.

As follows from the data of Table 1, an interaction of the intermetallides $LaNi_5$ and $SmCo_5$ with ammonia at 150°C (samples 1 and 6) for 25 h proceeds with formation (after removal of pressure) of solid solutions of hydrogen in metallic matrices of RT_5, that results in insignificant increase of the lattice constants constituting for $LaNi_5H_{0.5}$ $a = 5.015$ Å, $c = 3.990$ Å, for $SmCo_5H_{0.3}$ $a = 5.001$ Å, $c = 4.034$ Å. The specific surface of $LaNi_5H_{0.5}$ reaches 1.3 m^2/g, that is a bit higher in comparison with specific surface of phase which was obtained by a treatment of RT_5 by pure hydrogen (0.2-0.3 m^2/g).

The rise of the temperature of interaction of $LaNi_5$ with ammonia up to 200°C (sample 2) is accompanied by occurrence of metal nickel and intermetallic phase of not established composition (X-phase) on diffractograms of reaction products, alongside with a phase of a solid solution of hydrogen. On the data of a thermal analysis, X-phase decomposed with two endoeffects at 160 and 240°C. The interaction products are the high-dispersed powders with the advanced specific surface (6.8 m^2/g). The solid solution $LaNi_5H_{0.5}$ has the following parameters of a lattice: $a = 5.015$ Å, $c = 3.987$ Å.

Unlike $LaNi_5$ the intermetallic compound $SmCo_5$ in ammoniacal medium is more insensitive and at the temperature of 200°C (sample 7) interacts with ammonia only with formation of a solid solution of composition $SmCo_5H_{0.3}$ with the lattice constants $a = 5.090$ Å, $c = 4.069$ Å. The specific surface of reaction products makes 1.4 m^2/g.

At a temperature of 250°C at interaction of $SmCo_5$ with ammonia (sample 8) for 26 h the decomposition of hydride phase – the solid solution of hydrogen in intermetallic compound $SmCo_5$ - begins with an isolation of metallic cobalt and a formation of hydride phase of composition $Sm_2Co_7H_x$ with the smaller content of cobalt. On data of the X-ray phase analysis, the hydride phases have the following parameters of a lattice: for $SmCo_5H_{\sim 0.3}$ $a = 5.021$ Å, $c = 4.003$ Å; for $Sm_2Co_7H_x$ $a = 5.020$ Å, $c = 23.182$ Å (for comparison: Sm_2Co_7 crystallizes into a hexagonal lattice with the constants $a = 5.043$ Å, $c = 24.311$ Å). The specific surface of the interaction products of $SmCo_5$ with NH_3 at 250°C reaches 6.5 m^2/g, that in some tens of times is higher, than the specific surface of the products of hydride dispergation. This value of specific surface testifies about high dispersivity of a material.

At 250-300°C the process of decomposition of the intermetallide $LaNi_5$ under action of ammonia (samples 3-4) with formation of X-phase and metallic nickel bears the profounder pattern, and already at the temperature of 350°C (sample 5) the hydride phase $LaNi_5H_x$ practically disappears. The specific surfaces of the products of reactions carried out at the temperatures of 250-350°C are 1.7-2.5 m^2/g.

At the temperatures of 300-350°C at an interaction of intermetallic compound $SmCo_5$ with ammonia (samples 9-10) there is a further decomposition of hydride phase $SmCo_5H_x$. If at 300°C the content of $SmCo_5H_x$ and $Sm_2Co_7H_x$ carries equimolar character, at 350°C the hydride phase $Sm_2Co_7H_x$ prevails in a mixture. The lattice constants of the phases, obtained at 300°C, are $a = 5.004$ Å, $c = 4.064$ Å for $SmCo_5H_{\sim 0.3}$ and $a = 4.943$ Å, $c = 23.558$ Å for $Sm_2Co_7H_x$. The specific surface of interaction products of $SmCo_5$ with ammonia at 300-350°C forms 0.7-3.9 m^2/g.

The interaction of the intermetallic compound $SmCo_5$ with ammonia in the presence of NH_4Cl can be described by the reactions, the schemes (1-2) of which are shown below:

$$SmCo_5 + NH_3 \xrightarrow[150\text{-}200°C]{NH_4Cl} SmCo_5H_{0.3} + N_2 \qquad (1)$$

$$SmCo_5 + NH_3 \xrightarrow[250\text{-}350°C]{NH_4Cl} SmCo_5H_{\sim 0.3} + Sm_2Co_7H_x + Co + N_2 \qquad (2)$$

The similar processes occur and in the $LaNi_5 - NH_3$ system.

The particular attention deserves the fact of a growth of specific surface of nitriding products of $SmCo_5$ with a rise of the interaction temperature: from 0.3 m^2/g at 150°C till 6.5 m^2/g at 250°C, that testifies to significant increase of powder dispersity. This process, probably, is promoted by an accumulation of some of nitrogen in a lattice of intermetallic compound.

Acknowledgment

This work is supported by the Russian Basic Research Foundation (grant No. 01-03-33241).

4. References

1. Semenenko, K.N., Malyshev, V.P., Petrova, L.A., Burnasheva, V.V., and Sarynin, V.K. (1977) Interaction of $LaNi_5$ with hydrogen, *Izv Akad Nauk SSSR, Ser Neorg Mater* **17**, 2009-2013 (in Russian).

2. Kuijpers, F.A. (1973) RCo_5-H and related systems, *Philips Res Repts Suppl* N2, 1-102.
3. Fokin, V.N., Fokina, E.E., and Shilkin, S.P. (1994) Application of hydrogen of high purity for dispergation or embrittlement of magnetic materials, *Zh Prikladnoi Khim* **67**, 1372-1374 (in Russian).
4. Fokin, V.N., Shilkin, S.P., Fokina, E.E., and Mozgina, N.G. (1997) Synthesis of high-dispersed titanium nitride by interaction of titanium or its dihydride with ammonia, *Zh Obshch Khim* **67**, 705-708 (in Russian).
5. Fokin, V.N., Fokina, E.E., Tarasov, B.P., Korobov, I.I., and Shilkin, S.P. (2001) Hydrides $ScFe(Ni)_2H_x$: preparation and properties, *Int J Hydrogen Energy* **26**, 449-452.

TABLE 1. Conditions and investigation results of the interaction of RT_5 with ammonia

Sample No.	Initial RT_5	Conditions of synthesis			Phase composition of interaction products	Lattice Constants of interaction products		Specific surface of interaction products (m^2/g)
		T (°C)	Time (h)	P_{NH3} (MPa)		*a* (Å)	*c* (Å)	
1	$LaNi_5$	150	25	0.64	$LaNi_5H_{0.5}$	5.015	3.990	1.3
2	$LaNi_5$	200	25	0.64	$LaNi_5H_{0.5}$ +	5.015	3.987	6.8
					X-phase traces +			
					Ni traces			
3	$LaNi_5$	250	28	0.7	$LaNi_5H_{\sim0.5}$ +	4.981	4.024	2.5
					X-phase +			
					Ni	3.526	–	
4	$LaNi_5$	300	25	0.68	$LaNi_5H_{\sim0.5}$ +	5.023	3.983	1.7
					X-phase +			
					Ni			
5	$LaNi_5$	350	23	0.68	$LaNi_5H_x$ traces +			1.8
					X-phase +			
					Ni	3.523	–	
6	$SmCo_5$	150	25	0.64	$SmCo_5H_{0.3}$ +	5.001	4.034	0.3
					$Sm_2Co_7H_x$ traces			
7	$SmCo_5$	200	23	0.64	$SmCo_5H_{0.3}$ +	5.090	4.069	1.4
					$Sm_2Co_7H_x$ traces			
8	$SmCo_5$	250	26	0.62	$SmCo_5H_{\sim0.3}$ +	5.021	4.003	6.5
					$Sm_2Co_7H_x$ +	5.020	23.182	
					α-Co			
9	$SmCo_5$	300	23	0.56	$SmCo_5H_{\sim0.3}$ +	5.004	4.064	0.7
					$Sm_2Co_7H_x$ +	4.943	23.558	
					α-Co	2.504	4.073	
10	$SmCo_5$	350	25	0.64	$SmCo_5H_x$ traces +			3.9
					$Sm_2Co_7H_x$ +	4.929	23.323	
					α-Co	2.510	4.055	

DEVELOPMENT OF THE PHYSICOCHEMICAL BASIS FOR MATERIALS SCIENCE

L.V. BOBUKH, K.A. BOBUKH
National Metallurgical Academy of Ukraine,
4 Gagarin Avenue, Dnepropetrovsk, 49600 Ukraine

1. Introduction

Within the framework of the series of works [1, 2] the authors are engaged in development of the scientific concept of the energo-elemental essences of the matter. The state of the matter is described by the inseparably linked among themselves and changing in time physical properties - energies ***e*** and chemical properties - elements ***c***.

Such approach has allowed to present the mathematical (equation 1) and graphical (Fig. 1,2) expressions of the fundamental law of subsistence - the law of the matter motion - the law of the matter energo-elemental states changes.

2. Results and discussion

The systems are considered conjugation of qualitative and quantitative sets of elements and energie:

$$\begin{array}{c} e_1, e_2, e_3, \ldots, e_n \\ c_1, c_2, c_3, \ldots, c_n \end{array} \rightarrow \begin{array}{c} e'_1, e'_2, e'_3, \ldots, e'_n \\ c'_1, c'_2, c'_3, \ldots, c'_n \end{array} \rightarrow \ldots \rightarrow \begin{array}{c} e^\infty_1, e^\infty_2, e^\infty_3, \ldots, e^\infty_n \\ c^\infty_1, c^\infty_2, c^\infty_3, \ldots, c^\infty_n \end{array}$$

$$f(E\,;C) \qquad\qquad f'(E'\,;C') \qquad\qquad f^\infty(E^\infty\,;C^\infty)$$

where,

$E = (e_1;\ e_2;\ e_3;\ \ldots;\ e_n)$ - qualitative and quantitative set of energies that is inherent to the matter; $C = (c_1;\ c_2;\ c_3;\ldots;\ c_n)$ - qualitative and quantitative set of elements that inherent to the matter; f (E; C) - function of the matter state or function of the energy-element state system; Δf (E; C) - change of the matter state function in the result of the change of E and/or C.

Process of transition of the matter from one state into another f (E ; C) $\rightarrow$ $\rightarrow f'$ (E' ; C') $\rightarrow$ f^∞ (E^∞ ; C^∞) is perpetual, and the speed of this transition is described by equation:

$$V = \frac{\Delta f(E,C)}{\tau} \qquad (1),$$

where:

V - speed of the matter transition from one state into another;

τ - time during which the matter makes this transition from one state into another.

N. Veziroglu et al. (eds.), Hydrogen Materials Science and Chemistry of Metal Hydrides, 171–177.

In the practice of the science of materials to describe the state in which the system is in at the given moment of time, one uses diagrams (Fig. 1, 2), the plotting principles and use of which have been described in the works [1, 2].

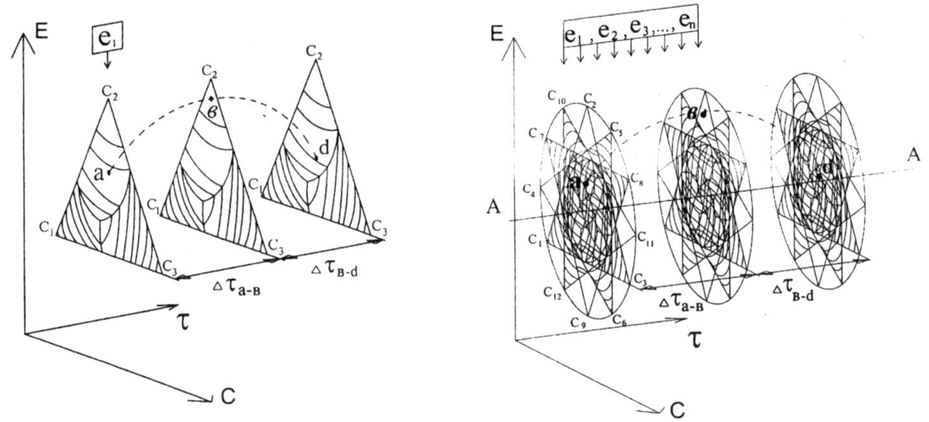

Fig. 1a, b. Diagrams of the state of the triple (a) and n-element (b) systems

Each point of the state diagram plane (Fig. la.) describes the real energy-element state for the given moment of time. If the system is isolated, then this state will exit arbitrarily long. For isolated systems: E, C=const; Δf=0; $\tau \rightarrow \infty$; $V \rightarrow 0$. Exploring the processes within the system at the nano-level we should speak of conditionally isolated systems. . For the metal system that consists of three elements c_1, c_2, c_3 to transfer from one state into another: a → b → d, it is necessary that there was a change in E and C values; the process of transition goes on during the time interval $\Delta\tau$. With the increase in the number of elements (c_1,... , c_n) and number of energy components (e_1, ... , e_n), the diagram becomes more complicated (Fig. lb). Thus, at each particular instant there exists a certain point a, b or d of the diagram that describes the system state determined by the certain relationship between E and C.

In case of E and/or C change (open system), the system transfers from one state into another and the process goes on in time. For open systems: E, C $\neq$ const; $\Delta f \neq 0$; τ and V have certain values. Description of processes of continuous transition of the open system from one state into another is convenient for examination with the help of the kinetic diagram given on Fig. 2a. Each plane of the kinetic diagram section is represented by the system diagram that is similar to the one shown on the figurel; any point (1, 2, 2') of this section plane describes the system state and is determined by the certain ratio of E and C at a certain moment of time. Time is the interval $\Delta\tau$ during which the biosystem is functioning (existing); for the human being it is his life time, from the moment of his conception $\tau_з$ to the moment of his death $\tau_{c..}$. If at some moment τ of the biosystem functioning its state is described by quantities E_1 and C_1 then the point 1 of the complex influence of the two quantities is within the volume of functioning - the biosystem lives. If, at the moment τ chemical and energy actions on the biosystem are described by quantities C_2 and E_2 (accordingly), then the point 2 of the complex influence of the two quantities also lays within limits of the area - the biosystem functions, but the degree of stability of the system in the point 2 is much lower than the degree of the system stability in the point 1. If, at the moment τ, chemical and energy actions are described by quantities C_3 and E_3 (accordingly), the point 3 of the complex influences of two quantities lays outside the area of functioning - the biosystem does not live.

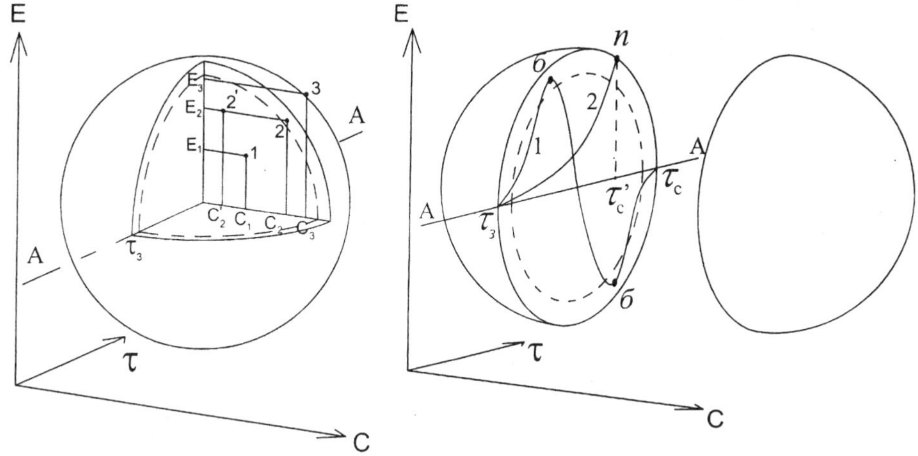

Fig. 2a, b. Diagrams of the matter state function change *f* (E; C) in time

The highest degree of the biosystem stability (the degree of approach to equilibrium) is proper for the points, which are laying on the line of equilibrium A-A (Fig. 2a, 2b). The least degree of the biosystem (the greatest remoteness from the line of equilibrium A-A) corresponds to the points, which are within the area between the dotted and continuous surfaces of the diagram.

Changes in the position of some point that describes the equilibrium of the system during a certain period of time $\Delta\tau = \tau_3 \div \tau_c$ is represented (Fig. 2b) on the cross-section plane that is perpendicular to the coordinate plane E-C of the kinetic diagram. If the biosystem is not effected by "exiting" factors of external environment (critical quantities $C_{кр}$, $E_{кр}$), then the period of the stable biosystem in equilibrium (lifetime) may be rather long $\Delta\tau = \tau_3 \div \tau_c$ (Fig. 2b). If diversion (line 1) from equilibrium has been observed in the biosystem under the influence of external factors (C and/or E) changes, it would mean that the biosystem has been twice in the non-equilibrium state (illness), but value $\Delta\tau$ remained the same: $\Delta\tau = \tau_3 \div \tau_c$ .If the biosystem, under the influence of changes in external factors $C_{кр}$ and/or $E_{кр}$ has revealed processes of diversion (line 2) from equilibrium (the system state is described by the point *n* , that is located outside the diagram), the biosystem has stopped its functioning for rather a short period of time $\Delta\tau' = \tau_3 \div \tau'_C < \Delta\tau$.

Action rendered by the changing in time physicochemical factors (E, C) of the surrounding medium on the physicochemical processes inside a certain system is described by Fig. 3. Any object of organic (biosystem) or inorganic (metal systems)

nature changes its state (levels the equilibrium, changes the position of equilibrium line A-A) following the change of values E and C of the surrounding medium. The changing in time equilibrium of the surrounding medium results in change of the inner equilibrium of each separate system 1-5 on the segment $\Delta\tau = \tau_3 \div \tau_c$. If changes in environmental characteristics of E and C are going on within the limits that are sufficient for leveling of inner physicochemical equilibrium of each separate object, such object maintains its characteristics and functions in equilibrium with the changed environmental characteristics. But if the surrounding medium is subject to considerable changes of E and/or C within a rather short period of time $\Delta\tau' < \Delta\tau$ (Fig. 2b.), and a separate biological system does not have enough time to level the inner physicochemical equilibrium following the abruptly changed physicochemical environmental characteristics, it terminates functioning as the uniform integral system./

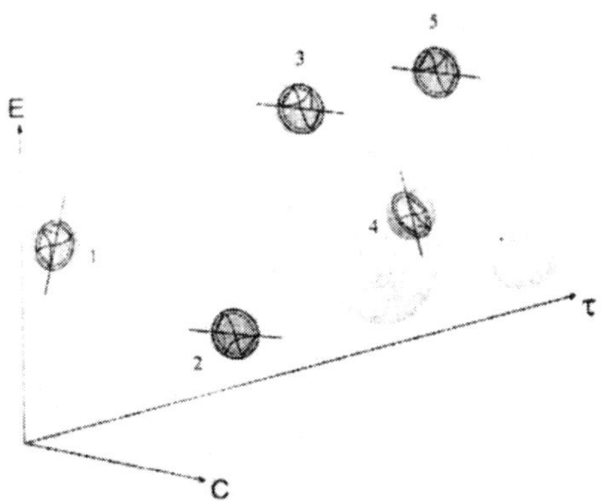

Fig.3. The cyclic change of functions of the environment state function f(E; C) in time and the system 1-5 that is maintain its equilibrium following the change in the energo-elemental features of the environment

Fig. 3 shows the change of the physicochemical environmental characteristics of the surrounding medium within the period comparable with the lifetime of each following generations of biosystems. In this case, it is necessary to say about leveling of the equilibrium state in each of the generations; with gradual transfer from the physicochemical characteristics inherent to the 1 system to the physicochemical characteristics inherent to 2, 3, 4, 5 systems, by digenesis, in harmony with the changed environmental characteristics.

Variety of possible versions of conjunctions that are forming and functioning in various, continuously changing energo-elemental environmental conditions explain the variety of forms and properties of the organic and inorganic nature systems. If forming and functioning of the systems takes place in similar at the nano-level energo-elemental conditions (systems 3, 5 Fig. 3), then the system properties at the macro-level will also be similar.

At the nano-level, transitions from one energo-elemental state into another are inherent in all without exception systems of the organic and inorganic matter. So, for systems of the organic and inorganic matter, conjunction of the qualitative and quantitative set of the same element (carbon) with various qualitative and quantitative sets of energies is exhibited in formation of various allotropic forms (diamond, graphite, fullerene):

$e_1, \dots, e_n$; c_1	→	$e'_1, \dots, e'_n$; c'_1	→	$e''_1, \dots, e''_n$; c''_1	→	$e'''_1, \dots, e'''_n$; $c'''_1, \dots, c'''_n$
f^d (E ; C)		f^g (E' ; C')		f^f (E'' ; C'')		$f^{f\,легир}$ (E''' ; C''')

Each function of the energy-element states f (E; C) is reflected by a certain allotropic form with certain characteristics. For example, carbon in the state f^f (E''; C'') is usually forming the hierarchical structure. This structure (fig.4a, [3]) is capable of "incorporating" (under certain changing in time energy-element environmental characteristics) oxygen, hydrogen, nitrogen [4], calcium [5] and other [6] biogenic elements.

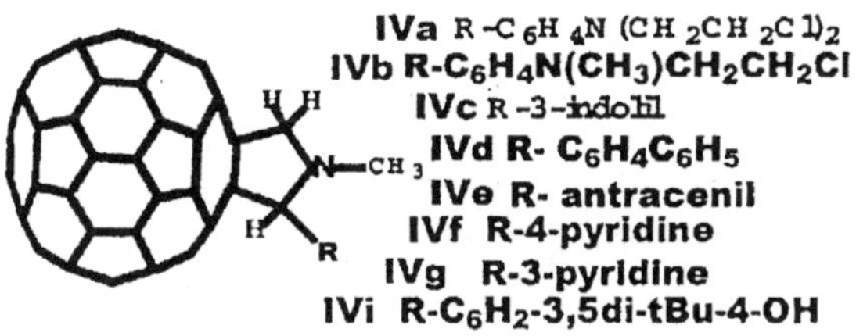

Fig. 4a. Scheme of the 5-substituted fulleropyrrolidines of the general formula IV a-i - organic derivatives of fullerene C_{60}

For the organic nature system considering the processes of the DNA formation and its transformation into a cell, an organism is of interest.

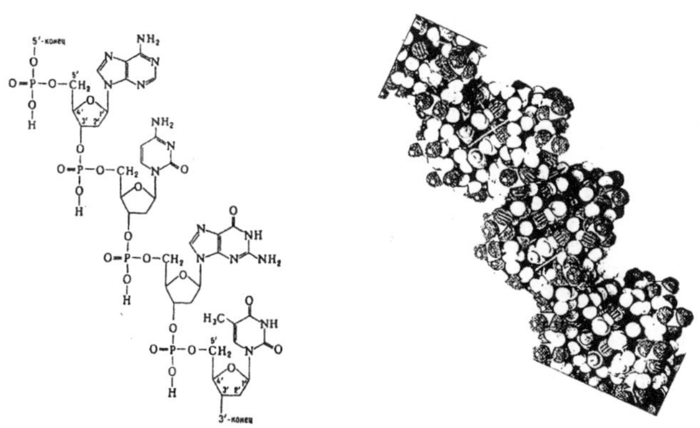

Fig.4 b. The structural formula and volumetric model of the DNA

The DNA (Fig. 4b) is conjunction of the qualitative and quantitative sets of elements C *(C-0-H-N*-nc*)* with the qualitative and quantitative set of energies E, that are intrinsic to the Earth conditions. E is the action caused by the Earth's revolution around the Sun and its own axis; thermal, ionizing, ultraviolet, optical radiations; the sound, the medium density, gravitation and other actions referred to as power-producing which are basis for formation and functioning of the given system. The developed energo-elemental approach allows to present the DNA as the function of the matter energo-elemental state

$$\begin{array}{|l|} e_1,\ e_2,\ e_3,\ \ldots,\ e_n \\ c_1,\ c_2,\ c_3,\ \ldots,\ c_n \end{array}$$

$$f^{DNA}(E\,;C)$$

And to explain the DNA formation process by way of self-organization of conjunction of the qualitative and quantitative set of elements C (Tables 1) with the qualitative and quantitative set of energies E that are intrinsic to the Earth conditions. Such a conjugation (DNA) is an open system that is functioning under condition of constant interchange by elements and energies within the environment (Fig. 3).

TABLE 1. Content of chemical elements in organisms, milligrams per 100 g of the dry substance (comparative data; 1996)

Chemical elements	Plants		Animals		Bacterium
	Sea	Land	Sea	Land	
C	34500	45400	40000	46500	54000
O	47000	41000	40000	18600	23000
H	4100	5500	5200	7000	7400
N	1500	3000	7500	10000	9600
Ca	1000	1800	150 – 2000	20 – 8500	510
Mg	520	320	500	100	700
Na	3300	120	400 – 4800	400	460
K	5200	1400	500 – 3000	740	11500
P	350	230	400 – 1800	1700	3000
S	1200	340	500 – 1900	500	530
Cl	470	200	500 – 9000	280	230
Si	150	20 – 500	7 – 100	12 – 600	18
Fe	70	14	40	16	25

Formation of the DNA and transformation of the DNA → cell → organism is described by of the matter from one energo-elemental state into another by using:

-scheme of transition

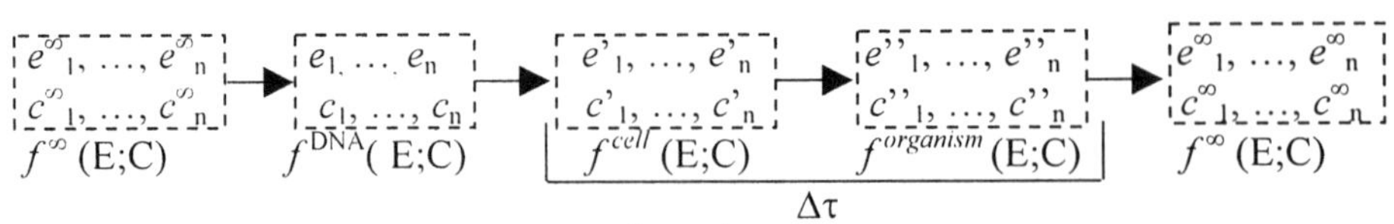

-equation $V = \dfrac{\Delta f(E,C)}{\tau}$;

- diagram of state – (Fig. 1, 2);

-schemes of the cyclic changes of the environment state function f(E;C) and systems of the organic matter (Fig. 3).

3. Conclusions:

1. We presented the concept of the unified energo-elemental essence of the systems of the organic and inorganic matter; mathematical and pictorial expressions of the fundamental law of life - the law of the change of the matter state.
2. We presented the qualitative picture of processes of the organic matter (DNA - cell - organism) self-organization as (the same as the systems of the inorganic matter) conjunction of qualitative and quantitative sets of elements and energies that are intrinsic to the Earth conditions; forming and functioning of this conjunction go on under the influence of the continuously changing energo-elemental states of the environment.
3. The determined regularities theoretically justify the processes of artificial formation and functioning of the inorganic and organic matter systems at the nano- and macro-levels.

4. References

1. Bobukh , L.V., Tomilin, I.A., Golovanenko, S.A., Svistunova, T.V. (1976) Termodynamic Analysis of the Processes of Carbide Formation and Intercrystalline Corrosion in Alloys of Nickel-Molybdenum System, *Metal Protection* **XII(4)**, 381-386.
2. Bobukh, L.V., Bobukh, K.A. (2001) General Regularities and Kinetics Diagram of Change in the Animate and Inanimate Nature System, *Reports of the National Academy of Science of Ukraine* **9**, 127-131.
3. Nuretdinov, I.A., Gubskaya, V.P., Yanilkin, V.V., Zverev, V.V. (2001) Synthesis, Structure and Properties of New Derivatives of Fullerenes, *VII International Conference "Hydrogen Materials Science & Chemistry of Metal Hydrides" Alushta-Crimea-Ukraine,* pp. 614-615.
4. Ying, Z.C.,. Hettich, R .L ., Compton, R. N., Haufler, R.E. (1996) *J. Phys. B : At . Mol. Opt .Phys* **29**, 4935-4942.
5. Zimmermann, U., Burkhardt, A., Martin, T.P. (1996) The Chemical Physics of Fullerenes 10 (and 5) Years Later. *Ed. By. W. Andreoni. Kluwer Acad. Publishers* 215-232.
6. Kraposhin, V.S., New Allotropic Form of Carbon – New Principle of Substance Structure Organization, (2000) *Steel* **1** 72-79.

SPONTANEOUS POLARIZATION AND ITS EFFECT ON PHYSICAL CHARACTERISTICS OF POTASSIUM, RUBIDIUM AND CAESIUM DIHYDROPHOSPHATES AND ARSENATES

S.Yu. ZAGINAICHENKO, Z.A. MATYSINA[#], D.V. SCHUR, V.A. CHUMAK[#]
Institute for Problems of Materials Science of NAS of Ukraine, 3, Krzhizhanovsky str., Kiev, 03142 Ukraine
[#]Dnepropetrovsk State University, 72, Gagarin ave., Dnepropetrovsk, 49000 Ukraine

Abstract

The statistical theory of paraelectrician-ferroelectrician phase transition in *KDP* crystals with H_{22} structure has been developed. It have been established the temperature dependences of free energy, thermodynamic potential, order parameters and energetic constants of interatomic interaction. The dependence of order parameter on temperature has been defined and it has been shown the conditions, when this dependence is close to transition of first kind. The temperature dependences of configuration heat capacity and dielectric susceptibility have been determined in the neighbourhood of Curie point. The Curie-Veiss law and the rule of negative two have been justified. The comparison of computer-aided design, theoretical plots with experimental data have shown their qualitative agreement.

1. Introduction

KDP ferroelectricians (ferroelectrics) are isomorphic crystals of dihydrophosphates and arsenates of potassium, rubidium, caesium KH_2PO_4, RbH_2PO_4, CsH_2PO_4, RbH_2AsO_4, CsH_2AsO_4 and their deuterated (either totally or partially) analogues. The *KDP* crystals investigation is of interest to physical engineers, these crystals are extensively used in laser technology. The physical properties of these crystals are described in monographs [1-5]. The thermodynamic theory of ferroelectricians of different structures is also given in these monographs.

The results of statistical theory of *KDP* crystals on the basis of molecular-kinetic concept are given below and theoretical results of an investigation provide an explanation and substantiation for their physical properties. It has been established that the phase transition from para- to ferrophase is the transition of ordering type [1-5] and the dipolar and deformational ordering occurs in *KDP* crystals. Previously the statistical theory of dipolar ordering was elaborated [6-8] for crystals of cubic structures [9] of perovskite $BaTiO_3$ (or Fe_3AlC) type and ReO_3 (or Cu_3N) type.

N. Veziroglu et al. (eds.), Hydrogen Materials Science and Chemistry of Metal Hydrides, 179–188.

At high temperatures the KH_2PO_4 crystals are in paraelectrical phase and have H_{22} tetragonal structure [2] with lattice parameters a, c (Figure 1a). With temperature decrease the displacement of potassium K and phosphorus P atoms takes place by and against the c-axis (Figure 1), the directional spontaneous polarization and deformation are established, the ferroelectrical and simultaneously the ferroelastical phase of orthorhombical structure with parameters a_1, a_2, c is formed (Figure 1b). In the neighbourhood of phase transition temperature the physical properties of KDP crystals undergo the distinctive changes [10-22] that manifest themselves as peaks, breaks, V-shaped minimums on the curves of temperature dependence of heat capacity, direct and inverse dielectric permittivity, thermal conductivity coefficient, spontaneous deformation degree, elasticity constants.

Our interests are in the determination of temperature dependence of polarization order parameter $\xi(T)$, proportional to the degree of spontaneous polarization and deformation, in the elucidation of conditions, when phase transition in *KDP* crystals will be found to be second-kind transformation, close to the first-kind change, in the evaluation of Curie temperature T_0 of transition between phases, polarizational order parameter dependence on the intensity of external electric field $\xi(E)$ or on the external oriented mechanical stress, in the assessment of configurational heat capacity and its temperature dependence C(T), in the estimation of temperature dependence of direct and inverse dielectric permittivity ε and susceptibility $\chi = 1 + \varepsilon$, in the testing of validity of Curie-Veiss law for $1/\chi$ quantity. To clear up all these questions the internal configuration energy and thermodynamic potential have been calculated and examined, the equations of thermodynamic equilibrium have been studied.

2. The equations of thermodynamic equilibrium. The equilibrium parameter of order

The calculation has been performed by the use of average energies method on the assumption of quadratic dependence of ordering energy on order parameters and in the approximation of pair interaction of the nearest atoms α, β = K, P, which are situated in lattice sites at a distances of

$$r_1' = a_1/2,\ r_1'' = a_2/2,\ r_2 = c/2,\ r_3 = \sqrt{a_1^2 + a_2^2 + c^2}\ 4,\ r_4' = \sqrt{a_1^2 + c^2}/2,\ r_4'' = \sqrt{a_2^2 + c^2}/2.$$

In this case by P is meant the structural unit from phosphorus atom, surrounded by tetrahedron from nearest atoms of oxygen. The hydrogen atoms, located in interstitial sites, are displaced at ordering perpendicularly to the c axis and don't make a contribution to the phenomenon of spontaneous polarization [2]. The directions of preferential displacements of base units are marked off by arrows, lattice constants a_1, a_2, c and distances r_1, r_2, r_3, r_4, for which the interactions of structural units are taken into account in calculations, are shown in the Figure 1. The crystal lattice is broken into two sublattices correspondingly for potassium atoms, that are displaced along c axis (rightward displacement) in ordered state, and for phosphorus atoms, that are displaced opposite to the c axis (leftward displacement) in ordered state. The order parameters ξ_1, ξ_2 are inserted for sublattices and they are defined by the difference of probabilities of right- and leftward displacements, that are opposite in sign. The order parameter of the whole crystal is equal to

$$\xi = \frac{1}{2}(\xi_1 - \xi_2). \qquad (1)$$

The computations of internal configuration energy u, Helmholtz free energy f for one lattice site and also of the equations of system thermodynamic equilibrium

$$\partial f / \partial \xi_1 = 0 \ , \ \partial f / \partial \xi_2 = 0 \qquad (2)$$

give the results

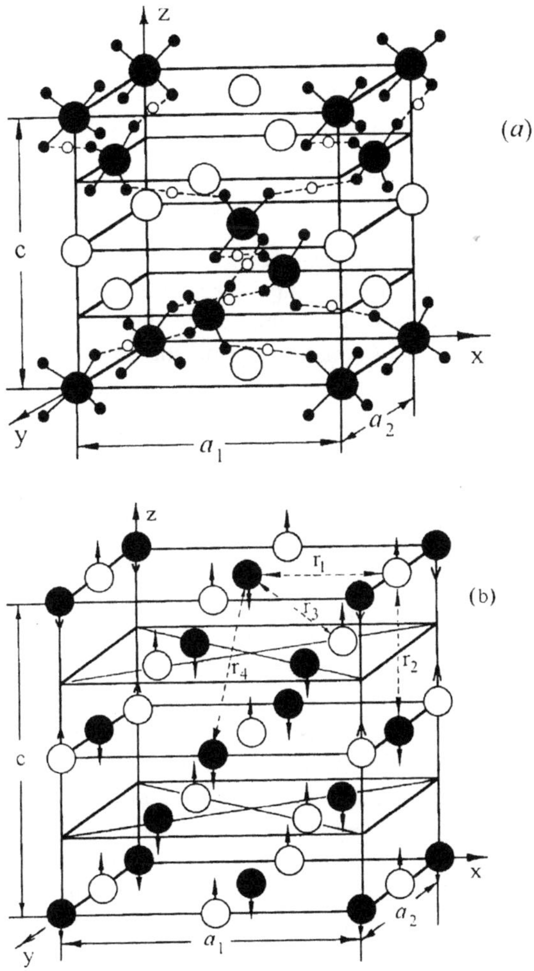

Figure 1. Elementary cell of crystal lattice of potassium dihydrophosphate KH_2PO_4.
(a) tetragonal structure of paraelectrical modification by West [6],
(b) orthorhombic structure of ferroelectrical modification by Frazer [7, 8] (oxygen and hydrogen atoms are not shown). The displacements directions of potassium K and phosphorus P atoms are indicted by arrows.

○ - potassium K atoms, ● - phosphorus P atoms,

• - oxygen O atoms, ∘ - hydrogen H atoms.

$$u = u(\xi_1, \xi_2) - u(0) = \frac{1}{2}\left[\omega + \frac{1}{4}\alpha(\xi_1 - \xi_2)^2\right]\xi_1\xi_2, \qquad (3)$$

$$f = f(\xi_1, \xi_2) - f(0) = \frac{1}{2}\left[\omega + \frac{1}{4}\alpha(\xi_1 - \xi_2)^2\right]\xi_1\xi_2 + \frac{1}{4}kT\left[(1+\xi_1)\ln\frac{1+\xi_1}{2} + (1-\xi_1)\ln\frac{1-\xi_1}{2} + (1+\xi_2)\ln\frac{1+\xi_2}{2} + (1-\xi_2)\ln\frac{1-\xi_2}{2} + 2\ln 2\right], \qquad (4)$$

$$\left.\begin{aligned} kT\ln\frac{1+\xi_1}{1-\xi_1} &= -2\xi_2\left[\omega + \frac{1}{4}\alpha\left(3\xi_1^2 - 4\xi_1\xi_2 + \xi_2^2\right)\right], \\ kT\ln\frac{1+\xi_2}{1-\xi_2} &= -2\xi_1\left[\omega + \frac{1}{4}\alpha\left(\xi_1^2 - 4\xi_1\xi_2 + 3\xi_2^2\right)\right], \end{aligned}\right\} \qquad (5)$$

where ω and α are energetic parameters, expressed in terms of energies of α, β = K, P pairs interaction; k is Boltzmann's constant, T is absolute temperature.

The equations (5) determine the temperature dependence of polarizational order parameters $\xi_1(T), \xi_2(T)$. These equations admit the solutions: $\xi_2 = -\xi_1$ and $\xi_2 = \xi_1$. With such substitutions in (5) one the equations transforms to another. Our interest is with the first solution

$$\xi_1 = -\xi_2 = \xi, \qquad (6)$$

which corresponds to the displacement of structural units K and P at ordering in the opposite directions. In this case the temperature dependence of polarizational order parameter $\xi(T)$ is determined by the following formula

$$kT\ln\frac{1+\xi}{1-\xi} = 2\left(\omega + 2\alpha\xi^2\right)\xi, \qquad (7)$$

on which the ordering temperature T_o is defined

$$kT_o = \omega. \qquad (8)$$

Figure 2 demonstrates the plots of temperature dependence of spontaneous polarization order parameter, constructed by formula (7) for different values of value α/ω. The comparison of calculation plots with experimental shows that in *KDP* crystals the $\xi(T)$ dependence is close to experimental for $\alpha>0$, for example at α/ω=0,2. In this situation the order parameter and spontaneous polarization at $T \to T_o$ have undergone the ever-change, but increase sharply near Curie temperature (Figure 2) and the phase transition is the second-kind transformation, close to the first-kind change.

3. Configuration heat capacity of crystal

With a knowledge of internal energy u (3) of the system, it is possible to determine the temperature dependence of heat capacity

$$C = \frac{\partial u}{\partial T} = \frac{\partial u}{\partial \xi}\cdot\frac{\partial \xi}{\partial T}, \qquad (9)$$

which in view of equation (7) can be written as

$$\frac{C}{k} = \frac{\left(1+2\frac{\alpha}{\omega}\xi^2\right)\xi \ln\frac{1+\xi}{1-\xi}}{2\left(\frac{kT/\omega}{1+\xi^2} - 1 - 6\frac{\alpha}{\omega}\xi^2\right)}. \tag{10}$$

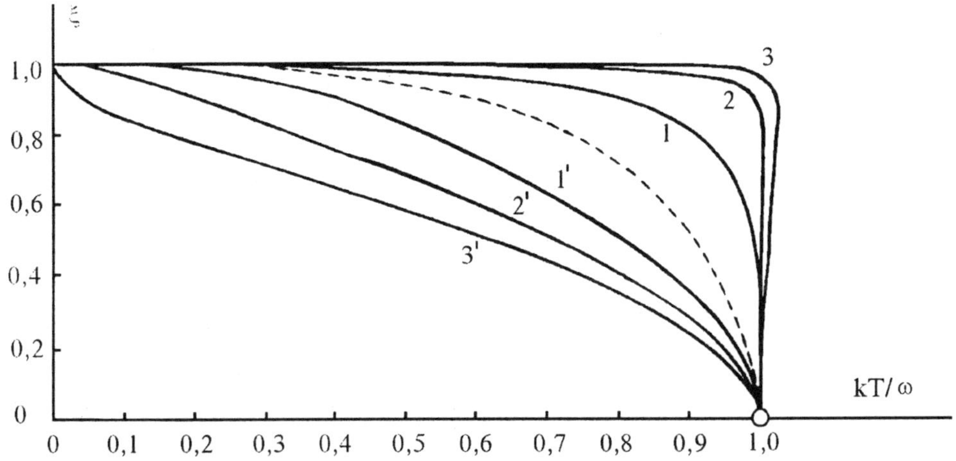

Figure 2. The calculation plots of temperature dependence of spontaneous polarization order parameter for different values of energetic parameter $\alpha/\omega=0$ (dotted curve), $\pm$ 0,2; $\pm$ 0,4; $\pm$ 0,6 (curves 1, 2, 3 for "+" sign and curves 1′, 2′, 3′ for the "-" sign).

The plots of the C(T) function in the field of phase transition, constructed by formula (10) for different α/ω values, are presented in Figure 3. In this case the order parameter is defined by formula (7). As indicated in Figure 3, this dependence is peak-shaped near Curie point and the greater is the α/ω value, the sharper and higher is the function peak. When this calculation result of Figure 3 is compared with that of experimental, it is apparent their qualitative correspondence.

4. Thermodynamic potential of crystal. The dependence of order parameter on intensity of electric field

In the case that investigated crystal is in the external electric field of intensity E and the oriented mechanical stress $\bar{\delta}$ acts on it, the crystal state is defined by thermodynamic potential. As this takes place, the degrees of spontaneous polarization P and deformation X are proportional to order parameter ξ:

$$P = \kappa\xi, \; X = \kappa'\xi, \tag{11}$$

where κ, κ' are coefficients of proportionality. The change of thermodynamic potential in design for one lattice site is equal to

$$\psi = \psi(\xi) - \psi(0) = -\frac{1}{2}\left(\omega + \frac{1}{2}\alpha\xi^2\right)\xi^2 + \\ + \frac{1}{2}kT\left[(1+\xi)\ln\frac{1+\xi}{2} + (1-\xi)\ln\frac{1-\xi}{2} + 2\ln 2\right] - \kappa E\xi - \kappa'\delta\xi. \tag{12}$$

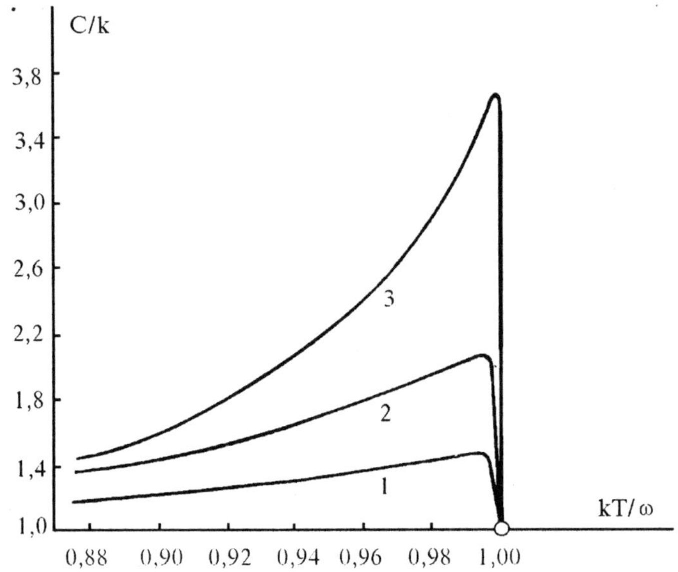

Figure 3. The calculation plots of temperature dependence of configuration heat capacity C(T) in the area of phase transition, constructed by equation (10) taking account of the ξ(T) dependence (7) for different values of energetic parameter $\alpha/\omega=0;\ 0{,}05;\ 0{,}1$ (curves 1, 2, 3 correspondingly).

The equation of thermodynamic equilibrium state

$$\partial\psi / \partial\xi = 0 \tag{13}$$

takes the form

$$-\left(\omega + 2\alpha\xi^2\right)\xi + \frac{1}{2}kT\ln\frac{1+\xi}{1-\xi} - \kappa E - \kappa'\delta = 0. \tag{14}$$

In the absence of impressed forces ($E = \delta = 0$), assuming $\xi \to 0$ in formula (14), in view of $\omega = kT_o$, we find the temperature dependence of polarizational order parameter near the Curie point T_o in the form

$$\xi = \sqrt{\frac{k\left(T - T_o\right)}{2\alpha}}. \tag{15}$$

From this relation it follows that in the absence of order at $T > T_o$ we have $\alpha > 0$. If $T < T_o$, so it is bound to be $\alpha < 0$. But for phase transitions, close to the first-kind transition between phases, the phase transformation temperature can slightly exceed the Curie temperature and therefore it is possible the case, when in ordered state in the neighbourhood of Curie point we shall have $T > T_o$ and $\alpha > 0$.

Assuming $\delta = 0$ in equation (14), we find the expression

$$\frac{\kappa}{\omega}E = \left(2\frac{\alpha}{\omega}\xi^2 + 1\right)\xi - \frac{1}{2}\left(1 - x\right)\ln\frac{1+\xi}{1-\xi}, \quad x = \frac{\left(T - T_o\right)}{T_o}, \tag{16}$$

defined the dependence of polarization order parameter on the intensity of external electric field ξ(E) for different temperatures. At $\xi \to 0$ close to Curie point and in terms of $\ln\frac{1+\xi}{1-\xi}\approx 2\xi$, we receive more simple and approximate expression instead of formula (16)

$$\frac{\kappa}{\omega}E = 2\frac{\alpha}{\omega}\xi^3 + x\xi. \qquad (17)$$

It follows from equations (16) and (17) that the $\xi(E)$ dependence is to be non-linear. Figure 4 shows the plots of this $\xi(E)$ dependence, constructed by formula (17) (full curves) and more precise formula (16) (dotted curves) for different temperatures and $\alpha/\omega=0{,}5$. This dependence $\xi(E)$ is non-linear, but become close to linear with increase in temperature T (value x). The portion of the curve 2′ from A to B corresponds to the unstable state and it is not realized. A qualitative agreement between theoretical curves of Figure 4 and experimental curves is evident.

5. Dielectric susceptibility

The formulae (16), (17) permit the determination of the dielectric susceptibility χ along the polar axis *c*

$$\varepsilon_0\chi = \partial P/\partial E = \kappa\partial\xi/\partial E, \qquad (18)$$

where ε_0 is electrical permitivity of vacuum, which is involved further in the κ coefficient. Taking into consideration the expression (16), we find formula

$$\frac{\kappa^2}{\omega}\frac{1}{\chi} = \begin{cases} -2x\left[1+\dfrac{1+x}{2\left(x-\dfrac{2\alpha}{\omega}\right)}\right] & \text{at } \xi\neq 0, \\ x & \text{at } \xi = 0 \end{cases} \qquad (19)$$

or approximately in view of equation (17)

$$\frac{\kappa^2}{\chi} = \begin{cases} 2k(T-T_0) & \text{при } \xi\neq 0, \\ -k(T-T_0) & \text{при } \xi = 0. \end{cases} \qquad (20)$$

From (20) it follows that close to Curie point the rule of negative two holds true and the Curie-Veiss law is valid. It follows from (19) that dependence $\frac{1}{\chi}(T)$ at $\xi\neq 0$ is non-linear.

Figure 5 shows the plots of temperature dependence of direct χ and inverse $1/\chi$ dielectric susceptibility, constructed by formula (19) for numerical value of energetic parameter $2\alpha/\omega= -0{,}15$. The theoretical calculation give the results in qualitative agreement with experimental results.

It should be also pointed out that the temperature dependence of direct and inverse elastic compliance can be estimated in a similar way as susceptibility. Such evaluations permit justification and explanation for the change of spontaneous deformation X at $T>T_0$, for the sharp decrease in elastic constant C_{66} and for the shear modulus G in the Curie point.

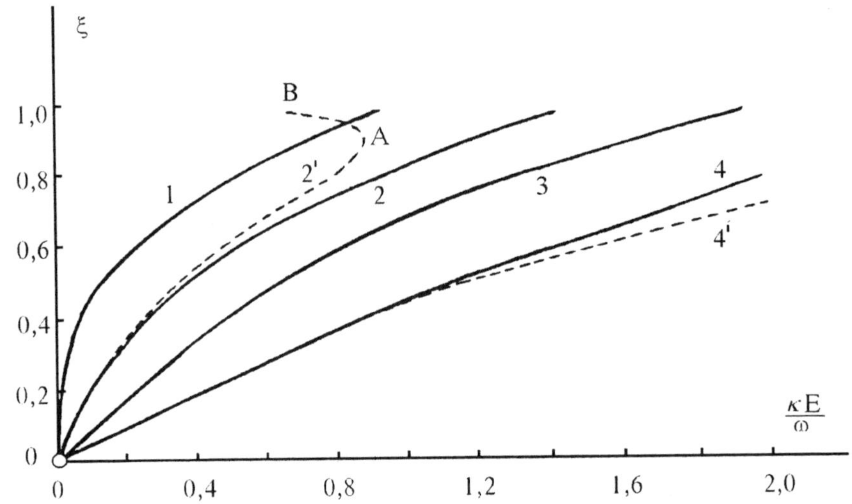

Figure 4. The calculation plots of order parameter, characterized the spontaneous polarization degree, for different temperatures ($x=(T-T_o)/T_o$) as a function of the intensity of external electric field $\xi(E)$, constructed by formulae (16) (full curves 1, 2, 3, 4) and (17) (dotted curves 2′, 4′) for $\alpha/\omega=0{,}5$ and for numerical values of quantity x=0; 0,5; 1; 2 (curves 1, 2, 3, 4 correspondingly).

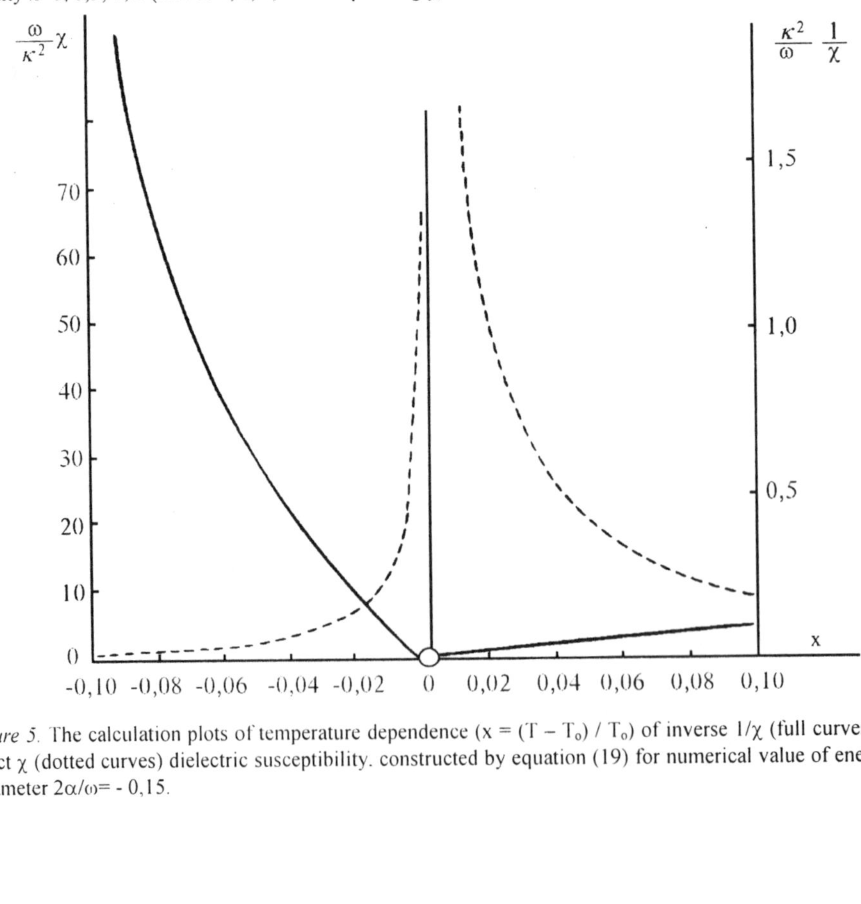

Figure 5. The calculation plots of temperature dependence ($x = (T - T_o) / T_o$) of inverse $1/\chi$ (full curves) and direct χ (dotted curves) dielectric susceptibility. constructed by equation (19) for numerical value of energetic parameter $2\alpha/\omega= - 0{,}15$.

6. Conclusions

Thus, the developed statistical theory of spontaneous polarization and deformation in *KDP* crystals with consideration for interpretation of the paraelectrician-ferroelectrician phase transition as of ordering type allows the elucidation and explanation for the dependence of order parameter on temperature (the nature of this dependence is close to the transition of first kind) and on the intensity of external electric field. This theory permits also the elucidation of special features of configuration heat capacity dependence on temperature. As described in the paper, this dependence is close to experimental just for the case of phase transition of first kind. Theory results allow the conclusions that the temperature dependences of direct and inverse dielectric susceptibility are such that the Curie-Veiss law (non-linear dependence of $1/\chi$ quantity on temperature) and the rule of negative two are true.

The obtained results of computer-aided design have been compared with literature experimental data for crystals with H_{22} structure, the comparison of revealed regularities shows their qualitative correspondence.

7. References

1. Kentsig, V. (1960) *Segnetoelektriki i antisegnetoelektriki*, M.: IL.
2. Smolenskiy, G.A., Bokov, V.A., Isupov, V.A., Krajnik, N.N., Pasynkov, R.E. & Shur, M.S.(1971) *Segnetoelektriki i antisegnetoelektriki*, L.: Nauka.
3. Smolenskiy, G.A. (Red.)(1985) *Fizika segnetoelektricheskih yavleniy*, L.: Nauka.
4. Yurkevich, V.E. (1988) *Fizika fazovykh perekhodov v segnetoaktivnykh tverdykh rastvorakh*, Rostov-na-Donu, RGU.
5. Rudyak, V.M. (1989) *Fizicheskiye svoistva segnetoelektricheskikh kristallov*, Kalinin: KGU.
6. Matysina, Z.A. (1995) Spontannaya polyarizatsiya i perepolyarizatsiya v kristallakh tipa perovskita. *Izv. vuzov. Fizika,* **5**, 8.
7. Matysina, Z.A. (1997) K teorii polyarizatsii i perepolyarizatsii v kristallakh tipa ReO_3, *Metallofizika i noveish. tekhn.*, **5**, 45.
8. Matysina, Z.A., Zaginaichenko, S.Yu., Chumak, V.A. & Schur, D.V. (1999) Sverkhprovodyaschee uporyadochenie v kristallakh $CsRbC_{60}$, *Metallofizika i noveish. tekhn.,* **9**, 78-82.
9. Gol'dshmidt, Kh.J. (1971) *Splavy vnedreniya*. M.: Mir.
10. Busch, G. (1933) *Helv. Phys. Acta.,* **11**, 239; **12**, 489.
11. Mason, W.P. (1939) *Phys. Rev.,* **55**, 775.
12. Bantle, W. (1942) *Helv. Phys. Acta.,* **15**, 373.
13. Arx, A. & W. Bantle, W. (1943) *Helv. Phys. Acta.,* **16**, 211.
14. Stephenson, C.C. & I.G. Hooly, I.G. (1944) *J. Am. Chem. Soc.,* **66**, 1397.
15. Von Arx, A. & Bantle, W. (1944) *Helv. Phys. Acta.,* **17**, 298.
16. Zwicker, B. (1946) *Helv. Phys. Acta.,* **19**, 523.
17. Beck, M. & Granicher, H. (1950) *Helv. Phys. Acta.,* **23**, 522.
18. Baumgartner, H. (1950) *Helv. Phys. Acta.,* **23**, 651.
19. Suemune, Y. (1966, 1967) *J. Phys. Soc. Japan,* **21**, 802; **22**, 735.
20. Strukov, B.A., Amin, M. & Kopzik, V.A. (1966) *Phys. Stat. Sol.,* **27**, 741.

21. Strukov, B.A. & Belov, A.A. (1992) *Izv. AN SSSR, Ser. fiz.*, **10**, 40.
22. Korotkov, L.N., Kravchenko, S.A., Gridnev, S.A. & Fedosyuk,R.M. (1998) *Izv. AN SSSR, Ser. fiz.*, **8**, 1598.

STRUCTURE OF $Ti_{0.9}Zr_{0.1}Mn_{1.5}V_{0.3}Ti_{0.2}D_{2.8}$

T.N.SMIRNOVA, S.V.MITROKHIN, V.A.SOMENKOV(1),
V.P.GLAZKOV(1), B.N.SAVENKO(2), V.N.VERBETSKY
Chemistry Department, Lomonosov Moscow State University, Vorobjovy gory, Moscow 119899, Russia
(1) RSC "Kurchatov institute", Institute for superconductivity and solid state physics, Kurchatov sq. 1, Moscow 123182, Russia
(2) Joint Institute for Nuclear Research, Frank laboratory of neutron physics, Joliot-Curie 6, Dubna 141980, Moscow region, Russia

Abstract

The structure of $Ti_{0.9}Zr_{0.1}Mn_{1.5}V_{0.3}Ti_{0.2}$ alloy and its deuteride was studied using methods of X-ray analysis and time-of-flight neutron diffraction (TOFND). The refinement of diffraction profiles was performed using Rietveld method. The refinement of the diffraction profile of the starting alloy showed, that the vanadium atoms are situated in 2(a) sites and titanium atoms in 6(h) sites. That introduction of hydrogen does not change the metal matrix structure and the hydriding is accompanied by isotropic increase of cell volume by 20%. The minimisation of R-factor showed that the best fit is achieved for the model of deuterium location in 3 sites, 24(l), $12(k)_1$ and $6(h)_1$. However, at low temperature approximately 0.11 deuterium atoms relocate from 24(l) to $12(k)_1$ position. The analysis of alloy and deuteride structure obtained in this work as well as the reference data allows to draw a conclusion, that the set of sites in hexagonal Laves phase structure which are occupied by deuterium is a stable one and does not depend on temperature or alloy composition.

1. Introduction

The prediction of hydrogen absorption properties of potential hydrogen storage materials is impossible without the knowledge of structural peculiarities of metal matrix and relevant hydride phase. In most number of works the structure of stoichiometric Laves phases and hydrides, such as $TiMn_2$ and $ZrMn_2$ was studied. These compounds are known to have a large homogeneity region. This region characterises the range of stable existence of nonstoichiometric compositions with deviations from stoichiometry ($B/A=2\pm x$). This stability is stipulated by the ability of one of the components to occupy the crystallographic positions of the other component. Earlier we showed that the addition of zirconium and vanadium to Ti-Mn alloys leads to some enlargement of the homogeneity region. Still the crystal chemistry of these phases is studied insufficiently. Structural investigations were only completed for $TiMn_{1.5}D_{3.13}$ [1], $ZrMn_{2.5}$ [2] and $TiMn_{1.12}V_{0.75}D_{1.9}$ [3].

N. Veziroglu et al. (eds.), Hydrogen Materials Science and Chemistry of Metal Hydrides, 189–195.

In present work we studied the structure of nonstoichiometric $(Ti_{0.9}Zr_{0.1})(Mn_{0.75}V_{0.15}Ti_{0.1})_2$ and its deuteride.

2. Experimental part

The structure of alloy and its deuteride was studied using methods of X-ray analysis and time-of-flight neutron diffraction (TOFND). The refinement of diffraction profiles was performed using Rietveld method. The main goal was to solve the structure of AB_2 Laves phase in which A-components is presented by two atoms ($A=Ti_{0.9}Zr_{0.1}$) and B-component – by three atoms ($B=Mn_{0.75}V_{0.15}Ti_{0.1}$), that is to determine the occupancy of A-sites by titanium and zirconium, and, secondly, to determine whether the distribution of manganese and vanadium atoms in B^1 and B^2 sites is random or ordered. All possible positions of atoms were considered while refinement procedure.

It should be noted, that the alloy structure is rather difficult to refine. This is due to the practically equal values of amplitudes of coherent diffusion of neutrons for titanium and manganese ($b_{Ti}=3.438\cdot10^{-12}$ cm, $b_{Mn}=3.73\cdot10^{-12}$ cm) which makes it impossible to distinguish these two atoms in the same position.

3. Results and discussion

The refinement of the diffraction profile of the starting alloy showed, that the vanadium atoms are situated in 2(a) sites and titanium atoms in 6(h) sites. Fig.1 and 2 show the X-ray and neutron diffraction patterns together with refined calculated profiles.

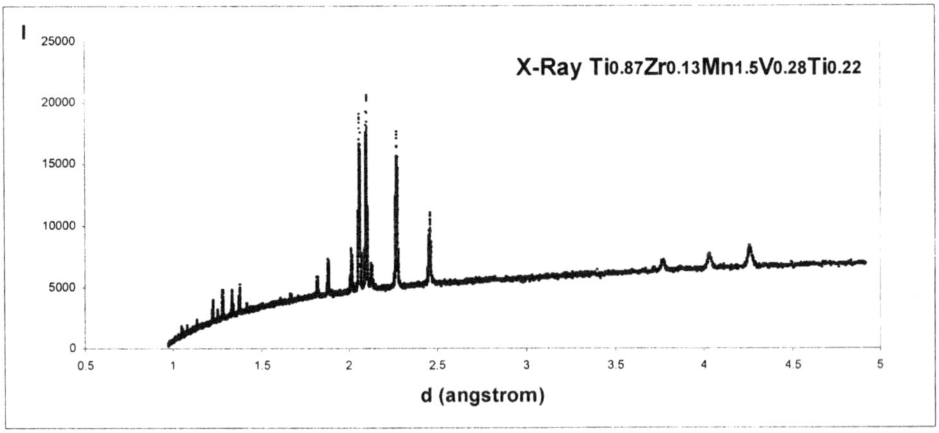

Figure 1. X-ray diffraction pattern of the starting alloy

The results obtained by X-ray and TOFND – cell parameters, atom coordinates, thermal factors, shown in Table 1 are in good agreement between each other. The refines composition also fits well to that obtained from electron-probe microanalysis.

The structure of deuteride phase was studied by TOFND. The obtained data shows that introduction of hydrogen does not change the metal matrix structure and the

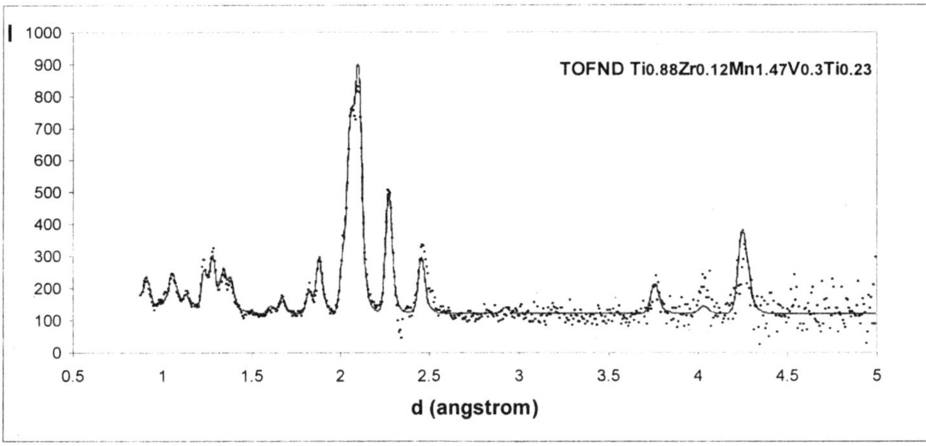

Figure 2. TOFND patterns of starting alloy

TABLE I. Structural parameters for $Ti_{0.9}Zr_{0.1}Mn_{1.5}V_{0.3}Ti_{0.2}$

Method	X-ray	TOFND
cell parameters (Å)	a=4,915 c=8,056 V=168.57	a=4,906 c=8,062 V=168.30
	A-component	
4f (1/3,2/3,z)	z=0.0639	z=0.059
Ti:Zr	0.87:0.13	0.88:0.12
thermal factor (Å^2)	0.5	1.5
	B-component	
2a (0,0,0)		
Mn:V	0.425:0.572	0.392:0.608
thermal factor (Å^2)	0.5	1.5
6h (x,2x,1/4)	x=0.830	x=0.827
Mn:Ti	0.85:0.15	0.85:0.15
Thermal factor (Å^2)	0.5	1.5
Refined composition	$Ti_{0.87}Zr_{0.13}(Mn_{1.49}V_{0.29}Ti_{0.22})$	$Ti_{0.88}Zr_{0.12}(Mn_{1.47}V_{0.3}Ti_{.23})$

hydriding is accompanied by isotropic increase of cell volume by 20%. So, the pattern exhibits the same diffraction maximums, which positions and intensities are somewhat changed due to the increase of cell parameters and contribution of deuterium atoms (Fig.3).

The comparison of the diffraction patterns obtained at 295 K for alloy and its deuteride in the same conditions shows the appearance of so-called "halo" in the case of deuteride phase (Fig.3). This can be connected with the formation of short range order in deuteride phase thus attesting some ordering in the range of interstitial distances

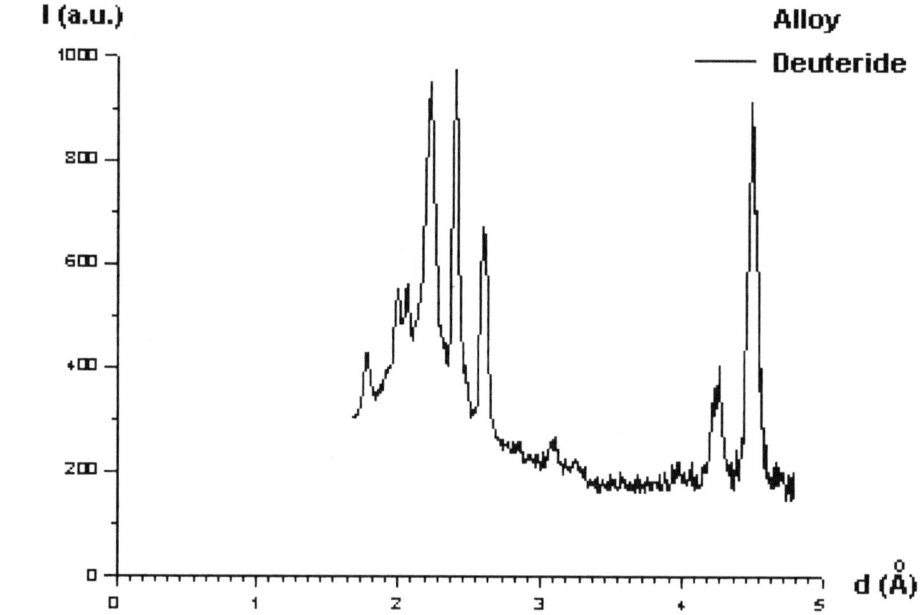

Figure 3. TOFND for deuteride phase with "halo"

Table 2. Structural parameters for $Ti_{0.9}Zr_{0.1}Mn_{1.5}V_{0.3}Ti_{0.2}D_{2.8}$

Atoms	Atom positions	Crystal parameters at different temperature: 295K Coord.	Occupancy	B (Å²)	17K Coord.	Occupancy	B (Å²)
Ti:Zr	**4f** 1/3,2/3,z	z=0.0721	0.89:0.11	1.5	z=0.0576	0.892:0.108	1.5
Mn:V	**2a** 0,0,0		0.387:0.613	1.5		0.392:0.608	1.5
Mn:Ti	**6h** x,2x,1/4	x=0.824	0.85:0.15	1.5	x=0.832	0.85:0.15	1.5
D_1	**24l** x,y,z	x=0.089 y=0.391 z=0.565	0.310	2.0	x=0.149 y=0.356 z=0.558	0.291	2.0
D_2	**$12k_1$** x,2x,z	x=0.472 z=0.607	0.230	2.0	x=0.499 z=0.593	0.294	2.0
D_3	**$6h_1$** x,2x,1/4	x=0.444	0.168	2.0	x=0.507	0.114	2.0
Ref. composition		$Ti_{0.89}Zr_{0.11}Mn_{1.47}V_{0.31}Ti_{0.22}D_{2.8}$			$Ti_{0.89}Zr_{0.11}Mn_{1.47}V_{0.3}Ti_{0.23}D_{2.79}$		
Cell parameters (Å)		a=5.228 c=8.553 V=202.7			a=5.232 c=8.540 V=201.8		

d =.3-2.4Å. It is known [4], that short range order is revealed in the pattern as a background modulation. Therefore, wide diffusion maximums arise in the patterns, besides the Bragg maximums. The preservation of the "halo" at low temperature shows, that no long range ordering of the interstitials atoms takes place.

The Laves phase hexagonal structure is characterised by 7 types of tetrahedral sites suitable for hydrogen occupation. They are shown in Fig.4. Since the alloy composition is nonstoichiometric one, it is rather difficult to determine the dimensions of these sites and to assume, which one will be preferable for hydrogen. Still, it is clear that two of the sites with maximum contribution of B-component 4(f) [AB_3] and 4(e) [B_4] are most unfavourable for hydrogen occupation. In fact, the attempt to locate deuterium atoms in these sites gave the zero or negative occupancy while refinement.

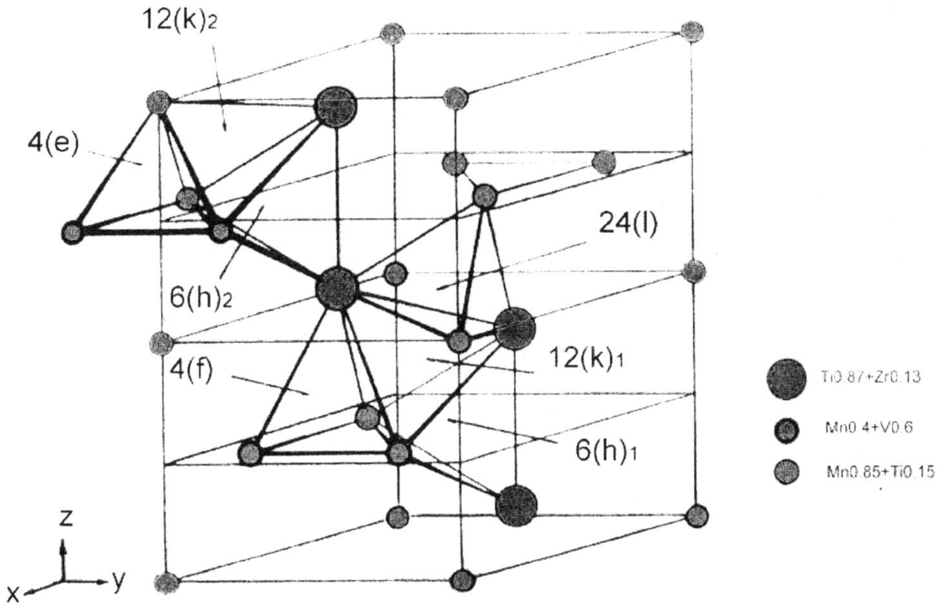

Figure 4. Interstitial sites in C14 Laves phase.

The minimisation of R-factor showed that the best fit is achieved for the model of deuterium location in 3 sites, 24(l), $12(k)_1$ and $6(h)_1$. However, at low temperature approximately 0.11 deuterium atoms relocate from 24(l) to $12(k)_1$ position. It is interesting that the same reordering at 80 K (~0.25D) was found for $TiMn_{1.5}D_3$ [1]. The structural parameters of deuteride phase are summarised in Table 2. The interatomic distances are presented in Table 3.

Table 3. Interatomic distsances for $Ti_{0.9}Zr_{0.1}Mn_{1.5}V_{0.3}Ti_{0.2}D_{2.8}$*

Atoms	d (Å)	Atoms	d (Å)
A-A	3.174	B_1-D_3	3.377
A-B_1	3.058	B_2-D_1	1.649
A-B_2	3.023	B_2-D_2	2.011
B_1-B_2	2.622	B_2-D_3	1.473
B_1-B_1	4.270	D_1-D_1	2.286
B_2-B_2	2.595	D_1-D_2	1.356
A-D_1	1.727	D_1-D_3	2.264
A-D_2	1.589	D_2-D_2	1.589
A-D_3	2.274	D_2-D_3	1.342
B_1-D_1	1.694	D_3-D_3	2.504
B_1-D_2	2.734		

* - $A=Ti_{0.89}Zr_{0.11}$, $B_1=Mn_{0.39}V_{0.61}$, $B_2=Mn_{0.85}Ti_{0.15}$

Table 4. Deuterium occupation of sites in the AB_2 hexagonal structures.

Deuteride	Sites and their occupancy							Ref.
	$[A_2B_2]$				$[AB_3]$		$[B_4]$	
	24l	$12k_1$	$6h_1$	$6h_2$	$12k_2$	4f	4e	
$ZrMoFeD_{2.6}$	0.275		0.63					[5]
$ZrVFeD_{3.6}$	0.312	0.275	0.530			0.1		[6]
$ZrVCuD_{4.0}$	0.342		1.0			0.175		[6]
$ZrVNbD_{5.4}$	0.25	0.792		0.917	0.05			[6]
$ZrMn_2D_{3.0}$	0.179	0.376	0.312	0.052				[7]
$ZrMn_2D_{3.04}$	0.211	0.372	0.369	0.079				[2]
$ZrMn_{2.5}D_{2.32}$	0.164	0.278	0.298	0.045				[2]
$ZrMn_{2.75}D_{2.12}$	0.14	0.247	0.286	0.045				[2]
$TiFe_{1.16}V_{0.84}D_{1.9}$	0.238			0.29				[8]
$TiV_{0.75}Mn_{1.12}D_{2.36}$	0.212	0.199	0.238			0.239		[3]
$(Hf_{0.7}Ti_{0.3})(Ni_{0.5}V_{0.5})_2D_{2.2}$	0.304			0.23				[9]
$ZrCoVD_{3.4}$	0.403	0.226	0.187	0.02				[10]
$TiMn_{1.5}D_{3.13}$	0.223	0.386	0.22	0.204				[1]
$Zr_{0.642}Mn_{0.358}(Mn_{1.519}Fe_{0.471})D_{1.52}$	0.167	0.056	0.233	0.0129				[11]
$Zr_{0.749}Mn_{0.251}(Mn_{1.487}Fe_{0.512})D_{4.08}$	0.236	0.388	0.527	0.468				[11]
$Zr(Cr_{0.6}Ni_{0.4})_2D_{3.3}$	0.257	0.334	0.128	0.382				[12]
$Zr(Cr_{0.6}Fe_{0.4})_2D_{2.5}$	0.202	0.329	0.05	0.359				[13]
$Zr(Cr_{0.6}Fe_{0.4})_2D_{2.97}$	0.227	0.327	0.09	0.372				[13]
$Zr(Cr_{0.25}Fe_{0.75})_2D_{2.2}$	0.156	0.274	0.027	0.318				[13]
$Zr(Cr_{0.25}Fe_{0.75})_2D_{2.5}$	0.177	0.288	0.033	0.323				[13]
$Ti_{0.89}Zr_{0.11}Mn_{1.47}V_{0.31}Ti_{0.22}D_{2.8}$	0.31	0.23	0.168					this work

4. Conclusions

The analysis of alloy and deuteride structure obtained in this work as well as the reference data allows to draw a conclusion, that the set of sites in hexagonal Laves phase structure which are occupied by deuterium is a stable one and does not depend on temperature or alloy composition (Table 4). In all cases the [A2B2] sites occupied by deuterium - 24(l), $12(k)_1$ $6(h)_1$ and $6(h)_2$, contain maximum number of A-component, which is active to hydrogen. However, equal occupation is not observed in any case.

Nevertheless, it is possible to make an assumption, that in the intermetallic hydrides, where a considerable part of hydrogen inactive B-atoms is substituted by hydride-forming metal, the occupancy factor for 24(l) site is higher.

5. Acknowledgement

This work was supported in part by RFBR Grants 99-02-16383, 00-15-97457 and 01-02-06024.

6. References

1. D.Fruchart, J.Souberoux and R.Hempelmann. J. Less-common metals, **99**, (1984), p.307.
2. L.Pontonnier, S.Miraglia and D.Fruchart. J. Alloys and Compounds 186, (1992), p.241.
3. H.Mayer, K.Alasafi and O.Bernauer. J. Less-common metals, **88, (**1982), p.L7.
4. Yu.Z.Nozik, R.P.Ozerov and K.Hennig. Structural neutronography. Moscow, Atomizdat, 1979, v.1.
5. V.A.Yartys, V.V.Burnasheva, N.V.Fadeeva, S.P.Solovjev and K.N.Semenenko. Russ. Kristallografiya, **27**, (1982), p.900.
6. V.A.Yartys, V.V.Burnasheva and K.N.Semenenko. Russ. Zhurnal neorganicheskoi khimii, **29**, (1984), p.615.
7. J.-J.Didisheim, K.Yvon, D.Shaltiel and P.Fisher. Solid State Comm. 31, (1979), p.47.
8. V.A.Yartys, S.V.Mitrokhin, V.N.Verbetsly and K.N.Semenenko. Russ. Zhurnal neorganicheskoi khimii, **37**, (1992), p.32.
9. E.Rönnebro, D.Noreus, M.Tsukahara and T.Sakai. J. Alloys and Compounds, **293-295**, **(**1999), p.169.
10. J.Souberoux, D.Fruchart and A.BirisAll-Compounds **293-295**, **(**1999), p.88.
11. G.Triantafillidis, L.Pontonnier, D.Fruchart, P.Wolfers and J.Soubeyroux. J. Less-common metals, **172-174**, (1991), p.183.
12. J.Joubert, M.Latroche, A.Percheron-Guegan and F.Bouree-Vigneron. J. Alloys and Compounds, **217**, (1995), p.283.
13. O.Canet, M.Latroche, F.Bouree-Vigneron and A.Percheron-Guegan. J. Alloys and Compounds, **210**, (1994), p.129.

PREPARATION AND PROPERTIES OF DEUTEROFULLERENES: NEW DATA

YU.M.SHUL'GA, B.P.TARASOV, V.N.FOKIN,
V.M.MARTYNENKO, D.V.SCHUR[(1)], G.A.VOLKOV,
V.I.RUBTSOV , G.A.KRASOCHKA, N.V.CHAPUSHEVA,
V.V.SHEVCHENKO
Institute of Problems of Chemical Physics of RAS, Chernogolovka, 142432 Russia
[(1)]Institute for Problems of Material Science Ukrainian National Academy of Science, Krzhyzhanovskogo Str.3, Kiev, 03142, Ukraine

1. Introduction

Many methods for preparing hydro- or deuterofullerenes are known (see, for example, [1]). Direct hydrogenation of solid fullerene was thought [1] to occur at fairly high temperature and pressure. It is obvious that composition of hydro(deutero)fullerene formed in interaction between solid C_{60} and gaseous H_2 or D_2 under constant pressure of the latter will depend on temperature in a complicated form. At first with increasing temperature one should expect the ratio x to increase (x=[H/C_{60}] or [D/C_{60}]), and then after some critical values of temperature (T_c) to decrease gradually. Decreasing x must occur due to thermal destruction of hydro- or deuterofullerene molecules formed during the reaction. In spite of simplicity of these considerations, we have not found any work on the study of $x=x(T)$ dependence for the given reaction, and on estimation of T_c value.

The decomposition of hydrofullerenes $C_{60}H_x$ is of interest from the viewpoint of using fullerene based materials as a medium for hydrogen accumulation. It was established by thermogravimetric analysis (TGA) [2-4] that the weight of the $C_{60}H_x$ begins to decrease noticeably at 723 K on heating in an inert medium. H_2 and fullerene were found in the products of thermal decomposition of hydrofullerenes. The complete reversibility of formation of $C_{60}H_2$ was established for the anodic oxidation of the C_{60} films supported on the graphite electrode [5]. The authors of the work [6] concluded that hydrogenation process is reversible. At the same time, it is known [7] that the products of $C_{60}D_{24}$ thermal decomposition at 823 K contain, along with fullerene molecules, an admixture identified by IR spectroscopy as a fullerene polymer. In addition, methane was found [4] in the gas phase products of $C_{60}H_{18.7}$ decomposition at temperatures above 823 K.

In this report we describe the study results on products after interaction between solid C_{60} and D_2 (p=2 MPa) at temperatures around T_c. The purpose of this work was also to study the products of thermal decomposition of deuterofullerite by mass spectroscopy. Deuterofullerite was chosen to distinguish the products of thermal decomposition from hydrocarbon admixtures, which can appear in the system due to desorption from cool walls and/or vacuum lubricants.

N. Veziroglu et al. (eds.), Hydrogen Materials Science and Chemistry of Metal Hydrides, 197–202.

2. Experimental

Fullerene C_{60} synthesized by us contained 99.8 wt.% of the main substance. Before deuteration fullerene was purified by vacuum sublimation to remove the solvent (toluene). A laboratory deuterium accumulator based on $La_{0.75}Mm_{0.25}Ni_5$ (Mm is a technical blend of rare earth metals) [3].

Fullerene was deuterated at P =2.5 MPa and T = 623-773 K in a standard high-pressure installation. The temperature was monitored with an accuracy of ±10 degrees and the accuracy of pressure monitoring was ±0.02 MPa. Samples were preliminarily degassed by heating to 500 K *in vacuo* (~ 1 Pa). Deuterium was supplied to a reactor autoclave directly from the accumulator heated to 373 K. In order to enhance efficiency, deuteration was conducted in the cyclic thermal mode, *i.e.*, the sample under the D_2 pressure was heated to 723 K, stored at this temperature for 1 h, the reaction mixture was cooled to ~310 K and the cycle was repeated. After three cycles, the gas phase was removed, and a fresh portion of D_2 was supplied from the accumulator. This deuteration mode was repeated five times.

Analysis of $C_{60}D_x$ sample composition has been carried out by oxygen burning ~4 mg sample at 1400K: $C_{60}D_x+(60+0.25x)O_2 \rightarrow 60CO_2+0.5xD_2O$. Carbon and deuterium content in the sample has been calculated using data on CO_2 and D_2O weights measurements.

IR absorption spectra were recorded on a Specord 75 IR spectrometer in KBr pellets prepared from 2 mg of the substance and 300 mg of KBr. X ray diffraction patterns were recorded using a DRON ADP-1 diffractometer with monochromatic Cu K_α radiation. Mass spectra of gases accompanying deuterofullerite were detected in the temperature range from room to 773 K on an MI 1201B mass spectrometer in the m/z interval from 4 to 90 (electron impact, 70 eV).

In experiments on thermal decomposition a deuterofullerite sample (16 - 40 mg) was placed in a quartz cell of a pyrolyzer connected through a fine adjustment valve with the supply system of the mass spectrometer. After the sample was annealed (373 K, 1 h), and ampule and supply system were evacuated to $\sim 2 \cdot 10^{-5}$ Pa. Gas sampling to a pre evacuated ampule was started from 373 K.

3. Results and Discussion

The absorption band at 2150 cm^{-1} attributed to stretching vibrations of the C-D bonds is the most intense in the IR spectrum of the studied samples [7]. Chemical analysis data (Table 1) show that the maximum value of x is reached at 673 K. Consequently the value of T_c is near this temperature.

Table 1. Deuteration temperature (T), composition (x, see text) and fcc lattice constant (a_0) for studied samples

Sample number	T, K	x	a_0, nm
2	623	18.6	1.456
3	673	29.3	1.487
4	723	28.5	*
5	773	27.5	*

*diffraction peaks are absent (fig.1)

Figure 1 shows diffractograms of initial C_{60} fullerite and deuterofullerite samples produced at different deuteration temperatures. As Figure 1 shows initial C_{60} fullerite has fcc lattice that is revealed with 16 well marked diffraction peaks over the given range in angles. a_0 parameter estimated by three the most intensive peaks {(111),(220),(311)} is 1.417 nm.

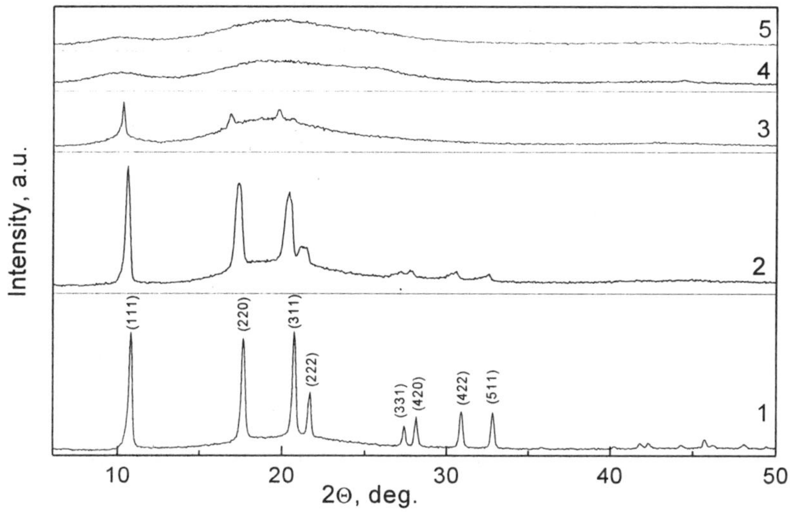

Fig 1. Diffractograms (Cu Kα) of initial fullerite (1) and deuterofullerites produced after interaction between gaseous D_2 (2 MPa) and initial fullerite at 623 (2), 673 (3), 723 (4) and 773 K(5).

Deuteration results in decreasing intensity of diffraction peaks and shifting them to the lower angles (Table 1). These effects enhance with increasing temperature. Diffraction peaks that can be attributed to *fcc* structure are not observed for the samples produced at 723 and 773 K.

Mass spectrum of the gas phase in the reactor-autoclave after interaction between C_{60} fullerite and deuterium at 773 K (before deuterium renovation) is given in Table 2.

One can see that besides the peak with m/z = 4 the intensive peak with m/z = 20 is present in the spectrum. This peak is attributed to CD_4^+ ion. Note that the peak with m/z = 20 have not been observed in the mass spectra of the residual gases in the mass spectrometer, in the gas entering from the laboratory cells and in the products of the preliminary degassing initial fullerite. The peak with m/z = 20 is absent in the mass spectrum of the gas phase if the process is carried out at 723K.

The mass spectrum of the gas phase decomposition products of deuterofullerite $C_{60}D_{18.6}$ heated to 773 K is presented in Table 3. This is below the temperature, which corresponds, according to TGA data, to the maximum rate of gas evolution (803 K). The residual pressure in the quartz vessel after preliminary annealing of the sample for 15 - 20 min at 373 K was at most $7 \cdot 10^{-3}$ Pa. After the end of the preliminary annealing, the quartz vessel was disconnected from the vacuum system and heated to a high temperature, which was accompanied by the pressure in crease to $(1-3) \cdot 10^2$ Pa. After storage at a required temperature for several minutes, the quartz vessel was connected through a needle valve with the ionization chamber of the mass spectrometer.

Table 2. Fragment from the mass spectrum (electron impact, 70 eV) of gas from the deuteration chamber.

m/z	I(rel.%)	Attribution
4	3	$[D_2]^+$
5	0	
11	0	
12	0.9	$[C]^+$
13	0.1	
14	8	$[CD]^+$
15	0.5	
16	8.5	$[CD_2]^+$
17	6	
18	34	$[CD_3]^+$
19	12	
20	100	$[CD_4]^+$

As can be seen in Table 3, the spectrum contains a rather intense peak of the CD_4+ ion with *m/z* = 20 along with the peak of D_2+ (*m/z* = 4). Peaks with *m/z* = 4 and 20 were not found in residual gases of the mass spectrometer and quartz vessel. These peaks are also absent in the spectra of the products of preliminary degassing of the deuterofullerite samples. However, if the sample pre-degassed at 373 K was heated not to 773 K but to 673 K, a peak with *m/z* = 20 appears in the spectrum and that with *m/z* = 4 disappeared. The total amount of the gas evolved from the sample at 673 K (~ 1 Pa) is much smaller than that evolved at 773 K (300 - 700 Pa). These values agree with the intensities of peaks in the mass spectra.

The gas phase above deuterofulleride at 773 K contains C_6D_6 along with CD_4. This is indicated by the peak with *m/z* = 84, being the most intense in the $60 \leq m/z \leq 100$ interval. For the decomposition of hydrofullerite $C_{60}H_x$, the peak of C_6D_6+ with *m/z* = 78 is the most intense in this mass spectral region.

Heating to 773 K does not change the diffraction pattern of the sample. The IR spectrum of $C_{60}D_{18.6}$, obtained under these conditions, exhibits a slight decrease in the peak intensity at 2150 cm^{-1}. Chemical analysis showed that the composition of the sample under such a treatment changed insignificantly and corresponded to the formula $C_{60}D_{18.1}$.

Thus obtained by authors data evidence that the optimum deuteration temperature is the temperature near 673 K (p=2 MPa) if the purpose is to prepare the samples with a maximum value of x. Increasing process temperature leads to destruction of the forming deuterofullerene. In particular destruction can be accompanied by failure of the close cage consisting of 60 carbon atoms. Deuteromethane evolution evidences it. Deuteration also results in failure of the initial fullerite lattice. It is revealed by decreasing intensities of diffraction peaks and shifting them to the lower angles (larger interplane distances). Diffraction peaks are not observed at 723 K and above. Our data also confirm the conclusions [4] about methane evolution on heating hydrofullerene to 773 K. The obtained results also demonstrate the possibility of formation of benzene C_6D_6 upon heating deuterofullerite to 773 K.

Table 3. Mass spectra of the gas phase above deuterofullerite heated to 673 and 773 K.

m/z	I (rel. units)		Assignment *
	T = 673 K	T = 773 K	
4	-	300	$[D_2]^{+\cdot}$
12	2	95	
13	-	5	
14	11	357	
15	3	33	
16	24	910	
17	10	460	
18	139	9100	$[CD_3]^+$, (CD_4)
19	22	1170	
20	280	13500	$[CD_4]^+$, (CD_4)
21	-	142	
22	-	7	
24	-	25	
25	-	3	
26	10	187	$[C_2D]^+$, (C_2D_2, C_2D_4, C_2D_6)
27	5	59	
28	72	3826	$[C_2D_2]^+$, $(C_2D_2, C_2D_4, C_2D_6, C_6D_6)$
29	8	141	
30	84	3800	$[C_2D_3]^+$, (C_2D_4, C_2D_6)
31	20	580	
32	323	8700	$[C_2D_4]^+$, (C_2D_4, C_2D_6)
33	13	357	
34	67	1850	$[C_2D_5]^+$, (C_2D_6)
35	18	267	
36	75	3667	$[C_2D_6]^+$, (C_2D_6)
37	-	43	
38	-	23	
39	2	3	
40	3	40	
41	4	10	
42	7	153	
43	5	13	
44	15	718	
45	-	17	
46	-	143	
47	-	10	
48	-	77	
49	-	17	
50	-	150	
51	-	33	
52	-	250	$[C_3D_8]^+$, (C_3D_8); $[C_4D_2]^+$, (C_6D_6)
53	-	7	
54	-	27	$[C_4D_3]^+$, (C_6D_6)
80	-	6	
81	-	3	
82	-	18	
83	-	17	
84	5	138	$[C_6D_6]^+$, (C_6D_6)
85	-	9	

*Ions are presented in brackets,and initial molecules are in parentheses.

4. Conclusion

With increasing deuteration temperature at first amount of deuterium absorbed by fullerite increases, then at temperatures above 723 K it begins to decrease. Decreasing absorbed deuterium content is accompanied by deuteromethane evolution that has been found experimentally at 773 K. It points to the destruction of the carbon cage at least in the part of C_{60} or $C_{60}D_x$ molecules.

The work has been carried out with financial support of RFBR (project 00-03-32106).

5. References

1. Goldshleger, N.F. and Moravskii, A.P. (1997) Fullerene hydrides: Preparation, properties, and structure, *Uspekhi khimii,* **66**, 353-375.
2. Tarasov, B.P., Fokin, V.N., Moravskii, A.P. and Shul'ga, Yu.M., (1996) Reactions of fullerenes with deuterium in the presence of palladium, *Russ.Chem.Bull.,* **45**, 1778-1779.
3. Tarasov, B.P., Fokin, V.N., Moravskii, A.P. and Shul'ga, Yu.M., (1997) Hydrogenation of fullerites in the presence of intermetallic compounds or metals, *Russ.Chem.Bull.,* **46**, 649-652.
4. Brosha, E.L., Davey, J., Garzon, F.H. and Gooesfeld, S., (1999) Irreversible hydrogenation of solid C-60 with and without catalytic metals, *J.Mater.Res* ., **14**, 1919-1923.
5. Nozu, R. and Matsumoto, O. (1996) Hydrogenation of C-60 by electrolysis of KOH-H2O solution, *J.Electrochem.Soc.*, **143**, 1919-1923.
6. Haufler, R.E., Canciecao, J., Chai, Y., Dyrne, N.E., Flanagan, S., Halley, M.M., O'Bien S.C., Pan C., Xiao Z., Billups W.E., Ciufolini M.A., Hauge R.H., Margrave, J.L., Wilson, L.J., Curl, R.F. and Smalley, R.E. (1990) *J.Phys.Chem.*, **94**, 8634-8639.
7. Shul'ga, Yu.M., and Tarasov, B.P. (1998) C-60 fullerite with a "stretched" fcc lattice, *JETP Lett.*, **68** , 253-256.

HYDROGEN INCORPORATION IN CVD DIAMOND FILMS

V.RALCHENKO[1*], A.KHOMICH[2], R. KHMELNITSKII[3], A. VLASOV[1]
[1)] General Physics Institute, 38 Vavilov str., Moscow 119991, Russia
[2)] Institute of RadioEng. & Electronics, 1 Vvedenskogo sq., 141190 Fryazino, Russia
[3)] P.N.Lebedev Physical Institute, 53 Leninsky prospekt, 111991Moscow, Russia

Abstract

Concentration of bonded hydrogen (C-H) in polycrystalline diamond films grown in microwave plasma, DC plasma and hot filament CVD reactors using CH_4/H_2 gas mixtures has been evaluated using IR spectroscopy. H content was found to be in the range of 30 ppm (optical quality thick films) to 20000 ppm (thin nanocrystalline films). As a consequence of decoration of defects and grain boundaries by hydrogen two effects were observed: (i) a decreases of thermal conductivity with concentration of incorporated hydrogen, (ii) a correlation of hydrogen and nitrogen impurities concentration in the diamond. High temperature annealing in vacuum results in breaking the C-H bonds followed by re-arrangement of defected regions and grain boundaries to graphitic (sp^2 – bonded) structures, that strongly increases the optical absorption of the material. It is observed that a H^+ ion implantation followed by annealing causes a formation of buried graphitized islands and even blisters rather than uniform continuous layer. Porous diamond layers prepared by oxidation of nanocrystalline films are shown to accumulate up to 20 at% hydrogen upon storage in ambient atmosphere.

1. Introduction

There is a growing interest to hydrogen interaction with carbon materials, in particular due to possible applications of nanotubes, porous graphite, or clustered carbon films for hydrogen storage technologies [1,2]. Less attention is paid to hydrogen incorporation in carbon materials with cubic diamond structure. In natural diamond the hydrogen occurs in concentrations that typically span a range of 500 to ca. 4000 ppm as determined by nuclear reaction analysis [3], and it seems to be uniformly distributed in the bulk. The C-H bonded hydrogen forms only a small part of the hydrogen present, the most hydrogen being, possibly, in methane and water forms [4]. The less H content (500-900 ppm) has been determined in synthetic diamond single crystals produced by high pressure – high temperature technique [3]. Currently polycrystalline diamond films of large size can be produced by chemical vapor deposition (CVD) technique that employs a hydrocarbon-hydrogen gaseous environment, so the deposited diamond may capture the hydrogen in substantial amounts. In the present paper the concentrations of bonded hydrogen (C-H) in as-grown and heat-treated polycrystalline and

N. Veziroglu et al. (eds.), Hydrogen Materials Science and Chemistry of Metal Hydrides, 203–212.

nanocrystalline diamond films is determined using IR spectroscopy. We consider the types of hydrogen bonding in diamond films of different structure perfection, hydrogen evolution upon heat treatment, and possibilities to increase hydrogen content in porous diamond. The thermal stability of H^+ ion implanted subsurface layer in diamond upon high temperature annealing is also examined.

2. Experimental

The diamond films were grown in methane/hydrogen mixtures on Si substrates using microwave plasma (MPCVD), DC plasma (DC-CVD) and hot filament (HFCVD) CVD systems. MPCVD diamond disks of 60 mm diameter and 0.15 to 1.1 mm thickness were prepared using a 5 kW microwave plasma reactor [5]. Deposition conditions were varied in a wide range to produce the films of different qualities. Typically the H_2 gas flow rate was 100 to 1000 sccm, methane content $[CH_4]:[H_2]$ of 2 to 5%, substrate temperature 700-800°C. Nitrogen was present only as a residual impurity in the feed gases. The Si substrates were chemically removed after deposition process to obtain free-standing films, that were laser cut to smaller pieces and mechanically polished. Thin (~10 μm) polycrystalline films were grown in a DC discharge as described elsewhere [6], while nanocrystalline films of 2-10 μm thickness were deposited in a HFCVD reactor [7].

Optical absorption/transmission measurements were carried out using UV-VIS and IR spectrometers ("Specord") in the wavelength range of 0.185-0.9 μm and 2.5-50 μm, respectively. The diamond annealing was performed at 100-750°C in air or at 850-1600°C in vacuum (10^{-5} torr). Optical spectra were taken from the samples after each successive annealing procedure. Thermal conductivity of MPCVD films was measured with a laser flash technique [5] perpendicularly to the film surface, and while the DC-CVD thin membranes were measured by a thermal grating technique [6] parallel to the film plane.

3. Results

3.1. IR ABSORPTION

Quantitative data on bonded hydrogen in diamond films can be obtained from the IR absorption features displayed in the range of 2800-3000 cm^{-1} [8]. The CH_x-stretching bands for the series of 11 μm thick diamond films deposited in DC plasma with variable methane content in feed gas is shown in Fig. 1. The sp^3-hybridized CH_2 groups (antisymmetric and symmetric stretches at 2850 and 2920 cm^{-1}, respectively), and the absorption band peaked in 2820-2836 cm^{-1} range (this is ascribed to hydrogen located at diamond grain surfaces), give the main contribution to the band intensity. The CH_3 sp^3-hybridized groups show a low intensity in translucent films [7,9].

The hydrogen in CVD diamond mostly is covalently bonded to carbon. It is believed that the hydrogen is chemisorbed mainly in disordered and even non-crystalline regions (grain boundaries, dislocations, voids surfaces, amorphous domains, etc). The concentration of CH_X groups in diamond usually increases with methane concentration in the gas phase. It depends also on the film texture, e.g. the hydrogen is easier captured in <111> textured films compared to <110> textured ones [10].

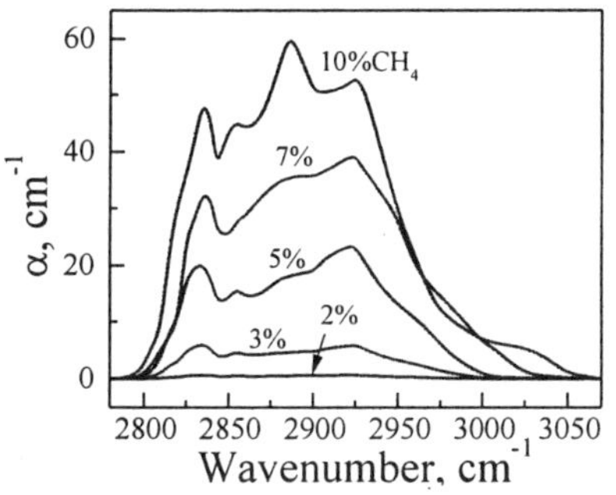

Fig. 1. The CH_x-stretching bands in 11 μm thick diamond films deposited in DC plasma with different methane percentage in feed gas.

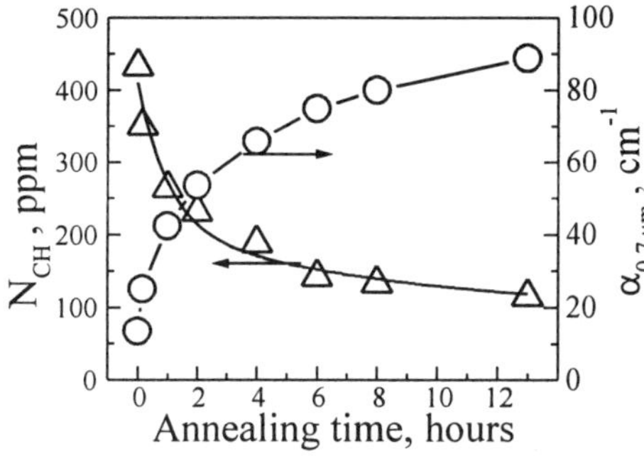

Fig. 2. The concentration of CH_x bonds N_{CH} (triangles) and absorption coefficient α at λ=0.7 μm (circles) in a "gray" diamond film vs. annealing time at T=1500°C.

In opaque samples with high non-diamond (a-C, graphite) content, the absorption due to sp^2-hybridized CH_x groups at 2880-2890 cm^{-1} was observed as well (Fig.1, spectra for 7% and 10% CH_4). The examination of the black diamond films with transmission electron microscopy (TEM) [11] revealed a very high density of planar defects, which frequently intersect forming disordered regions. In those low quality films the hydrogen is supposed to decorate a variety of the defects, the CH_x band intensity is much higher, and the IR bands at 2880-2890 cm^{-1} appear or become dominant (Fig.1).

We calculated the concentration of bonded hydrogen, N_{CH}, using the relationship [13] N_{CH} (ppm) = 1.7·A, where A (cm^{-2}) is integral intensity of CH_x stretching mode vibrations. This corresponds to C-H oscillator strength of $0.87 \cdot 10^{21}$ cm^{-1}. Different values of the C-H oscillator strength equal to $1.0 \cdot 10^{21}$ cm^{-1} [8] and $0.7 \cdot 10^{21}$ cm^{-1} [14] have also been reported. The H concentration in the range of $(0.05 - 400) \cdot 10^{19}$ cm^{-3} (30–20000 ppm) is determined in our samples. More specifically it was in the range of 30-10000 ppm for thick MPCVD films, 120-10000 ppm for DC plasma grown films, and 5000-20000 ppm for nanocrystalline thin films.

Annealing of CVD diamond films in vacuum has been studied by IR and UV-visible optical absorption, Raman spectroscopy and TEM [12,15,16]. It was found that hydrogen atoms at the internal grain surfaces and/or intergranular hydrogenated inclusions became mobile at T>1000°C. At least 25% of the total amount of hydrogen can be in the unbound state in some of the as-grown CVD films. The hydrogen de-bonding is shown to begin at temperatures of about 1300°C, and this process is accompanied by internal degradation (darkening) of the films.

The change in concentration of CH_x bonds with annealing time at T=1500°C is shown in Fig. 2 for a 200 μm MPCVD diamond film. The H content decreases from 440 ppm to 120 ppm for 12 hours annealing, the de-bonding process slowing down with time. Simultaneously a strong, an order of magnitude, increase of absorption coefficient (measured at λ=0.7 μm) is observed (Fig. 2). The redistribution and loss of hydrogen during the high-temperature annealing destabilizes the internal grain boundaries leading to formation of amorphous carbon and/or graphite layers up to 20 nm thick at the grain boundaries [16]. These transformations occur easier in the more defective CVD diamond films.

3.2. THERMAL CONDUCTIVITY

The pure natural single crystal diamonds of type IIa ("nitrogen-free") exhibit thermal conductivity (TC) of 20-25 W/cmK at room temperature, the highest value for any material known. The best quality thick (>0.2 mm) diamond films show TC values within the same range [15,17-19]. Variations in thermal conductivity are related to differences in phonon scattering rate. Besides intrinsic phonon-phonon scattering, various defects like vacancies, impurities, dislocations, grain boundaries, voids and non-diamond inclusions contribute to thermal resistance [20]. The direct effect of hydrogen impurity atoms on TC of diamond films is little [20]. However, since the hydrogen has a tendency to decorate the defects [17,19,21] its concentration can be a good indicator of imperfections in CVD diamond [19].

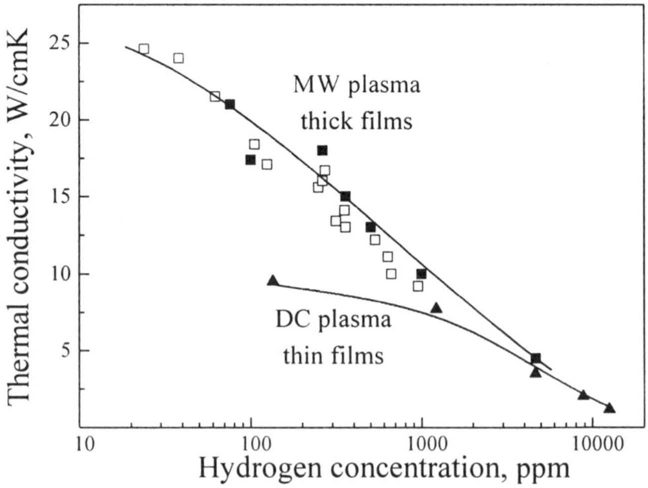

Fig. 3. Thermal conductivity versus H concentration in thick films grown by MPECVD (full squares) and in thin films grown in a DC plasma (triangles). For comparison the data for MPECVD films taken from Ref. [19] are also shown (open squares).

An inverse correlation of TC and bonded hydrogen content as determined from IR absorption was previously reported for CVD diamond by other authors [18,19]. Here we show (Fig. 3, full squares) the TC versus H (ppm) plot for our thick (0.2-1.1 mm) films grown by MPCVD. The deposition parameters, such as methane content, hydrogen flow rate, gas purity, and substrate temperature, were variable for this collection of samples. The hydrogen content in these films varied within two orders of magnitude. The TC value decreases from 21 to 4.5 W/cmK with H concentration increase from 70 to 4500 ppm. Interestingly, that the present data are in a good agreement with the results of Coe and Sussmann [19] who characterized the thick films produced also by MPCVD, but probably at conditions different from ours. Their data recalculated by us from integrated

C-H absorption intensity (in cm^{-2}) to the concentration (in ppm) using a relation 1 cm^{-2} = 1.7 ppm [13] are shown in Fig.3 (open squares) for comparison purpose. We note, that a similar functional relationship TC vs hydrogen content was found also by McNamara et al. [18], however with much wider scattering in TC data.

At low (<1000 ppm) H concentrations the TC for thin (ca. 10 μm) DC plasma deposited films is a factor of 2 lower than for thick MPCVD films. This can be due to smaller grain size (2-7 μm) in the DC plasma films. Because of large total area of grain boundaries (GB) in the thin films the H content may be not enough to decorate all defects and GB, thus nonhydrogenated defects could contribute significantly to thermal resistance in this case. At the higher H content (>1000 ppm) the intragranular defects rather than GB become the main source of thermal resistance. Note, that the IR absorption data give the hydrogen concentration averaged over the sample thickness. Actually hydrogen is distributed non-uniformly, and a thin, 20-50 μm, fine-grained layer adjacent to substrate is an order of magnitude enriched with H [20-22], this being in qualitative agreement with the higher density of the defects and GB there.

3.3. HYDROGEN-NITROGEN CORRELATION

Not only hydrogen but also nitrogen impurity concentration increases with abundance of the extended defects as was reported for a limited collection of diamond films [9]. Those defects very likely offer lattice sites with lower incorporation energy both for H and N (nitrogen is present mostly in substitutional form). On the other hand, the presence of nitrogen in plasma promotes formation of twins (and possibly other defects) [23]. Then, the hydrogen can be captured at nitrogen-induced defects, resulting, again, in apparent correlation between the nitrogen and hydrogen impurities.

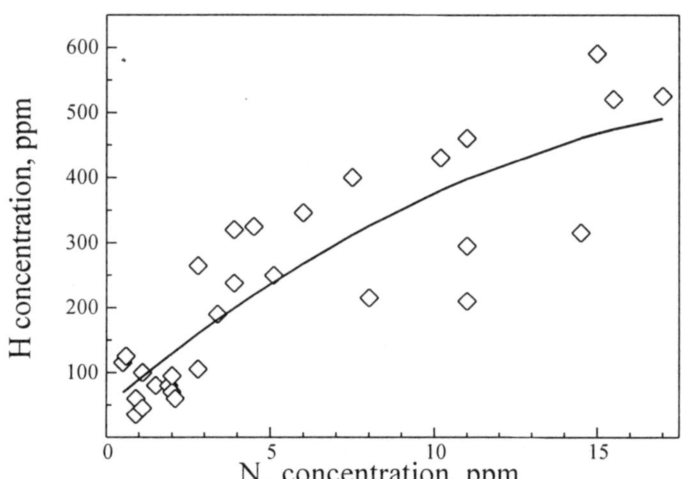

Fig. 4. Interrelation between concentrations of nitrogen (in substitutional position) and hydrogen (C-H bonded) impurities in MPECVD translucent diamond films as determined from UV and IR absorption spectra, respectively.

Figure 4 shows an interrelation between nitrogen and hydrogen impurities as established for a batch of 29 translucent diamond samples. Each data point on the plot represents a separate diamond wafer, or, in a few cases, pieces cut from different portions of one mother wafer. The substitutional nitrogen concentration was determined from intensity of UV absorption band at 270 nm [9]. Hydrogen concentration increases approximately from 50 ppm to 500 ppm with nitrogen content variation from 0.5 to 17 ppm. The H/N ratio for those translucent films lies in the range of 30-100. For low quality opaque films (not shown here) this ratio was 2-3 times larger because, probably, only a minor part of the defects have nitrogen-related origin in this case. We conclude that the low hydrogen concentrations ($\ll$100 ppm) can hardly be obtained if nitrogen is trapped in CVD diamond at the level of a few ppm or higher.

3.4. POROUS NANOCRYSTALLINE DIAMOND

To accommodate a significant quantity of adsorbed hydrogen the diamond material should be porous. Diamond powder of ca. 5 nm size produced by detonation technique [24] and CVD diamond with regular 3-D porosity of submicron scale (200-400 nm) prepared by a replica method using a synthetic opal template [25] are examples of such porous systems with specific surface area of 350 m^2/g, and 10 m^2/g, respectively. The porosity is still too low to be of interest for the gas storage. We prepared highly porous diamond layers by etching the fine-grained films in oxygen atmosphere, this treatment resulting in preferential oxidation of grain boundaries, defects and non-diamond carbon regions.

The as-grown very smooth nanocrystalline diamond films of 0.5-10 μm thickness consisted of 100-200 nm polycrystalline conglomerates as revealed by TEM [7]. A superperiodical structure of microtwin lamellas along the <111> plane was observed with 1-1.5 nm period. Defect-free regions in the diamond single crystals were about 10 nm in size. The etching in air at 630-650°C removes not only the non-diamond carbon, but also the diamond with high density of defects, leading to significant reduction of the absorption near the band gap edge. Typically the transmission of free-standing diamond films in the far IR grew from 70% up to 85-92% as a result of the oxidation. The interference fringes analysis shows that such porous films have the effective refractive index of 1.5-1.7 corresponding to volume fraction of the diamond of 60-40 %. The diamond peak in Raman spectrum has become more pronounced and its contribution strongly increased with respect to a-C band after oxidation. Polycrystalline conglomerates were transformed into broom-like units consisting of stacks of highly flattened crystallites with maximum linear size about 10 nm.

The IR absorption band due to CH_x groups in as-grown film with main peak at 2836 cm^{-1} band (Fig. 5, spectrum *a*) almost vanishes after oxidation (spectrum *b*). After hydrogenation in microwave H plasma and storage for 5 day in ambient atmosphere the CH_x stretching band shape differed significantly from the one for the as-grown film by higher contribution of the sp^3 CH_3-groups and absence of 2836 cm^{-1} band (Fig.5, *c*). The integral intensity of CH stretching band is 15 times higher than in as-grown film. The long-term keeping of the sample in ambient air leads to further increase of CH absorption band. After 180 days the concentration of bonded hydrogen grew up to 10-20 at.% as calculated according to [13]. The spectral shape of CH-band in the aged sample is close to that observed for hydrogenated ultradispersed diamond powder [27].

Upon the heating of the 180-day-aged-sample in air a decrease of CH_x band intensity starts at 150-180°C, and the heating to 280-300°C was sufficient to remove all the adsorbed hydrogen (Fig.6). These latter temperatures are somewhat lower than those (300-380°C) reported [28] for oxygen-assisted dehydrogenation process on diamond powder surfaces. Note, that the hydrogen desorption from CVD diamond films in vacuum starts at much higher temperatures of 800-900°C [29]. After repeated cycle of oxidation, plasma hydrogenization and storage in air the CH_x band recovered to the high intensities, indicating that the surface reactions involved are reversible. The process of the slow increment of bonded hydrogen in porous nanocrystalline diamond aged in ambient (humid) atmosphere is not well understood. It might include the interaction of hydroxyl groups with diamond surface [27,30], or a presence of hydrogen in primarily IR inactive state [31] in the near-surface region [32] immediately after H plasma treatment, followed by a gradual C-H bonding at room temperature.

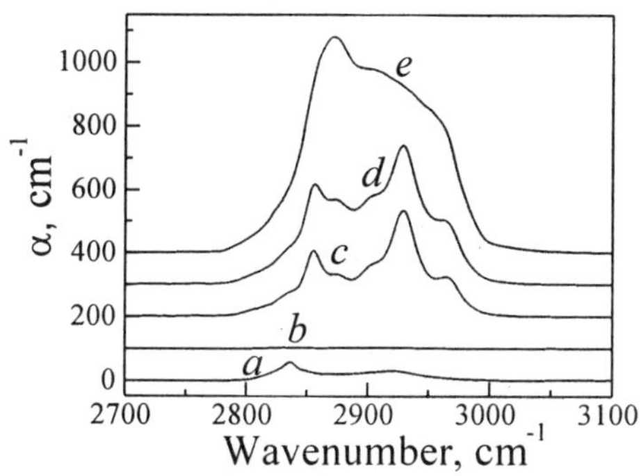

Fig. 5. CH_x stretching absorption spectra for 5μm thick HF CVD nanocrystalline diamond film: as grown film (a); after oxidation in air at 645°C (b); after hydrogen treatment in MW plasma and subsequent aging ambient atmosphere for 5 (c), 50 (d) and 180 (e) days. The spectra are 100 cm^{-1} off-set vertically for clarity.

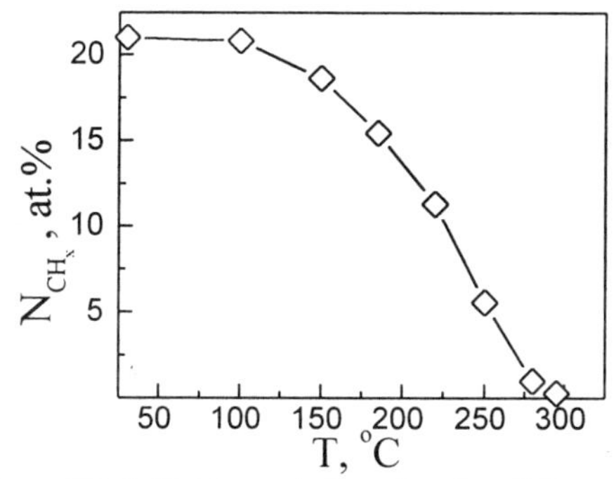

Fig. 6. The concentration of surface C-H bonds in MW plasma-hydrogenated porous nanocrystalline diamond film aged for 100 days in ambient atmosphere upon annealing in air in steps for 1 hour at different temperatures.

3.5. HYDROGEN IMPLANTATION

To obtain a high volumetric H concentration, though in a very thin (of the order of 200 nm) buried layer, H^+ implantation in diamond with ion energy of 350 keV and doses in the range $(2\div12)\cdot10^{16}$ cm^{-2} was performed. The growth side of polished 0.5 mm thick films was irradiated at room temperature, then the sample was annealed in a graphite furnace at temperatures up to 1700°C for 1 hour in a vacuum of 10^{-5} Torr. According to a Monte-Carlo simulation the maximum H concentration at the dose $D=5\cdot10^{16}$ cm^{-2} is $2.8\cdot10^{21}$ cm^{-3} (≈1.5 at%) achieved at the depth of 1.75 μm.

A buried graphitized layer is formed by structural re-arrangement of the highly disordered material upon the annealing. In contrast to the case of implantation of other light ions like He^+ [33] or D^+ [34] the hydrogen implantation followed by annealing typically creates islands of graphitic layer rather than the continuous layer. Round dark graphitized islands with lateral size of 20-50 μm formed on the sample annealed at 1460°C (Fig. 7). At doses $>8\cdot10^{16}$ cm^{-2} a blistering effect, that is the formation of (flat) hydrogen bubbles inside diamond, was observed at T>1460°C.

The phenomenon of blistering in diamond [34,35] is similar to the one well known for hydrogen- and helium-implanted single-crystal silicon [36,37]. The implanted hydrogen atoms are believed to form associations with radiation defects, which break upon annealing. The disordered regions are no longer stabilized by hydrogen and they re-arrange to graphite clusters. The liberated H atoms diffuse, recombine to H_2 molecules, are accumulated within the buried transformed layer, and eventually blisters are created under the rising intrinsic pressure. We found that the blisters form when local H concentration reaches 2.5%, while no evidence of the blisters was observed at 1.5%. In comparison with H^+ implantation in single crystal diamond [34] the number and size of blisters in CVD diamond are smaller at the dose and annealing temperature given. This might be due to vacancies diffusion to grain boundaries in polycrystalline material. Typically grain boundaries (GB) look light (Fig. 7) indicating a less graphitization degree (if any) in those regions. The highly transparent regions where graphitization does not take place can be found not only in the vicinity of GB but also within some crystallites. Thus the reactivity and high mobility of hydrogen, as well as the spotty graphitization upon annealing are specific features of H-implantation in diamond.

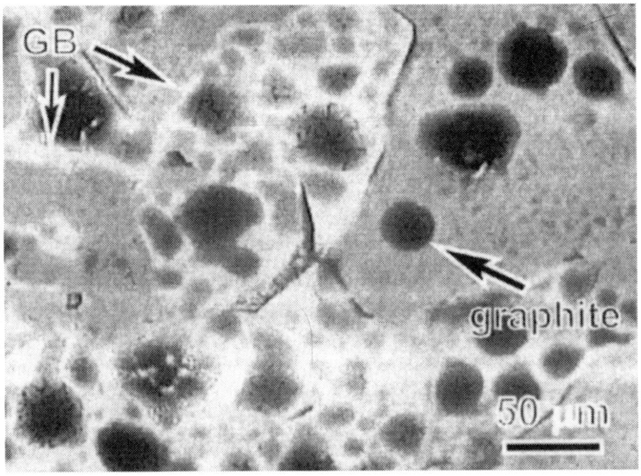

Fig. 7. Optical micrograph of H^+ implanted diamond film (E=350 KeV, dose $8 \cdot 10^{16}$ cm^{-2}) annealed at 1460°C. The image is taken in transmitted light. The grain size ≈100 μm. Grain boundaries GB with white contrast, and graphite islands (dark spots) are indicated by arrows.

4. Conclusions

The abundance of bonded hydrogen (C-H) in CVD diamond films grown in CH_4/H_2 mixtures and having the thickness in the range of a few microns to about 1 mm was studied using IR spectroscopy. The H concentration was found to span from 30 ppm for high quality thick films produced with microwave plasma to 20000 ppm for nanocrystalline thin films produced by HFCVD. As the hydrogen has a tendency to decorate the defects it is a good indicator of imperfections in CVD diamond, in particular, thermal conductivity is found to monotonically decrease with concentration of incorporated hydrogen, in a good agreement with the results of other authors [19].

The observed correlation between concentrations of hydrogen and nitrogen impurities in diamond can be explained by a similar reason if the nitrogen is responsible for formation of a part of extended defects.

The break of C-H bonds occurs at high temperature (>1300°C) annealing in vacuo. This process slows down in time as was monitored for more than 10 hours annealing process. The hydrogen loss causes a re-arrangement of defected regions and grain boundaries to graphitic (sp^2 – bonded) structures. Yet, the final state of liberated hydrogen is unknown.

Highly porous diamond layers were prepared by etching of thin nanocrystalline films in air. Those films with 40-60% porosity being treated in H microwave plasma demonstrated a significant increase of bonded hydrogen concentration up to 20 at% upon subsequent storage in ambient atmosphere.

H^+ ion implantation in diamond with ion energy of 350 keV and doses up to $12\cdot10^{16}$ cm^{-2} followed by annealing revealed the laterally non-uniform, spotty graphitization and even the blistering in the buried disordered layer, this being the specific feature of H implantation compared to other elements.

Acknowledgements - The authors are thankful to V.A. Dravin for performing the ion implantation, I.I. Vlasov for optical microphotograpy, I.G. Teremetskaya and V.P. Varnin for supplying the HFCVD nanocrystalline diamond films. This work was supported in a part by Russian Foundation for Basic Research, project No. 01-02-16826.

5. References

1. Dresselhaus, M.S., Williams, K.A., and Eklund, P.C. (1999) Hydrogen absorption in carbon materials, *MRS Bulletin* **24**, 45-50.
2. Dillon, A.C., Heben, M.J. (2001) Hydrogen storage using carbon adsorbents: past, present and future, *Appl. Phys. A* **72**, 132-142.
3. Sellshop, J.P.F. (1992) Nuclear probes in the study of diamond, in J.E. Field (ed.), *The properties of natural and synthetic diamond*, Academic Press, London, pp. 81-179.
4. Maclear, R.D., Connell, S.H., Doyle, B.P., Machi I.Z., Butler, J.E., Sellschop, J.P.F., Naidoo, S. R., and Fritsch, E. (1998) Quantative trace hydrogen distributions in natural diamond using 3D-micro-ERDA microscopy, *Nucl. Instrum. and Meth. in Phys. Research B*, **136-138**, 579-582.
5. Ralchenko, V., Sychov, I., Vlasov, I., Vlasov, A., Konov, V., Khomich, A., and Voronina, S. (1999) Quality of diamond wafers grown by microwave plasma CVD: effects of gas flow rate, *Diamond Relat. Mater.*, **8**, 189-193.
6. Graebner, J.E., Ralchenko, V.G., Smolin, A.A., Obraztsova, E.D., Korotushenko, K.G., and Konov, V.I. (1996) Thermal conductivity of thin diamond films grown from d.c. discharge, *Diamond Relat. Mater.*, **5**, 693-698.
7. Khomich, A.V., Polyakov, V.I ., Perov, P. I., Varnin, V.P., Teremetskaya, I.G., Balakirev, V.G., and Obraztsova, E.D. (1996) Surface chemical effects on the electrical and optical properties of thin nanocrystalline diamond films, in D.K. Gaskill, C.D. Brandt and R. J. Nemanich (eds.), *III-Nitride, SiC and Diamond Materials for Electronic Devices. MRS Symposium Proceedings*, Pittsburg, PA, USA, **423**, pp. 723-728.
8. Cheeraet, E. and Deneuville, A. (1992) IR characterization of diamond films on Si substrates, *Diamond Relat. Mater.* **1**, 584-587.
9. Nistor, S.V., Stefan, M., Ralchenko, V., Khomich, A., and Schoemaker, D.. (2000) Nitrogen and hydrogen in thick diamond films grown by microwave plasma enhanced chemical vapor deposition at variable H_2 flow rates, *J. Appl. Phys.*, **87**, 8741-8746.
10. Hag, S., Tunniciliffe, D.L., and Savage, J.A. (1994) Optical characterization of textured microwave CVD diamond, *Diamond Relat. Mater.*, **3**, 593-597.
11. Nistor, L., Van Landuyt, J., and Ralchenko, V. (1999) Structural aspects of CVD diamond wafers grown at different hydrogen flow rates, *Phys. Stat. Sol. (a)*, **174**, 5-9.

12. Khomich, A., Ralchenko, V., Nistor, L., Vlasov, I., and Khmelnitskii, R. (2000) Optical properties and defect structure of CVD diamond films annealed at 900-1600C, *Physica Status Solidi (a)*, **181**, 37-44.
13. Coudberg, P. and Catherine, Y. (1987) Structure and physical properties of plasma-grown amorphous hydrogenated carbon films, *Thin Solid Films*, **146**, 93-107.
14. Erz, R., Dotter, W., Jung, K., and Ehrhardt, H. (1995) Investigation of boron and hydrogen concentrations in p-type diamond by infrared spectroscopy, *Diamond Relat. Mater.*, **4**, 469-472.
15. Khomich, A.V., Ralchenko, V.G., Vlasov, A.V., Khmelnitskiy, R.A., Vlasov, I.I., and Konov, V.I. (2001) Effect of high temperature annealing on optical and thermal properties of CVD diamond, *Diamond Relat. Mater.*, **10**, 546-551.
16. Nistor, L., Ralchenko, V., Vlasov, I., Khomich, A., Khmelnitskii, R., Potapov, P., and Van. Landuyt, J. (2001) Formation of amorphous carbon and graphite in CVD diamond upon annealing: A HREM, EELS, Raman and optical study, *Phys.Stat.Sol.(a)*, **186**, 207-214.
17. Woerner, E., Wild, C., Muller-Sebert, W., Locher, R., and Koidl, P. (1996) Thermal conductivity of CVD diamond films: High-precision, temperature-resolved measurements, *Diamond Relat. Mater.*, **5**, 688-692.
18. McNamara Rutledge, K.M., Scruggs, B.E., Gleason, K.K. (1995) Influence of hydrogenated defects and voids on the thermal conductivity of polycrystalline diamond, *J. Appl. Phys.*, **77**, 1459-1462.
19. Coe, S.E., Sussmann, R.S. (2000) Optical, thermal and mechanical properties of CVD diamond, *Diamond Relat. Mater.*, **9**, 1726-1729.
20. Graebner, J.E., Mucha, J.A., and Baiocchi, F.A. (1996) Sources of thermal resistance in chemically vapor deposited diamond, *Diamond Relat. Mater.*, **5**, 682-687.
21. Maclear, R.D., Butler, J.E., Connell, S.H., Doyle, B.P., Machi, I.Z., Rebuli, D.B., Sellschope, J.P.F., and Sideras-Haddad, E. (1999) The distribution of hydrogen in polycrystalline CVD diamond, *Diamond Relat. Mater.*, **8**, 1615-1620.
22. Pimenov, S.M., Kononenko, V.V., Ralchenko, V.G., Konov, V.I., Gloor, S., Luthy, W., Weber, H.P., and Khomich, A.V. (1999) Laser polishing of diamond plates, *Appl. Phys. A.*, **69**, 81-88.
23. Butler, J., private communication.
24. Staver, A.M., Gubarev, N.V., Lyamkin, A.I., and Petrov, E.A. (1985) Ultrafine diamond formation mechanism under detonation synthesis, *Fizika Goreniya I Vzryva*, **20**, 100-105.
25. Gordeev, S.K., Nanoporous and nanofragmental carbon composite materials, *in Nanostructured Carbon for Advanced Applications*, ed. G. Benedek et al. Kluwer, Dordrecht, 2001, 71-88.
26. Zakhidov, A.A., Baughman, R.H., Iqbal, Z., Cui, C., Khayrullin, I., Dantas, S., Marti, J., and Ralchenko, V.G. (1998) Carbon structures with three-dimensional periodicity at optical wavelengths, *Science*, **282**, 897-901.
27. T. Jiang, K. Xu. (1995) FTIR study of ultradispersed diamond powder synthesized by explosive detonation, *Carbon*, **33**, 1663-1671.
28. Ando, T., Yamamoto, K., Ishii, M., Kamo, M., and Sato, Y. (1994) Oxidation of diamond surfaces studied by FTIR, TPD and TPR spectroscopies, in S.Saito, K. Yamamoto, M. Ishii, M. Kamo, and Y. Sato (eds.), *Advances in Diamond Science and Technology*, MYU, Tokyo, pp.431-434.
29. Schulberg, M.T., Fox, C.A., Kubiak, G.D., and Stulen, R.H. (1995) Hydrogen desorption from chemical vapor deposited diamond films, *J. Appl. Phys.*, **77**, 3484-3490.
30. Ji, S., Jiang, T., Xu, K., and Li, S. (1998) FTIR study of the adsorption of water on ultradispersed diamond powder surface, *Appl. Surf. Sci.*, **133**, 231-238.
31. Goss, J.P., Jones, R., Heggie, M.I., Ewels, C.P., Briddon, P.R., and Oberg, S. (2002) Theory of hydrogen in diamond, *Phys. Rev. B*, **65**, 115207.
32. Looi, H.J., Pang, L.Y.S., Molloy, A.B., Jones, F., Foord, J.S., Jackman, R.B. (1998) An insight into the mechanism of surface conductivity in thin film diamond, *Diamond Relat. Mater.*, **7**, 550-555.
33. Gippius, A.A., Khmelnitsky, R.A., Dravin, V.A., and Tkachenko, S.D. (1999) Formation and characterization of graphitized layers in ion-implanted diamond, *Diamond Relat. Mater.*, **8**, 1631-1635.
34. Gippius, A.A., Khmelnitsky, R.A., Dravin, V.A., and Khomich, A.V. (2001) Defect-induced graphitisation in diamond implanted with light ions, *Physica B*, **308-310**, 573-576.
35. Tong, Q.-Y., Gutjahr, K., Hopfe, S., Gossele, U., and Lee, T.-H. (1997) Layer splitting process in hydrogen-implanted Si, Ge, SiC and diamond substrates, *Appl. Phys. Lett.*, **70**, 1390-1392.
36. Kozlovskiy, V. V., Kozlov, V. A., and Lomasov, V. N. (2000) Modification of semiconductors by proton beams, *Semiconductors*, **34**, 123-140.
37. Fichtner, P.F.P., Koschny, J.R., Kling, A., Trnkaus, H., Yankov, R.A., Mucklich, A., Skorupa, W., Zawislak, F.C., Amaral, L., da Silva, M.F., and Coares, J.C. (1998) Nucleation and growth of platelet bubble structures in the He implanted silicon, *Nucl. Instrum. and Meth. in Phys. Research B*, **136-138**, 460-464.

HYDROGEN INFLUENCE ON THE NANOSIZED INTERMETALLICS MICROSTRUCTURE FORMATION IN Fe-Mo SYSTEM

T.Yu.KISELEVA, A.A.NOVAKOVA, E.I.HAIT
Department of Physics, Moscow M.V.Lomonosov State University, 117234 Vorobiovy gory, Moscow, Russia

Abstract

Different nanosized compositions of the Fe-Mo system, obtained by coupled reduction of mechanical mixture of iron and molybdenum oxygen containing compounds in hydrogen flow, are studied. Low temperature phase formation that accompanied the Fe and Mo fine particles reduction is studied by Mossbauer spectroscopy and X-ray diffraction. The hydrogen influence on the disordering of formed intermetallic compounds have been revealed.

1. Introduction

Chemical reduction of precipitated metal oxygen containing compounds in hot hydrogen flow seems to be one of the perspective and common used techniques of nanocrystalline metal powders production [1-4]. It allows getting the homogeneous nanocrystalline state with narrow particles size distributions. In this paper we present a detail analysis of the Fe-Mo system powder compounds microstructure formation that take place during the coupled reduction of fine dispersed Fe-hydroxide and H_2MoO_4 in continuous hydrogen flow at 700°C. As was shown recently [2-4] the such two-metal-components reduction resulted in metal fine particles metallization and their subsequent alloying with formation of nanocrystalline phases and intermediate compounds.

We imply several factors that affect the microstructure parameters. They are the nanoscale initial interacting reagents; the nonequilibrium conditions of H_2 flow at reaction-chamber and the thermodynamical potential of the 3d-Fe and 5d-Mo transition metals interaction.

Mixtures of nascent fine-grained oxygen containing compounds are very reactive. For small particles there increases the role of surface atoms, characterized by less saturated bonds. This results in facile bonding to other atoms, recrystallisation and coagulation of particles. Mixing and easy diffusion interaction, caused by surface activity during reduction by hydrogen, leads to the phase formation at lower temperatures than that implied by equilibrium phase diagram. In this case formation results in nanosized compositions consistituing many phases, and structure of nascent phases (especially near interfaces) can be disordered or even amorphous.

N. Veziroglu et al. (eds.), Hydrogen Materials Science and Chemistry of Metal Hydrides, 213–220.

By the way, in Fe-Mo system one can expect an amorphous phase formation in narrow concentration range of intermetallic compounds formation [5] as well as the extension of the primary bcc solid solutions. All of that is possible due to the negative interface enthalpies ΔH=-7 (in kJ per mol for solved Mo in Fe) and enough atomic size difference between constituent atoms (r_{Fe}=0.126 nm and r_{Mo}=0.150 nm according Goldsmith) [6].

The purpose of the present work is to elucidate the hydrogen role as the active participant of the low temperature alloying and nanocrystalline Fe-Mo compounds microstructure formation.

2. Experimental

Chemically dispersed nanocrystalline Fe-Mo compounds with different Mo concentration were obtained by means of coupled reduction of fine dispersed FeOOH and H_2MoO_4 in continuous H_2 flow at the temperature 700°C [1-2]. This temperature have been chosen in view of time temperature ranges of Fe and Mo metallization and to get the narrow distribution of reduced particles sizes. The starting concentrations of Fe and Mo are given in the Table I.

TABLE I. The starting elements concentrations in the samples

Sample	**Fe,** at.%	**Mo,** at.%
1	100	0
2	98	2
3	87	13
4	72	28
5	53	47
6	0	100

X-ray diffraction (XRD) and Mossbauer Spectroscopy (MS) were used for the structural study of synthesized samples (grain sizes, phase composition, concentrations of diluted elements in solid solutions). XRD was performed at Rigaku DX/Max θ-2θ diffractometer with Cu Kα radiation. The ^{57}Fe Mossbauer spectra were received at room temperature using constant acceleration spectrometer with ^{57}Co(Rh) radiation source. The Mossbauer spectra hyperfine parameters with reconstruction of distribution functions of the hyperfine interaction parameters of the ^{57}Fe nuclei were analyzed by means of MSTools software. As usual, all the isomer shift values are given with respect to the metallic Fe at room temperature.

3. Results and discussion

The equilibrium phase diagram of the Fe-Mo system (fig.1) shows the occurrence of the intermetallic compounds Fe_2Mo(λ) and Fe_7Mo_6(μ), which are formed by peritectic reactions at 950°C and 1370°C, respectively [7]. The Fe_2Mo(λ) phase is homogeneous in the concentration range 33,7 at% Mo; it has a hexagonal structure of the $MgZn_2$ type.

The intermetallic compound $Fe_7Mo_6(\mu)$ with rhombohedral structure is homogeneous in the ranges 39-44 at.% Mo. Under equilibrium conditions, iron and molybdenum have limited mutual solubility. Below 800°C, the solubility of Mo in α-Fe is no more than 4 at.%. The solubility of Fe in Mo below 1235°C is no more than 6 at.% .

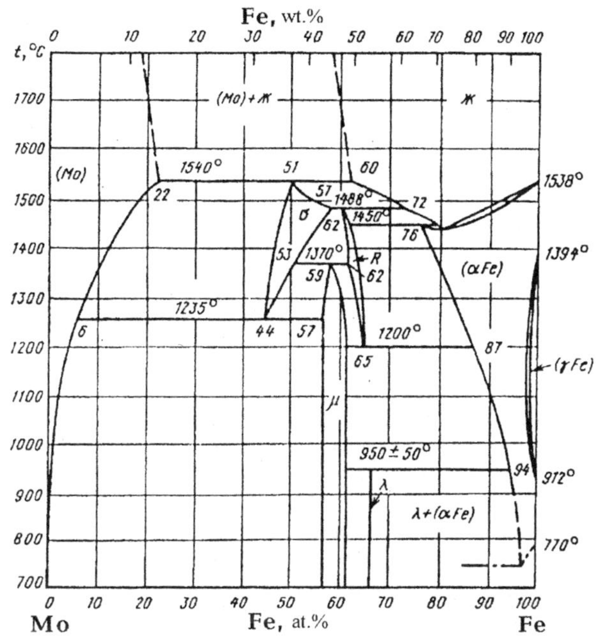

Figure 1. Equilibrium phase diagram of Fe-Mo system

Analysis of the obtained Mossbauer spectra and X-ray powder diffraction patterns of Fe-Mo compositions with different Mo concentration shows that the iron and molybdenum metallization that occurs in the course of reduction in hydrogen flow is accompanied by alloy formation, and the course of alloy formation changes with an increase of the molybdenum content in the samples. The reduced particles sizes (fig.2), calculated from diffraction peaks profile analysis, depend on the mutual initial compounds concentration and reach their minimum of 18 nm at the middle Mo concentration close to the 28 at %. The X-ray powder diffraction patterns of the samples show, in addition to the well resolved structural lines of the nascent crystalline bcc-phases of Fe and Mo, rather diffuse "halo" in the angle ranges 40-65°. The strongest halo is observed for the sample containing 28 at.% Mo. Such halo are often claimed to arise from the amorphous phase. The above angle range also contain the lines that arise from the intermetallic compounds $Fe_2Mo(\lambda)$, $Fe_7Mo_6(\mu)$ and bcc-Mo(Fe) solid solutions. In any case it seems to be the coexistence of several dispersed and not well diffractionally distinguished phases with most of lines overlapping. Analysis of the Mossbauer spectra hyperfine structure allows us to differentiate the regions of the intermetallic compounds formation and to elucidate the reason of the emergence of halos in the X-ray powder diffraction patterns.

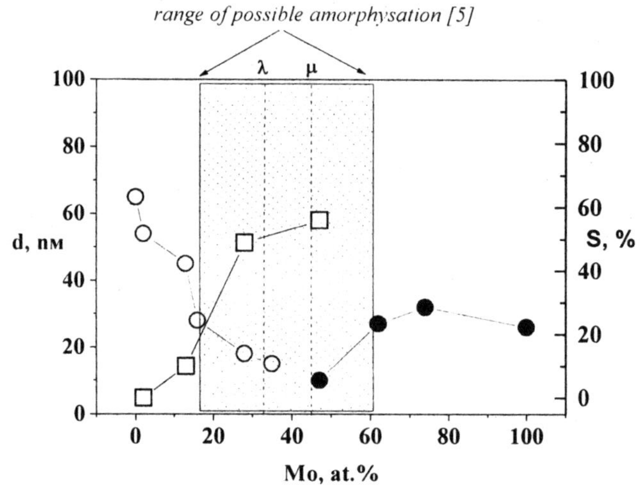

Figure 2. Concentration dependencies of the grain sizes (d, nm) of the b.c.c. solid solution for the Fe-Mo compounds and total quantity (as mossbauer subspectrum area S, %) of intermetallic phases in these compounds

Fig.3 shows the Mossbauer spectra of several Fe-Mo compounds with different Mo concentrations. Spectra of iron–rich samples show paramagnetic subspectra, in addition to strong magnetically split components corresponding to reduced Fe in different environments of Mo atoms because of the α-Fe(Mo) solid solution formation. The paramagnetic subspectra correspond to the intermetallic compounds and to the solid solutions of iron in molybdenum (Mo(Fe)). The amorphous phase of Fe-Mo system is paramagnetic [5]. Its room-temperature Mossbauer spectrum is a doublet with the quadruple splitting Δ= 0.3 mm/s and isomer shift δ=-0.18 mm/s. These values strongly differ from the Mossbauer parameters of the intermetallic phases $Fe_2Mo(\lambda)$ (Δ=-0.22 mm/s, δ= -0.25 mm/s), $Fe_7Mo_6(\mu)$ (Δ= 0.27 mm/s, δ= -0.27 mm/s) [5,8]. Increasing of the Mo concentration leads to an increase in the relative intensity of paramagnetic components, and the spectrum of the Mo-rich sample lacks magnetically split components.

Note that parameters of hyperfine magnetic splitting in the subspectra corresponding to the α-Fe(Mo) solid solution also change with the molybdenum content increase in the samples: spectral intensity deteriorates, and asymmetry becomes more pronounced, which corresponds to the appearance of new components caused by the changes in the nearest iron environment. The Mo content in the α-Fe(Mo) solid solutions was assessed by analyzing the distribution functions P(H) of the effective magnetic fields derived from the corresponding subspectra (Fig.4) and agree with the results of diffraction peaks profile analysis: the maximum Mo solubility in α-Fe extend the narrow equilibrium regions (3,4% [7]) up to 4,6 at.%. The fine structure of the P(H) distribution function near the smallest values of the hyperfine fields showed the appear of the partial amorphysation of the grain surface.

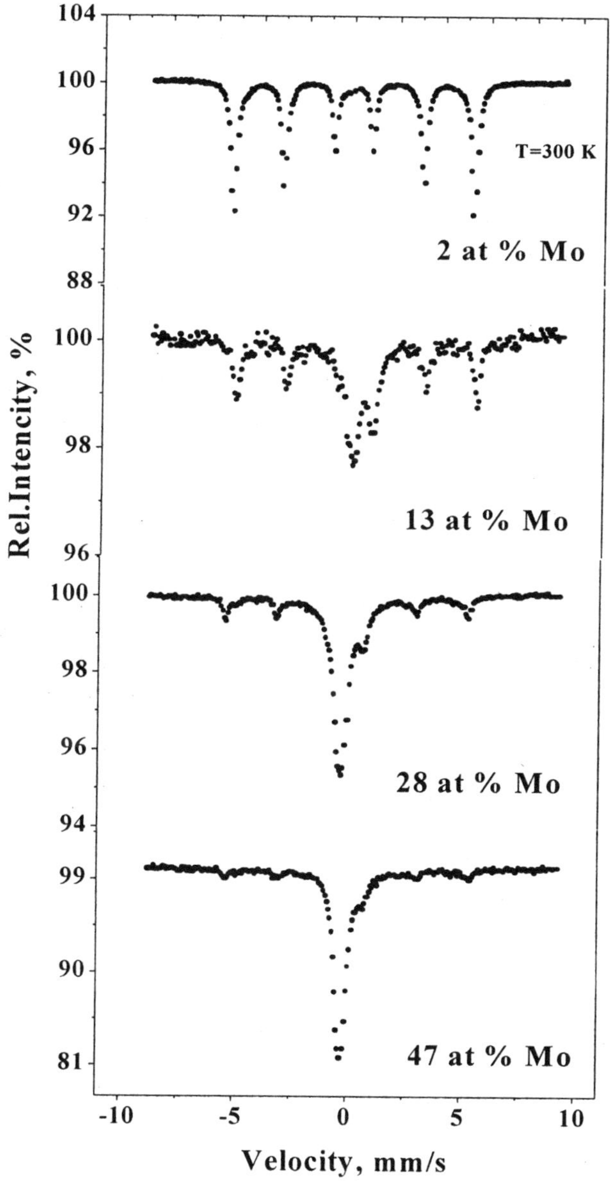

Figure 3. Room-temperature Mossbauer spectra of Fe-Mo nanocrystalline compositions

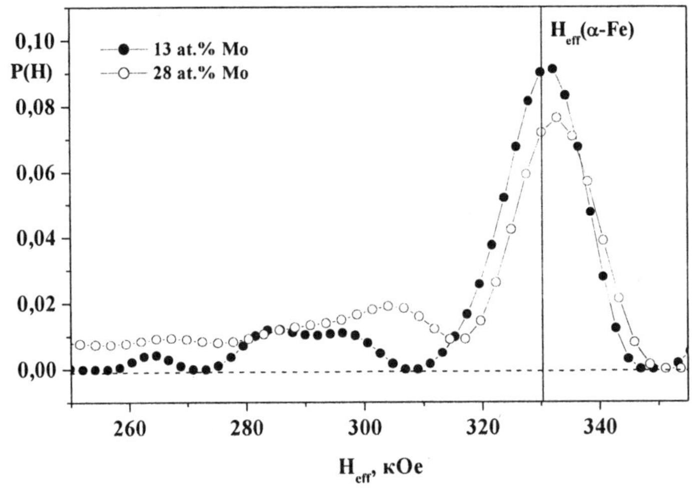

Figure 4. Distribution function P(H) of effective hyperfine magnetic fields for ferromagnetic components of the Fe-Mo compositions Mossbauer spectra.

The more interesting seems the Mossbauer results of the intermetallic compounds formation in that nanosized compositions. The corresponding central nonmagnetic part appears gradually in the spectra with increase of the Mo concentration in the samples. It consists of several subspectra with parameters of quadrupole splitting and isomer shift corresponding to intermetallic and even amorphous phases known in these systems and bcc Mo-solid solutions. All these intermetallic phases may be considered as disordered because their subspectra have linewidths extended from experimental value 0.28 up to 0.36 mm/s.

In the narrow concentration range closed to the 28 at. % Mo, where the smallest crystallite size have been reached and amorphysation was thermodynamically enabled the maximum quantity of disordered intermetallic phase was observed. The Mossbauer spectra analysis allows us to reveal and differentiate contributions of small amount (1-2%) of the amorphous and crystalline phases, because their magnetic and Mossbauer parameters are well resolved. To separate the spectral contributions of formed intermetallic phases we performed the analysis of distribution functions P(Q) of quadruple splitting and P(δ) isomer shifts for corresponding spectral components. The hyperfine parameters of the central paramagnetic component were calculated separately to magnetically resolved component. P(Q) profile showed that the most probable intermetallics structure corresponding to the equilibrium phase diagram coexists with another stoichiometry phases and even amorphous phase, which have the less probability. Analysis of normalized distribution function P(δ) (fig.5) allowed revealing the iron atoms chemical state when the local surrounding of iron in the intermetallic phase is changed.

At the Mo concentration of 13 at. % a broad single peak covers all the region of both intermetallic and amorphous phases existence. The most likely phases to be formed at this concentration are λ-Fe_2Mo and amorphous one. The distribution function for the samples with high concentration of Mo (47%) have a narrow single peak with maximum corresponding to μ-phase formation and another small peak corresponding to amorphous-phase. Considerable shift of the main peak to the negative isomer shift values as well as an increase of its intensity could be observed in the picture of the sample with middle concentrations 28 %. The increase of the isomer shift negativity may be explained by the higher electron density near iron atoms. This is believed to be the result of possible hydrogen trapping. Hydrogen does not dissolve in the solid phase of Fe and Mo at room temperature. Nevertheless its weak solubility in binary Fe-Mo system (especially at the temperature of reducing 700°C) seems to be very probable. Hydrogen atoms incorporated in the structure should add their electrons to the electron density near iron atoms and correspondingly the isomer shift parameter decreases.

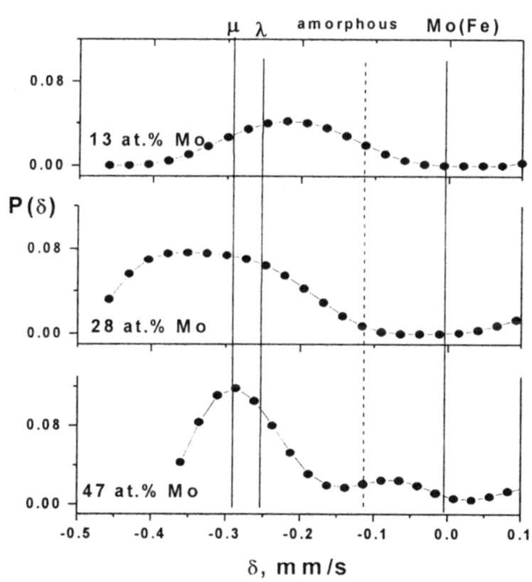

Figure 5. Isomer shift distribution functions P() for paramagnetic components of Fe-Mo Mossbauer spectra.

4. Conclusions

The coupled reduction of fine dispersed oxygen containing Fe-Mo compounds in continuous H_2 flow at the temperature 740°C results in different partially disordered and even amorphous intermetallic compounds and solid solutions with extended concentration formation. This phenomenon of phase formation at lower temperatures that implied by equilibrium phase diagram is due to the energetic contribution of the high density of interfaces formed upon the reduction process. The grain sizes, type and quantities of forming phases depend on the initial concentration of constituent elements in obtained compositions.

We observed the narrow concentration range of disordered intermetallic phase formation. Partial and short time hydriding of the forming intermetallic phase crystallites in the narrow concentration range closed to the most possible λ and μ-phase formation leads to the amorphysation of the grains surface and their strongest refinement.

5. Acknowledgements

We would like to thank Drs. V.V.Lyovina and D.V.Kuznetsov (Moscow State Steel and Alloys Institute) for preparing the sample compounds. *This work was supported by the Russian Foundation for Basic Research (Grant 99-02-16135).*

6. References

1. Ryishonkov, I., Levina, V.V., Samsonova T.V., , (1995) The method of iron powder producing from the acid dilution, *Patent №2038195*, Russia, №18, 1-33.
2. Kuznetsov, D.V., Dzidziguri, A.L., Lyovina, V.V., Sidorova, E.N., Ryizonkov, D.I. (1999) X-ray studies of ultradispersed compounds based on Fe and Mo, *Mater. Sci. Transactions*, **8,** 44-52 .
3. Novakova, A.A., Kiseleva, T.Yu., Lyovina, V.V. (2000) Specific Features of Phase Formation in nanosized Fe-W Compositions, *Russian Journal* of *Inorg. Chemistry*, **45,** 8, 1267-1272.
4. Novakova, A.A., Kiseleva, T.Yu., Lyovina, V.V., Kuznetsov, D.V., Dzidziguri, A.L. (2001) Low temperature formation of nanocrystalline Fe-W and Fe-Mo compounds, *J.Alloys&Compounds* , **317-318,** 423-427.
5. Sumiyama, K., Ezawa, H., Nakamura, Y., (1987), Magnetic properties of metastable bcc and amorphous $Fe_{1-x}Mo_x$ alloys produced by vapor quenching, *J.Phys.Chem.Sol*ids, **48,** 30, 255.
6. Bakker, H, (1998), *Enthalpies in Alloys*, Trans.Tech.Publications Ltd., Switzerland, 1-78.
7. Lyakishev I.P. (eds.) (1997), *Diagrams of binary metallic systems*, V.2, Moscow, Mashinostroenie, Russia, 567-568.
8. Kumar, T.S.S., Mallaya, R.M. , Deshpande, V.M., (1983) Mossbauer study of oxides and intermetallics of the systems Fe-W, Fe-Mo and Fe-Mo-W, *Proc. Of the Nuclear Physics and Solid State Physics Symposium,* Misore, 26C, 351-2.

PHASE STATES OF HYDROGEN IN METAL PALLADIUM

N.V.PISKUNOV, Y.T. SINIAPKIN, V.M.KULGAVCHUK,
N.A.PROTOPOPOV
Russian Federal Nuclear Center - All-Russia Research Institute of Experimental Physics (RFNC-VNIIEF)
37 Mir Ave., Sarov, Nizhny Novgorod Region, 607190 Russia

Abstract

The experimental relation of integral energy of atomic bounding in palladium and relative atomic concentration of hydrogen, absorbed in it is considered (it was published in the Engineering and Physical Journal in 1961). There are found intervals of hydrogen concentrations in palladium, where hydrogen exists mainly as the shielded complexes, in a structure of palladium hydride molecules in a molecular kind. It is shown, that the reduction of integral energy of connection at the initial segment of the relation can happen in the result of destruction of Pd_2^+ molecular complexes, the experience non-elastic collision with the shielded complexes of hydrogen. Growth of energy of connection at the terminal phase of the relation can be stipulated by the process of molecular hydrogen formation from the earlier absorbed atomic hydrogen. The examples of implementation of the obtained results of the research of so called cold nuclear fusion, heterogeneous catalysis at the d-metals are presented and structural materials hydrogen degradation.

In 1961 the results of measurement of the relation between the integral energy of the atomic bounding E in palladium and relative nuclear concentration C of the hydrogen absorbed in it were published in [1-3]. N.A.Protopopov and V.M.Kulgavchuk got these results with the help of their own technique [3-5].

It was obvious that the obtained relation contains new information on phase condition of hydrogen in palladium. However, much to our regret, it was not possible to find this information in time and to use it in the research effectively in an appropriate way.

It is a rare case in practice of scientific surveys, when 40 years later practically the same authors have to address the data, which were not interpreted and used properly in their due time. There are several incentive reasons for that. And first of all, these are appearing more than 10 years ago publications on co called cold nuclear fusion (CNF) and unsolved problems of the nature of active centers, catalysis mechanism, on the role of the holes in d-shells of atoms, etc. still existing

N. Veziroglu et al. (eds.), Hydrogen Materials Science and Chemistry of Metal Hydrides, 221–228.

for more than 150 years in research of heterogeneous catalysis on transient d-metals.

On the basis of the published data [1-3] the present report gives new information on phase composition of hydrogen in metallic palladium, as well as some results on implementation of these data in the research of CNF, heterogeneous catalysis at transient d-metals and structural materials hydrogen degradation.

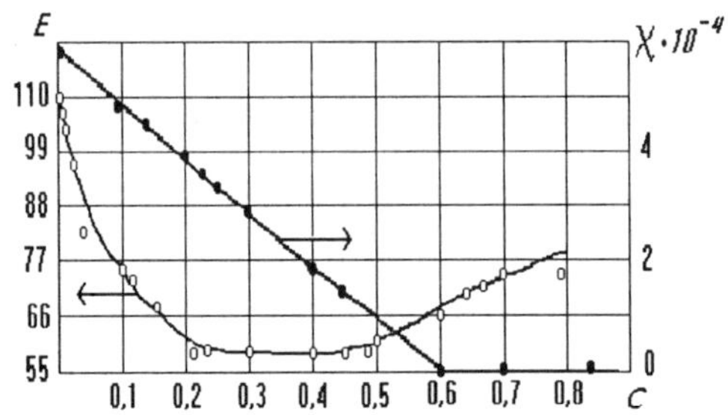

Figure 1. Hydrogen concentration $C = \dfrac{H}{Pd}$ (atomic ratio) dependence of sublimation (integral) energy E (kcal/mol) and paramagnetic susceptibility χ (mol^{-1}) of Pd–H system.

Fig. 1 shows a curve of experimental relation of E=f(C). It is necessary for solution of this task and is based on reproduction of the borrowed curve from [1], but updated considering the data of [2,3], the thermal effect of the palladium hydride formation, and the dependency of paramagnetic susceptibility of palladium χ on hydrogen concentration C borrowed from [6].

As it is shown on Fig. 1 the curve dependence E=f(C) has four characteristic sections that correspond to the hydrogen concentration intervals: from 0 up to 0.1; from 0.1 up to 0.2; from 0.2 up to 0.5 and from 0.5 up to 0.8.

Let us study each of the mentioned intervals.

In C interval from 0 up to 0.1 the average bounding energy ***E*** almost linearly decreases with the increase of C. The energy $\Delta E = 110 - E(C)$ taken from the gas of collective (valence) palladium electrons increases following almost the same law.

As it is shown from the curve, the measured value of this energy that falls on one atom of hydrogen, ε_λ at the considered interval makes up the following value:

$$\varepsilon_\lambda \approx \frac{\Delta E}{\Delta C} \approx \frac{110 - 77}{0,1} \approx 330 \text{ kcal/mol} \quad (1)$$

or ≈ 14,5 eV for each Pd atom.

This value coincides well with the energy of the hydrogen atom ionization (≈ 13,6 eV [7]).

Such concurrence of indicated values lets us suppose that in the considered interval of concentration C hydrogen is present in palladium in the foam of shielded complexes $(H^+)_{sh.}$, and this complies with earlier obtained data from [8,9].

Thus it is necessary to take into account that the process of ionization of hydrogen atoms energetically does not change the state of the electron gas, since neutral atom of hydrogen being dissolved in palladium brings into palladium kinetic energy equal to 13.6 eV, and that is very close to the value obtained with the help of formula (1).

Based on the obtained data on hydrogen phase state on the considered interval of its concentration it is possible to make definite significant conclusions in the field of CNF.

For this purpose one of the co-authors of the paper (N.A.Protopopov) derived equation in Thomas-Fermi [9] approximation for calculation of shielding energy or energy, used for one shielded complex formation $(D^+)_{sh}$, and taken away from electron gas:

$$\varepsilon_\lambda = \int_{5\lambda}^{0} \rho(r) \cdot \varphi(r) \cdot 4\pi r^2 dr \tag{2}$$

Density of shielding charge ρ(r) and shielding potential φ(r) in this approximation are determined with the following formulas [9]:

$$\rho(r) = -e \cdot \frac{1}{4\pi r \lambda^2} \cdot \exp^{-r/\lambda} \tag{3}$$

$$\varphi(r) = \frac{e}{r} \cdot \exp^{-r/\lambda} \tag{4}$$

After substitution of (3) and (4) into (2) the obtained integral have solution and ε_λ could be determined with formula

$$\varepsilon_\lambda \approx \frac{e^2}{2\lambda}$$

From here it is possible to determine λ :

$$\lambda \approx \frac{e^2}{2\varepsilon_\lambda} \tag{5}$$

In CNF experiments numerical value of λ is very important.

It is known [10] that at λ = (0,03 – 0,035) Å neutrons of light nuclei should be formed on reaction:

$$D + D \rightarrow {}^3He(0{,}82\ MeV) + n\ (2{,}45\ MeV) \quad \text{or}$$
$$D + D \rightarrow T(1{,}01\ \text{MэB}) + p(3{,}02\ \text{MэB})$$

With further interaction of p and D and n =0.75 MeV output.

Calculations, presented on formula (5) and using obtained data (figure 1), show that at ε_λ =13,6 eV λ ≈ 0,5 Å.

Thus, on the considered interval of hydrogen concentrations in palladium from 0 to 0.1, formation condition of light nuclei synthesis neutrons is not fulfilled.

Nevertheless, taking into account that Coulomb barrier of positive charge repulsion of reactive particles is the only obstacle for light nuclei fusion, and the only way to weaken this barrier is their shielding with electrons, experiments on CNF using shielding effect in this or other hydrogen concentrations intervals will be more productive.

It is necessary to mention that hydrogen existence in palladium in the considered concentrations interval in the form of shielding complexes is not sole condition.

It is known [6] that in palladium crystal lattice hydrogen at absorption is located in octahedral interstice, quantity of which corresponds to the quantity of palladium atoms. In this case electrons fill holes in 4d-shells of palladium atoms.

It is known [6,11] that average holes quantity in these palladium atoms shells, falling on one palladium atom, is equal to 0.36.

To compare with this value for palladium face-centered cubic lattice we calculated the number of covalently-coupled pairs of palladium atoms from among atoms, located at the centers of adjacent lattice faces, falling on one palladium atom, which is equal to 0.37.

Such coincidence of these values led us to supposition that palladium in **α**-phase consists of molecular complexes Pd_2^+, covalently-coupled pared of atoms, in one of which hole in 4d-shell is localized.

On that ground we supposed that shielded complexes $(H^+)_{sh}$, diffusing into palladium, will collide with Pd_2^+ complexes and destroy them, on equation, for example:

$$(H^+)_{sh} + Pd_2^+ \rightarrow PdH + Pd\ (+\varepsilon_h) \qquad (6)$$

This equation corresponds to well-known fact of β-phase palladium formation, i.e. palladium hydride at hydrogen absorption, which is started at $C \geq 0{,}001$.

Thus, on the considered concentration interval in palladium exists both in the form of shielded complex, which on the total charge represents neutral hydrogen atom, and in the structure of palladium hydride molecule.

Our supposition Pd_2^+ on molecular complexes existence in palladium allows to make important conclusion in the field of heterogeneous catalysis on transition d-metals and structural materials hydrogen degradation.

Complex with the hole in 4d-shell has significant free energy (as estimated, not less than the second ionization potential) and, locating on the surface of metal palladium, could be the source of chemical reaction at non-elastic collisions with adsorbate molecules from gas phase at catalysis, i.e. work as active center.

Formation of the complexes increases the integral interatomic binding energy in transition d-metals, thus providing the needed physical and mechanical properties.

The complexes' decay under the impact of high-mobile $(H^+)_{sh}$ complexes is accompanied by Pd ß-phase formation, that has a higher value of the constant lattice and lower physical and mechanical properties than the initial Pd α-phase. Such decay includes all the stages of construction materials' (transition metals) hydrogen degradation: creepage and demolition under external loading.

Process of non-elastic destruction of Pd_2^+ molecular complexes continues on the interval C from 0.1 up to 0.2 (figure 1) according to the equation (6).
The second generation of Pd_2^+ complexes is formed from free palladium atoms. They are also destroying like the first ones. Similar complexes of the third, forth, etc. generations are formed and destroyed. In this case the next generation of Pd_2^+ complexes has smaller energy value of covalent constraint forces, than the previous one.
Pd_2^+ complex decomposition in the hydrogen concentration interval from 0 up to 0.2 is accompanied by decrease of energy of covalent constraint forces of palladium atoms almost up to zero. At that total energy of atomic bounding in palladium decreases from 100 up to 59 kcal/mol.
This sudden decrease of integral energy of atomic bounding in palladium on this interval could be explained by the following.
It is known [11] that hydrogen diffusion coefficient at temperature of 300 K is $\approx 10^{-6}$ cm^2/sec, and hydrogen settled life time in octapairs is $\approx 10^{-9}$sec. This means that if palladium saturation with hydrogen up to C=0.1 takes place at 1000 sec, shielded hydrogen complexes suffer on average $3 \cdot 10^{34}$ $(0{,}5 \cdot 1000 \cdot 10^9 \cdot 6 \cdot 10^{23})$ collisions at this time, including with $2{,}16 \cdot 10^{23}$ $(0{,}36 \cdot 6 \cdot 10^{23})$ molecular complexes, destroying them.
From these data for palladium obtained dependence E=f(C) (figure 1) allows to obtain new data on structure and values, forming total energy of atomic bounding in palladium:

$$E = E_c + E_m = 51 \text{ kcal/mol} + 59 \text{ kcal/mol} = 110 \text{ kcal/mol}$$

The following equality is also valid:

$$\varepsilon = \varepsilon_c + \varepsilon_m$$

Numerical values ε, ε_c and ε_m are determined with the following way:

$$\varepsilon = \frac{110 \cdot 26{,}3 \cdot 10^{21}}{6 \cdot 10^{23}} \approx 4{,}8 \text{ eV}$$

$$2\,\varepsilon_c = \frac{51 \cdot 26{,}3 \cdot 10^{21}}{0{,}36 \cdot 6 \cdot 10^{23}} = \frac{51 \cdot 26{,}3 \cdot 10^{21}}{2{,}16 \cdot 10^{23}} \approx 6{,}2 \text{ eV}$$

$$\varepsilon_c \approx 3{,}1 \text{ eV}$$

$$\varepsilon_m = \frac{59 \cdot 26{,}3 \cdot 10^{21}}{6 \cdot 10^{23}} \approx 2{,}6 \text{ eV}$$

Approximate equality ε and $\varepsilon_c + \varepsilon_m$ confirms our suppositions.
It is very important to estimate value of relative hydrogen concentration in palladium, at which the process of formation and destruction of Pd_2^+ molecular complexes is stopped and zero value of holes quantity in 4d-shells of palladium atoms is obtained.
Using the formula for determination of sum of infinitely decreasing geometric progression terms:

$$S = \frac{a_1}{1-q}$$

at $a_1 = 0,36$ and $q = 0,5$ it was determined that the process of palladium hydride formation processes in normal conditions is finished at C= 0,72, which corresponds to the data presented in [6] and obtained at investigation of palladium superconductivity phenomenon. It somewhat differs from C = 0,6 value, obtained earlier at investigation of paramagnetic susceptibility and presented in [6] and figure 1. We believe that this is stipulated by insufficient sensitivity of measuring method of this parameter, which allows to record three holes generation only in 4d-shells of palladium atoms.

Really, sum of numbers 0.36, 0.18 and 0.09 is equal to 0.63, which is very close to 0.6.

That is why one can see that if there are no detected holes in palladium atoms, the do not exist in the form of statistically distributed on sample atoms, as it is usually accepted, but the are localized in the volume of molecular complexes only.

Moreover, one can count that for holes formation in palladium atoms 4d-shells energy distortions are necessary. There are no energy distortions in case of insulated atoms and in atoms of metal palladium, but they appear in the process of Pd_2^+ molecular complexes formation.

Thus, hydrogen in palladium exists in the form of shielded complexes and in palladium hydride molecules on the considered interval C from 0.1 up to 0.2 like on the previous interval C from 0 up to 0.1 (figure 1).

On the interval C from 0.2 up to 0.5, based on performed calculations of hydrogen concentration boundary limit, at which the process of holes filling in palladium atoms 4d-shells is stopped, and curve trend of E = f(C) dependence (figure 1), the process of palladium hydride formation on the reaction (6) is still dominating. However, both the first and further generations of Pd_2^+ complexes, having different decreasing energy values of covalent constrained forces, participate in this process. In this case curve trend of E = f(C) dependence on this section shows that the processes of formation of further generations of Pd_2^+ complexes and decomposition of the previous ones, taking into account reaction (6) thermal effect, equal to 9.6 kcal/mol, are energetically balanced.

Integral energy of atomic bounding in palladium on the given interval of hydrogen concentration is determined mainly by metal constrained forces, equal to 59 kcal/mol or ≈2.6 eV per each Pd atom.

On the interval C from 0.5 up to 0.8, as one can see from the curve trend of E = f(C) dependence (figure 1), significant growth of integral energy of atomic bounding in palladium E from 59 up to 78 kcal/mol takes place.

Taking into account that 50% of palladium crystal lattice interstices remain unoccupied, and based on the data [6] at C=0.6 and based on the data [11] and our data at C=0.72 the process of palladium hydride formation on reaction (6) and holes filling in 4d-shells of palladium atoms stops, one can suppose that from the moment C>0,5 absorbed hydrogen is spent at other processes.

We have calculated increase of atomic boundary specific energy in palladium, falling on one hydrogen molecule, using the data of E = f(C) dependence, presented on figure 1, on the interval ***C*** from 0.5 up to 0.8 with the following formula:

$$\varepsilon_\lambda \approx \frac{\Delta E}{0{,}5\Delta C} \approx \frac{78-59}{0{,}5\cdot 0{,}3} \approx 125\,\text{kcal/mol}$$

or ≈ 5,5 eV per one Pd atom.

Obtained result well coincides with energy value of hydrogen molecule formation (4.77 eV per one H atom [7]).

Based on this we suppose that on the considered interval of hydrogen concentration in palladium the formation process of molecular hydrogen from atomic one, absorbed earlier with palladium, is dominating.

In connection with this we have made another supposition that surplus heat generation, really observed at CNF investigation in the conditions, similar to the conditions of the performed experiment, result of which is published in [1-3] is stipulated by thermal effect, accompanying the formation process of molecular hydrogen from atomic one. Obtained results are reported on International conference on CNF, held in September 1999, Dagomys, Krasnodar region.

Table of symbols

e – electron charge, eV; λ – shielding distance (distance from nucleus, on which its potential is decreasing in exp times), Å; ε_h – heat effect of palladium hydride formation reaction, kcal/mol; $\mathbf{E_c}$ – covalent constrained forces component, kcal/mol; $\mathbf{E_m}$ – metal constrained forces component, kcal/mol; ε – total energy of atomic bounding in palladium, falling on one Pd atom, eV; ε_c – energy of covalent constrained forces of atom pare in Pd_2^+ complex, falling on one Pd atom, eV; ε_m – energy of metal constrained forces, falling on one Pd atom, eV; **S** - sum of infinitely decreasing geometric progression terms; $\mathbf{a_1}$ – value of the first progression term; **q** – progression denominator. Indices: **h** – heat; c – covalent; **m** – metal; **sh**– shielded.

References

1. Protopopov, N.A. and Kulgavchuk, V.M. (1961) The pulse method for determination of the sublimation energy of metals and alloys, *Engineer-Physical Journal*, 4-7, 84-89.
2. Protopopov, N.A. and Kulgavchuk, V.M. (1961) Determination of the sublimation energy of the Pd-H system by the pulse method, *Engineer-Physical Journal*, 4-9, 102-103.
3. Protopopov, N.A. and Kulgavchuk, V.M. (1960) Generator of high-power pulse currents, *Journal "Experimental Instruments and Technique"*, 1, 85-89.
4. Protopopov, N.A. and Kulgavchuk, V.M. (1961) To the theory on the mechanism of current pause and shock waves emerging while heating a metal by pulses of high-density electric current, *Journal of Technical Physics*, 31-5, 557-564.
5. Kulgavchuk, V.M. (1962) Study of the kinetics of the Pd-H system heating with the pulse method, *Journal of Physical Chemistry*, , 36-8, 1713-1716.
6. Geld, P.V., Ryabov, R.A. and Mokhracheva L.P. (1985) *Hydrogen and Physical Properties of Metals and Alloys. Hydrides of Transition Metals,* Nauka, M.
7. Drits, M.E. (1985) *Elements Properties,* Reference book, Metallurgy, M.
8. Galaktionova, N.A. (1967) *Hydrogen in Metals*, Metallurgy, M.
9. Kittel, Ch. (1967) *Quantum Theory of Solid Bodies*, Nauka, M. , 136-137.
10. Burrows, A. (1989) *Physical Review B*, 40-5, 3405-3408.
11. Alefeld, G. And Felkl, I. (1981) *Hydrogen in Metals*, Mir, M.

THE INVESTIGATION OF HYDROGEN ABSORPTION AND ELECTROCHEMICAL PROPERTIES OF ALLOYS Ti-Zr-Ni-V-Mn WITH LAVES PHASES IN NONSTOICHIOMETRIC REGION

T.A. ZOTOV, O.A. PETRII, V.N. VERBETSKY
Chemical Department, Moscow State University,
Russia, Moscow, 119992 Leninskiye Gory

Abstract

The electrochemical properties and hydrogen storage performance of $Ti_{0.45}Zr_{0.55}Ni_{0.85}V_{0.45}Mn_x$ (x = 0.1, 0.18, 0.23, 0.33, and 0.53) and $Ti_{0.45}Zr_{0.55}Ni_{0.7}V_{0.45}Mn_x$ (x = 0.08, 0.35, 0.71, 0.9, 1.15 and 1.33) alloys with a different amount of nickel and manganese were investigated. The equilibrium hydrogen pressure of alloys increased as the amount of manganese increased. All of these alloys were found to have a single phase with a C14 Laves phase structure by x-ray diffraction analysis and electronic microscopy. The relationship between discharge performance, the equilibrium hydrogen pressure of alloys and alloy's lattice parameters is also discussed. The $Ti_{0.45}Zr_{0.55}Ni_{0.85}V_{0.45}Mn_{0.23}$ alloy with a nonstoichiometric composition has a maximum discharge capacity (363 mAh/g with the discharge current density 100 mA/g).

1. Introduction

Nickel-metal hydride (Ni-MH) batteries widely adopted today for their high energy characteristics and better environmental compatibility. However, Laves phase AB_2-type Ti, Zr-based MH-electrodes relatively poor developed as distinct from rare-earth metals AB_5–type based alloys. High hydrogen capacity of Ti, Zr-based Laves phase alloys and low equilibrium decomposition pressure of hydride phase, enables to use them as a material for manufacturing of MH - electrode for Ni-MH batteries.

Many investigations dealt with the study of stoichiometric (AB_2) [1-3] and over-stoichiometric (AB_{2+n}) [4, 5] Ti, Zr-based Laves phase alloys. However, Laves phase in Ti-Mn and Zr-Mn systems have wide areas of homogeneity ($TiMn_{1.4}$ - $TiMn_2$, $ZrMn_{1.9}$ - $ZrMn_{2.9}$) [6, 7]. Thus, it is interesting to study how the changes in the stoichiometric composition of IMC in the Laves phase homogeneity region for Ti-Zr-Ni-V-Mn system affects the hydrogen absorption and electrochemical characteristics. For these purposes, in this work two series of alloys were prepared. The alloy composition was chosen for the following conditions:

(a) there was a constant Ti:Zr ratio;
(b) constant vanadium content;
(c) constant nickel content in each series of the alloys;

N. Veziroglu et al. (eds.), Hydrogen Materials Science and Chemistry of Metal Hydrides, 229–234.

(d) variable content of manganese - $Ti_{0.45}Zr_{0.55}Ni_{0.85}V_{0.45}Mn_x$ (where x = 0.18, 0.23, 0.33, and 0.53) and $Ti_{0.45}Zr_{0.55}Ni_{0.7}V_{0.45}Mn_x$ (where x = 0.08, 0.35, 0.71, 0.9, 1.15 and 1.33).

In addition, we also discuss the relationship between the discharge performance and characteristics of alloy structure and hydride properties.

2. Experimental Details

Alloys were prepared by arc melting of initial metals under argon atmosphere. With the purpose of homogenization of alloys samples were exposed to annealing at temperature of 900^0C for 240 hours with subsequent quenching in cold water.

The structure and phase composition of alloys were investigated by means of X-ray, electronic microscopy and probe analysis.

In order to investigate the hydrogen absorption properties of alloys, pressure-composition-temperature (PCT) curves were measured using Siewert's type apparatus. About 5 g of solid uncrushed alloy sample was taken for each absorption-desorption experiment. For the activation, sample was exposed to the hydrogen atmosphere at a pressure of about 50 atm. for 12 hours. Then sample was dehydrided by heated about 300-400 ^{0}C for 10-15 minutes. The PTC data were measured after two cycles of activation process.

The MH-electrode for electrochemical experiments was prepared by mixing alloy powder and copper powder at weight ratio IMC : Cu = 1 : 4. The mixture was cold pressed to a pellet with weight of 0,1 g and diameter 7 mm at compacting pressure of 8.1 ton/cm^2.

The measurement of electrochemical capacity was carried out in a glass three-electrode electrochemical cell using the metal hydride alloy powder electrode as a working electrode, platinum were as a counter electrode and a hydrogen electrode as a reference electrode. The electrolyte was 6 M KOH solution. The charge-discharge was controlled by the potential of working electrode with respect to reference electrode with an manual galvanostatic apparatus. . The discharge cutoff potential was +300 mV referenced to the standard hydrogen reference electrode.

The preliminary activation was made by boiling in 6 M KOH solution at 115 ^{0}C before the first polarization of a MH-electrode. Sample was charged by current density 100 mA/g IMC for 6 hours. After charging, the half-cell with MH-electrode was bubbling by argon gas during resting time about 15 minutes. The discharge capacity was investigated at discharge current densities about 100 and 200 mA/g IMC.

3. Results and discussion

The alloys $Ti_{0.45}Zr_{0.55}Ni_{0.85}V_{0.45}Mn_x$ (x = 0.1, 0.18, 0.23, 0.33, and 0.53) and $Ti_{0.45}Zr_{0.55}Ni_{0.7}V_{0.45}Mn_x$ (x = 0.08, 0.35, 0.71, 0.9, 1.15 and 1.33) were identified as having a single-phase structure according to electronic microscopy. The crystal structures of the each alloys were confirmed a hexagonal C14-type Laves phase by x-ray powder (XRD) diffraction analysis.

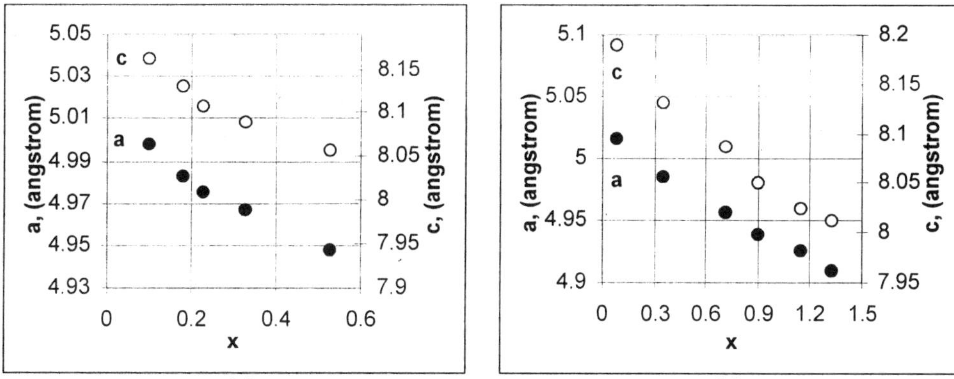

(a) (b)

Figure 1. Lattice parameters of $Ti_{0.45}Zr_{0.55}Ni_{0.85}V_{0.45}Mn_x$ (a) and $Ti_{0.45}Zr_{0.55}Ni_{0.85}V_{0.45}Mn_x$ (b).

Lattice constants were calculated from the XRD data. Lattice constants of each alloy are presented in Fig 1. Lattice constants decreased linearly as the content of Mn in the alloys increased (Fig 1). It mean s that all of these alloys were in the aria of Laves phase homogeneity.

TABLE 1. The alloys composition, hydrogen absorption and electrochemical properties.

Alloys composition	Stoichiometry	Equilibrium pressure P_{eq}, atm	Hydrogen content mass%	Discharge capacity ,mAh/g	
				i_{dis}=100 mA/g	i_{dis}=200 mA/g
$Ti_{0.45}Zr_{0.55}Ni_{0.85}V_{0.45}Mn_{0.1}$	$AB_{1.4}$	0.01	1.78	232	160
$Ti_{0.45}Zr_{0.55}Ni_{0.85}V_{0.45}Mn_{0.18}$	$AB_{1.48}$	0.015	1.83	313	296
$Ti_{0.45}Zr_{0.55}Ni_{0.85}V_{0.45}Mn_{0.23}$	$AB_{1.53}$	0.025	1.99	363	325
$Ti_{0.45}Zr_{0.55}Ni_{0.85}V_{0.45}Mn_{0.33}$	$AB_{1.63}$	0.04	1.87	314	300
$Ti_{0.45}Zr_{0.55}Ni_{0.85}V_{0.45}Mn_{0.53}$	$AB_{1.83}$	0.28	1.79	260	250
$Ti_{0.45}Zr_{0.55}Ni_{0.7}V_{0.45}Mn_{0.08}$	$AB_{1.23}$	-	-	21	8
$Ti_{0.45}Zr_{0.55}Ni_{0.7}V_{0.45}Mn_{0.35}$	$AB_{1.5}$	0.10	1.69	160	97
$Ti_{0.45}Zr_{0.55}Ni_{0.7}V_{0.45}Mn_{0.71}$	$AB_{1.86}$	0.25	1.82	296	250
$Ti_{0.45}Zr_{0.55}Ni_{0.7}V_{0.45}Mn_{0.9}$	$AB_{2.05}$	-	-	315	255
$Ti_{0.45}Zr_{0.55}Ni_{0.7}V_{0.45}Mn_{1.15}$	$AB_{2.3}$	-	-	258	210
$Ti_{0.45}Zr_{0.55}Ni_{0.7}V_{0.45}Mn_{1.33}$	$AB_{2.48}$	-	-	77	75

The PCT-investigations were carried out for all alloys of the $Ti_{0.45}Zr_{0.55}Ni_{0.85}V_{0.45}Mn_x$ series (x = 0.1, 0.18, 0.23, 0.33, and 0.53) and only two alloys $Ti_{0.45}Zr_{0.55}Ni_{0.7}V_{0.45}Mn_x$ series (x = and 0.53).Hydrogen absorption properties of these alloys are presented in the Table 1. From those, it is obvious that the hydrogen equilibrium pressure and the hydrogen storage amount of these alloys increase until the maximum, as the amount of Mn increases and than, particularly diminished. And besides, hydrogen equilibrium pressure of some IMC hydrides was much lower 1 atm., that makes enables to use these IMC as a material for manufacturing MH-electrodes.

Figure 2 shows the discharge curves at 25 ^{0}C for the $Ti_{0.45}Zr_{0.55}Ni_{0.85}V_{0.45}Mn_x$ (x = 0.18, 0.23, 0.33 and 0.53) and $Ti_{0.45}Zr_{0.55}Ni_{0.7}V_{0.45}Mn_x$ (x = 0.08, 0.35, 0.71, 0.9, 1.15 and 1.33) alloys, respectively.

It was established that after the preliminary activation the discharge capacity of the MH-electrodes rise their maximum during 1 –3 cycles. The preliminary activation processes were operated during 90 min by the hot immersion treatment in 6M KOH solution.

In the case of $Ti_{0.45}Zr_{0.55}Ni_{0.85}V_{0.45}Mn_x$ (x = 0.1, 0.18, 0.23, 0.33 and 0.53) alloys, the higher nickel type, the discharge capacities were larger then those of $Ti_{0.45}Zr_{0.55}Ni_{0.7}V_{0.45}Mn_x$ (x = 0.08, 0.35, 0.71, 0.9, 1.15 and 1.33) alloys.

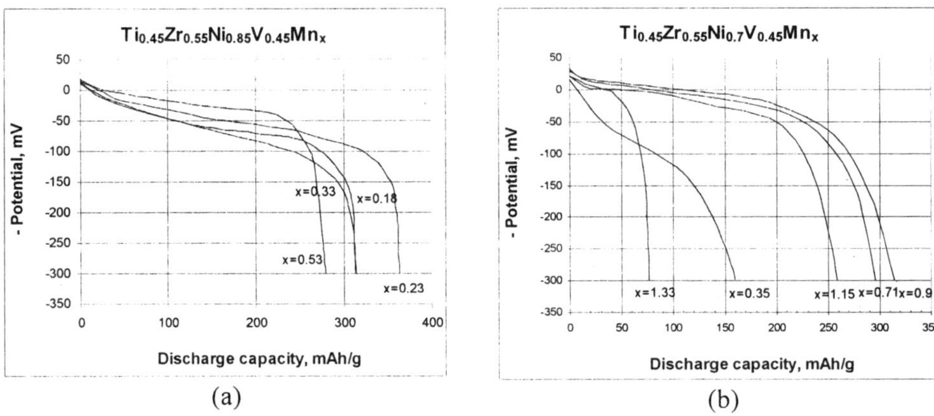

Figure 2. Discharge curves of $Ti_{0.45}Zr_{0.55}Ni_{0.87}V_{0.45}Mn_x$ (x = 0.18, 0.23, 0.33 and 0.53) alloys (a) and $Ti_{0.45}Zr_{0.55}Ni_{0.68}V_{0.45}Mn_x$ (x = 0.35, 0.53, 0.71, 0.9, 1.15, 1.33 and 1.47) alloys (b) at 25 ^{0}C and the discharge current density of 100 mA/g.

Also, it was obtained, that exist a maximum of discharge capacity, when the manganese content was increased in samples. Values of the discharge capacities at 25 ^{0}C and discharge current densities 100 and 200 mA/g for two series of alloys were shown in the Table 1. The dependence discharge capacities of alloys from the stoichiometric Laves phase A and B contents ratio (n in AB_n) was presented in the Fig. 3. The discharge capacity arises until maximum value when the stoichiometric coefficient increases and decreases at further increasing the stoichiometry.

The explanation of this phenomenon can be caused by two ways.

(a) There is a maximum hydrogen storage capacity of the alloys with various manganese content. The hydrogen storage capacity increases from the fixed limit as the stoichiometric factor (n in AB_n) increase in the Laves phase homogeneous region; it lead to increasing discharge capacity. However, subsequent increasing stoichiometric factor in alloys, caused by the manganese addition, gave rise to decrease of lattice constants (Table 1, Fig. 1), and therefore both the lattice volume and the size of tetrahedral interstities, the size of hydrogen occupancy sites, decreased. As a result, it seems that the increase of the hydrogen equilibrium pressure and the decrease of the hydrogen storage amount are due to the lattice shrinkage and the change of chemical affinity for hydrogen of metals (Mn is not a hydride forming element).

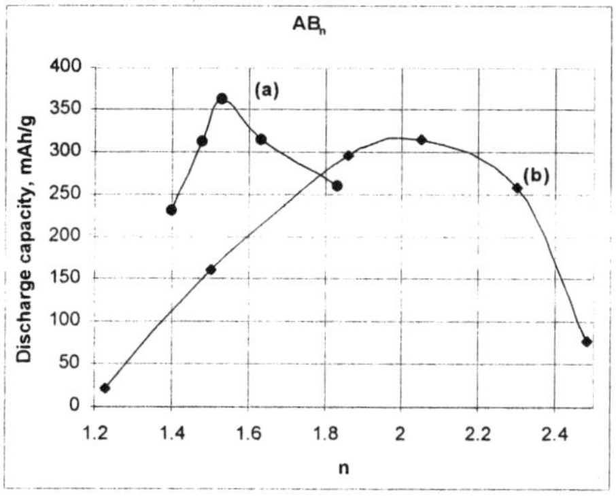

Figure 3. The dependence of discharge capacities of alloys on the stoichiometric A and B contents ratio, where (a) are $Ti_{0.45}Zr_{0.55}Ni_{0.85}V_{0.45}Mn_x$ alloys and (b) are $Ti_{0.45}Zr_{0.55}Ni_{0.7}V_{0.45}Mn_x$ alloys.

(b) From another hand, the maximum of discharge capacity can de occur at another stoichiometric ratio, then maximum of the hydrogen storage capacity. The discharge capacity of alloys with equilibrium pressure more then 1 atm. tend to decrease when the hydrogen equilibrium pressure increase. Because the MH-electrode manufactured from the IMC with hydrogen equilibrium pressure more then 1 atm. begin to emit hydrogen after a power will cut off.

Analyzing the discharge capacity data at current densities 200 mA/g (Fig 1), we can conclude, that rate capability of two series of alloys increases with the manganese content and stoichiometric coefficient increase. And besides, the rate capability better of the series with higher nickel content.

The maximum discharge capacity of $Ti_{0.45}Zr_{0.55}Ni_{0.7}V_{0.45}Mn_x$ alloys with lower nickel content has a $Ti_{0.45}Zr_{0.55}Ni_{0.7}V_{0.45}Mn_{0.9}$ alloy with the stoichiometric composition $AB_{2.05}$. But in case higher nickel content $Ti_{0.45}Zr_{0.55}Ni_{0.85}V_{0.45}Mn_x$ series of alloys maximum discharge capacity has a $Ti_{0.45}Zr_{0.55}Ni_{0.85}V_{0.45}Mn_{0.23}$ alloy with the nonstoichiometric composition $AB_{1.53}$.

4. Conclusions

The structure, electrochemical properties and hydrogen storage performance of $Ti_{0.45}Zr_{0.55}Ni_{0.85}V_{0.45}Mn_x$ (x = 0.18, 0.23, 0.33, 0.53 and 0.7) and $Ti_{0.45}Zr_{0.55}Ni_{0.7}V_{0.45}Mn_x$ (x = 0.08, 0.35, 0.71, 0.9, 1.15 and 1.33) alloys with a C14 Laves phase structure were investigated.

The equilibrium hydrogen pressure of alloys increased as the amount of manganese increased from the study of hydrogen storage performance of $Ti_{0.45}Zr_{0.55}Ni_{0.85}V_{0.45}Mn_x$ (x = 0.18, 0.23, 0.33, 0.53 and 0.7) alloys.

It was found, that the discharge capacity of two series of alloys has a maximum then stoichiometric ratio increase.

The maximum discharge capacity for the lower nickel series has $Ti_{0.45}Zr_{0.55}Ni_{0.7}V_{0.45}Mn_{0.9}$ alloy (315 mAh/g), and for the higher nickel content series has $Ti_{0.45}Zr_{0.55}Ni_{0.85}V_{0.45}Mn_{0.23}$ alloy (363 mAh/g).

5. References

1. Yu J.-S., Lee H.-H. J. (1999) A new activation process for a Zr-based alloy as a negative electrode for Ni/MH electric vehicle batteries, *J. of Electrochem. Soc* **146**, № 12, 4366-4370.
2. Kim J.S., Paik C.H. (1998) Corrosion behavior of $Zr_{1-x}Ti_xV_{0.6}Ni_{1.2}M_{0.2}$ (M=Ni, Cr, Mn) AB_2-type MH-electrodes. *J. of Power Sources* **75**, 1-8.
3. Lee H.-H., Lee K.-Y. (1997) The hydrogenation characteristics of Ti-Zr-V-Mn-Ni C14 Laves phase alloys for metal hydride electrodes. *J. Alloys and Comp* **253-254**, 601-604.
4. Kim D.-M., Lee S.-M. (1998) Electrochemical properties of over-stoichiometric $ZrMn_{1-x}V_xNi_{1.4+y}$ alloys with C15 Laves phase, *J. Electrochem. Soc.* **145**, № 1, 93-97.
5. Kim D.-M., Jeon S.-W. (1998)A study of the development of a high capacity and high performance Zr-Ti-Mn-V-Ni hydrogen storage alloy for Ni-MH rechargeable batteries. *J. Alloys and Comp* **279**, 209-214.
6. Waterstrat R.M. (1961) The Ti-Mn system. *Trans. Amer. Inst. Mineral, Metal* **225**, 687-695.
7. Schlesinger M.E. (1999) The Mn-Zr (manganese-zirconium) system. *J. Of Phase Equilibria* **20**, №1, 79-81.

STUDY OF PHYSICO-CHEMICAL PROCESSES ON CATALYST IN THE COURSE OF SYNTHESIS OF CARBON NANOMATERIALS

D.V. SCHUR, Z.A. MATYSINA, S.Yu. ZAGINAICHENKO
Institute of Hydrogen and Solar Energy, P.O. Box 150, Kiev, 03150 Ukraine
Dnepropetrovsk State University, 49000 Ukraine
Institute for Problems of Materials Science of Ukrainian Academy of Sciences, 3, Krzhizhanovsky str., Kiev, 03142 Ukraine

The catalysts based on metals of iron group (Fe, Ni, Co) and their alloys are used at the different methods of production of carbon nanomaterials. The selection of optimal regimes of these materials production calls for the fundamental consideration of processes on the catalyst.

An investigation of distribution and thermally activated redistribution of interstitial carbon atoms about the volume and surface of crystalline films has been carried out. These crystals have bcc lattice and various types of free facets. The kinetic curves of interstitial atoms redistribution under changes of temperature have been studied. The correlation has been made between theoretical calculations and experimental data for the graphitization of Fe-C, Ni-C, Co-C alloys and Fe-C alloy with V, Cu, Ti, Co or Ni impurity. The energetic conditions and possibility of macroscopical surface segregation of interstitial atoms have been established.

1. Introduction

The production of carbon nanotubes by the laser ablation, arc vaporization of graphite, catalytic pyrolysis of hydrocarbons and CO disproportionation assumes the use of metallic catalysts, on which the carbon nanotube comes into being and continue its growth.

The nature of catalyst and electronic structure of its surface dictate in many respects the moment of carbon nanotubes nucleation and the morphology of their growth. But growth morphology of carbon nanotubes defines basically the type of tubes and their chemical and physical properties (electrical and heat conduction, strength and others). The field of tubes application depends on their type and properties.

The base of catalysts, used at present in actual practice, are metals of iron group (Fe, Co, Ni) and their alloys with different additions. These additions will assist both in dissolution of carbon atoms in the catalyst volume and in creation of sufficiently strength surface film, favourable to the synthesis of nanotubes or its inhibitory.

The change of physico-chemical properties with the introduction of one or another addition and then properly carried out heat treatment can vary the efficiency of catalyst. This can be a consequence of the redistribution of atoms, different components of alloy, the formation or destruction of order in their mutual arrangement.

The atoms redistribution over the sites and interstitial sites of crystal lattice with free surface can stimulate the processes of surface macroscopic segregation of one of

N. Veziroglu et al. (eds.), Hydrogen Materials Science and Chemistry of Metal Hydrides, 235–241.

components, forward and inverse liquation of segregation. The surface segregation of interstitial atoms is used in practice for the obtaining of certain surface physical characteristics of catalysts. The knowledge of processes rates can prove to be very important.

This paper describes a theoretical investigation of the impurity distribution of interstitial C atoms on the surface and in the volume both of the films and massive crystals AB. The thermally activated processes of interstitial atoms redistribution among the volumetric and surface film interstices are investigated theoretically and also results from the calculation of the dependences of the equilibrium concentrations of carbon in the film interstitial sites on the alloy composition, temperature, atomic order and energetic parameters of interatomic interaction are listed below.

2. Interstitial carbon atoms at the volumetric and surface interstices

The monomolecular model of the surface and the principle of detailed equilibration are used in calculations. The subject of inquiry is the film of an ordering alloy AB with the body-centred cubic structure that consists of n atomic layers with two free surfaces. It is supposed that interstitial atoms C are located in interstitial sites of one type (e.g., octahedral) with the different coordination on the surface and in the volume, i.e. the film has the volumetric V and the surface S interstitial sites.

The temperature change from the value T_1 to T_2 disturbs the thermodynamic equilibrium and initiates the redistribution of C atoms. This theory includes the premise that the redistribution of slow A, B atoms does not take place at lattice sites.

The solution of the kinetic equations for the $V \rightleftarrows S$ transitions shows the exponential dependence of ν_V, ν_S concentrations of interstitial atoms on time t

$$\begin{aligned} \nu_V(t) &= \Delta\tilde{\nu}_V \, exp\frac{-t}{t_*} + \tilde{\nu}_V \ (T_2), \\ \nu_S(t) &= \Delta\tilde{\nu}_S \, exp\frac{-t}{t_*} + \tilde{\nu}_S \ (T_2), \end{aligned} \tag{1}$$

where

$$\tilde{\nu}_V(T) = \left[1 + \frac{1}{n} exp\frac{v_S - v_V}{kT}\right]^{-1}, \qquad \tilde{\nu}_S(T) = \left[1 + n exp\frac{v_V - v_S}{kT}\right]^{-1} \tag{2}$$

are the equilibrium concentrations of interstitial atoms C at the V and S interstices; $\Delta\tilde{\nu}_V = \tilde{\nu}_V(T_1) - \tilde{\nu}_V(T_2)$; $\Delta\tilde{\nu}_S = \tilde{\nu}_S(T_1) - \tilde{\nu}_S(T_2)$ are the changes of equilibrium concentrations $\tilde{\nu}_V, \tilde{\nu}_S$ at the $T_1 \rightarrow T_2$ temperature change, v_V, v_S are the energies of C atom in the V and S interstices, which are dependent on the alloy structure and composition, atomic order and the v_{AC} and v_{BC} energetic parameters of the interatomic interaction of AC and BC pairs, v_B is the C atom energy in the vertex of potential barrier at the $V \rightleftarrows S$ transitions;

$$t^* = \frac{\tau}{4}\left[exp\frac{-(v_V - v_B)}{kT} + n exp\frac{-(v_S - v_B)}{kT}\right]^{-1}, \qquad \tau \approx 10^{-3}\,s \tag{3}$$

is the relaxation time of the redistribution process of the interstitial atoms that escape the potential barrier v_B at the $V \rightleftarrows S$ transitions.

The $\nu_V(t), \nu_S(t)$ dependences are monotonic functions in accordance with (1). The investigation of equations (1) shows that plots (Fig. 1) of $\nu_V(t)$ and $\nu_S(t)$ dependences have a greater rate of change if $v_V < v_S$ in comparison with the same if $v_V > v_S$. Therefore, from the curve slope one might suppose the value and the sign of energetic parameters v_V, v_S and the kind of this dependence (strong or weak) dictates the interaction character of AC, BC pairs.

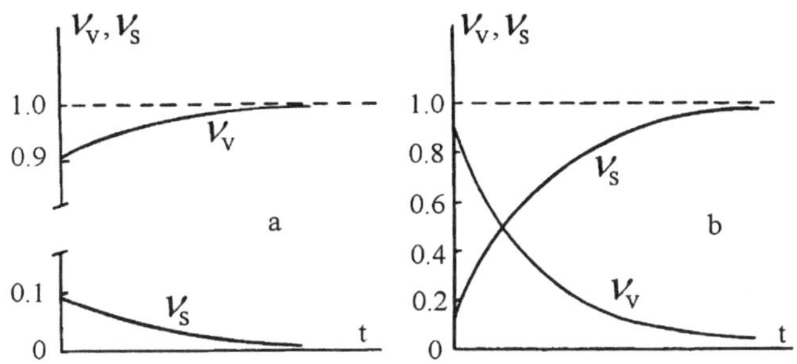

Fig. 1. Time dependence of concentrations of interstitial atoms in interstices of two types V, S in response to a sharp temperature decrease of metal thin film (n=10)
a) attraction character of interacting forces of atomic pairs AC ($v_V > v_S$),
b) repulsion character of interacting forces of atomic pairs AC ($v_V < v_S$).

3. Distribution of interstitial atoms at the interstices of three types

In certain cases, i.e. when the nodal plane (111) is the free face, the potential surface of the crystal lattice provides two types of surface interstitial sites S_1, S_2 with a different depth of the potential well (Fig. 2). Kinetic equations permit the establishing of the time dependence of the $\nu_V, \nu_{S_1}, \nu_{S_2}$ concentrations for the $V \rightleftarrows S_1$, $V \rightleftarrows S_2$, $S_1 \rightleftarrows S_2$ transitions and the extremal dependence of interstitial atoms concentrations $\nu_S(t)$ on time may be shown for interstitial sites with average depth of potential well (Fig. 3).

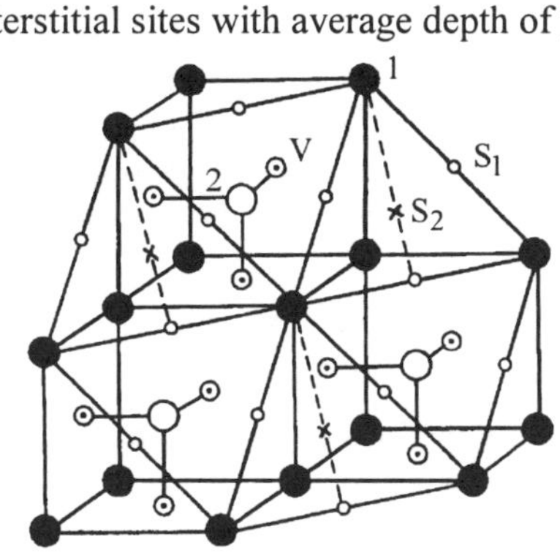

Fig. 2. Face (111) of crystal with bcc structure and upper surface layer

● - sites of first type, ○ - sites of second type,
⊙ - volume interstices V, o , × - surface interstices S_1, S_2.

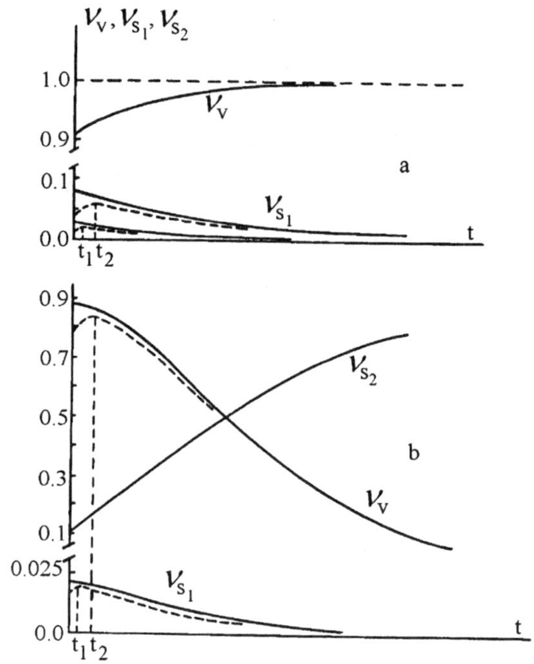

Fig. 3. Possible view of time dependences of interstitial atoms concentrations in interstices of three types V, S_1, S_2 in response to a sharp temperature decrease of metal thin film (n=10).
Dotted lines show these dependences in presence of maximum.
a) interacting forces of atomic pairs AC have attraction nature $(v_V > v_{S_1} > v_{S_2})$,
b) interacting forces of atomic pairs AC have repulsion nature $(v_V < v_{S_1} < v_{S_2})$.

The extremal dependence of concentrations (Fig. 1) on time don't take place with a distribution of interstitial atoms only in interstitial sites of two types. Therefore, as a result of experiments the presence of a maximum at the dependence $\nu_S(t)$ suggests the distribution of interstitial atoms in interstitial sites of three types.

The equilibrium concentrations (2) of interstitial atoms C, the rate or time of relaxation (3) of interstitial atoms redistribution are dependent on the *a*, b concentrations of A, B alloy components, on the long range order η in the A, B atoms distribution at the sites of crystal lattice and on the temperature T. These dependences may be elucidated if the dependences of v_V, v_S, v_B energies on *a*, b, η, T can be determined.

In the particular cases: 1) b = 0; 2) b $\ll 1$; 3) $\eta = 0$; 4) $\eta \ll 1$ the formulae for the relaxation time assume the forms

$$t_A^* = \frac{\tau}{4}\left(n\,exp\frac{v_{AC}}{kT} + exp\frac{v'_{AC}}{kT} \right)^{-1} \qquad \text{if} \quad b = 0\ , \tag{4}$$

$$t^* = t_A^*\left(1 + b\frac{n\alpha\, exp\dfrac{v_{AC} + v'_{AC}}{kT} - \beta}{n\,exp\dfrac{v_{AC} + v'_{AC}}{kT}} \right) \qquad \text{if} \quad b \ll 1\ , \tag{5}$$

$$t_o^* = \frac{\tau}{4}\left(n\,exp\frac{a v_{AC} + b v_{BC}}{kT} + exp\frac{-a v'_{AC} - b v'_{BC}}{kT}\right)^{-1} \quad \text{if} \quad \eta = 0\ , \tag{6}$$

$$t^* = t_o^*\left[1 - \frac{1}{2}\eta\frac{3n(\alpha - 2\beta)exp\frac{a v_{AC} + b v_{BC}}{kT} + (3\alpha - 7\beta)exp\frac{-a v'_{AC} - b v'_{BC}}{kT}}{n\,exp\frac{a v_{AC} + b v_{BC}}{kT} + exp\frac{-a v'_{AC} - b v'_{BC}}{kT}}\right] \quad \text{if}\ \eta \ll 1, \tag{7}$$

where

$$\alpha = (v_{AC} - v_{BC})/kT, \qquad \beta = (v'_{AC} - v'_{BC})/kT\ , \tag{8}$$

and the $v_{AC}, v_{BC}, v'_{AC}, v'_{BC}$ are the interaction energies of atomic pairs AC, BC being correspondingly at the 0,5d and 0,7d distances, where d is the lattice parameter.

The formula (4) represents the relaxation time for the pure metal A. The formula (5) permits to ascertain the exerted influence of an impurity B to the metal A on the relaxation time and it follows from this dependence that the B impurity increases the time of relaxation and reduces the rate of redistribution of H atoms among the interstitial sites in the case of

$$n\alpha\ exp\frac{v_{AC} + v'_{AC}}{kT} > \beta. \tag{9}$$

As appears from the above, doping and suitable selection of B component can change the relaxation time in the necessary direction.

The formula (6) defines the relaxation time of a disordered alloy AB of any composition. It follows from that formula that the dependence $t_o^*(a)$ can have a maximum at the composition

$$a^* = \left[ln\frac{v'_{AC} - v'_{BC}}{n(v_{AC} - v_{BC})} - 1\right] \Big/ \left(\frac{v_{AC} + v'_{AC}}{v_{BC} + v'_{BC}} - 1\right). \tag{10}$$

The knowledge of the energetic parameters for each specific disordered alloy gives the possibility of selecting the interval of concentrations with the most suitable value of the relaxation time that can permit either the acceleration or the slowing down of the redistribution process of the interstitial atoms C along the film interstices.

The formula (7) permits the obtaining of more specific information about the influence of the long range order on the redistribution process of interstitial atoms C. It follows from (7) that atomic order η decreases the relaxation time and, consequently, speeds up the process of redistribution of interstitial atoms C under the conditions

$$3n(\alpha - 2\beta)exp\frac{a(v_{AC} + v'_{AC}) + b(v_{BC} + v'_{BC})}{kT} > 4\alpha - 7\beta. \tag{11}$$

Knowledge of the energy parameters and the temperature (that must be below the Kurnakov temperature) permits the estimation of the influence of order on the relaxation time.

The formulae, determined the influence of B admixture to the metal A on the equilibrium concentrations of interstitial atoms C, at b << 1 have a form

$$\tilde{v}_V(T) = \tilde{v}_V^A\left[1 - b(\alpha + \beta)\tilde{v}_S^A\right], \quad \tilde{v}_S(T) = \tilde{v}_S^A\left[1 + b(\alpha + \beta)\tilde{v}_V^A\right], \tag{12}$$

where v_V^A, v_S^A are the such concentrations for metal A. As is obvious from (12), the B admixture decreases the volumetric concentration ν_V and increases the surface concentration ν_S of interstitial atoms C under condition $\alpha+\beta>0$. Therefore, one can regulate the process of segregation and liquation by addition and selection of B admixture.

Thus, if the values of energetic parameters for each alloy are known from independent experiments, for example, by the estimated values of the diffusion activation energies, then by means of formulae (4)-(7) one can predetermine the influence of composition, atomic order and temperature on the value of the equilibrium concentration of interstitial atoms and the relaxation time for the process of temperature redistribution of C atoms among the interstices in AB alloy. By suitable selection of alloy composition and specific atomic order in it (by means of thermal annealing), one can purposefully change the t^* value with a view to attain equilibrium quickly or, on the contrary, to freeze a definite non-equilibrium state. And that in turn will exert an influence equally upon the forming of one or another physical and chemical properties of the alloy surface and interaction of surface with carbon atoms during the process of synthesis of carbon nanomaterials.

Theoretical results for equilibrium concentrations (2), (12) of interstitial atoms, for the kinetic curves (1) and for the relaxation time (4)-(7) have been compared with experimental data concerning the graphitization of Fe-C, Co-C, Ni-C alloys and Fe-C alloy with the addition of Cr, V, Co, Ni, Ti metals [8]. The agreement of theory and experimental data has been satisfactory.

A knowledge of diffusion activation energies of interstitial atoms allow the evaluation of energetic parameters entered into theoretical equations. For example, with the assumption of diffusion by the vacancy mechanism we have for A-C alloy

$$\nu_V-\nu_S=\nu_{AC}=-Q, \tag{13}$$

where Q is the diffusion activation energy of interstitial atoms.

4. Conclusions

The comparison of theoretical and experimental plots for the volume concentration of carbon in Co-C, Fe-C, Ni-C demonstrate that the relationship between activation energies corresponds to data of independent experiments, i.e. $Q_{Ni} > Q_{Co} > Q_{Fe}$. In this case the values of activation energies are low in comparison with experiments [5-7], but our Q values are closer to experiment than in the paper [8].

Furthermore, as $Q_V < Q_{Cr} < Q_{Fe} < Q_{Ti} < Q_{Co} < Q_{Ni}$, then by our calculations V or Cr addition to Fe must decrease the surface concentration of carbon, but Ti, Co or Ni impurity must increase it. This conclusion is in complete agreement with experimental results: it is well known that addition of V or Cr to Fe inhibit and of Ti, Co or Ni to Fe accelerate the graphitization of FeMe-C alloys, where Me=H, Cr, Ti, Co, Ni [4].

The comparison of theoretical kinetic curves (1) with experimental have been performed for Fe-C, Co-C, Ni-C alloys according to the plots of the dependence of graphitization onset temperature on the rate of alloy cooling. This allows the statement that carbon atoms at their low surface concentration are arranged on free face of crystal in interstitial sites of one type and as their surface concentration is increased these atoms begin to occupy the interstitial sites of different type. This is evident from the fact of extremum appearance on the kinetic dependences of onset temperature of Co-C, Fe-C alloys graphitization on the rate of alloy cooling with increasing carbon concentration on the crystal surface.

The assessment of relaxation time in accordance with the values of energetic parameters, determined by the energy of diffusion activation, shows its strong dependence on the latter in correspondence with formulae (4)-(7). This fact substantiates the conclusion about the possibilities of regulation and control over kinetic processes by the B admixture to the metal A and their Q_A, Q_B quantities differ greatly.

The presented dependences for the equilibrium concentrations of impurity and the relaxation on time of redistribution of impurity atoms can be used for creation of polycrystalline films with resistant surface properties conserved in a wide temperature range. This is especially important for the production processes of carbon nanotubes by the method of catalytic pyrolysis of hydrocarbons and CO disproportionation.

5. References

1. Bolshov, L.A. and Natarpovich, A.P. (1977) *Uspeh.fiz.nauk* **122**(1), 125.
2. Klimenko, V.A. and Terentyev, G.E. (1983) *Uspeh.fiz.nauk* **5**(6), 1090.
3. Masharov, S.I. and Rybalko, A.F. (1984) *Fiz.met. i metalloved.* **58**(2), 213.
4. Lewis, F.A. (1967) *The Palladium-Hydrogen System*, Academic Press, New York.
5. Alefeld, G. and Volkl, J. (1978) *Hydrogen in Metals* **1-2**, Springer-Verlag, Berlin.
6. Suda, S. (1987) *Int. J. Hydrogen Energy* **12**(5), 323.
7. Nemoshkalenko, V.V., Chopovskaya, E.V. and Sheludchenko, L.M. (1993) *Int. J. Hydrogen Energy* **18**(10), 843.
8. Salli, I.V. and Lnyanoi, V.N. (1973) *Carbon on the Surface of Interstitial Solid Solutions*, Naukova dumka, Kiev.

THE INFLUENCE OF HYDROGEN-THERMAL TREATMENT ON THE DENSIFICATION AND STRUCTURE OF NITRIDES AT SINTERING

R.A. MOROZOVA, I.A. MOROZOV, T.V. DUBOVIK,
V.M. PANASHENKO, A.I. ITSENKO, N.V. DANILENKO
Institute for Problems of Material Science of NAS Ukraine,
3 Krzhyzhanovsky st., Kiev, 03142 Ukraine

1. Abstract

Positive effect of hydrogen-thermal treatment (HTT) of AlN, Si_3N_4 and BN powders on sintering was established. HTT of the powders influences on its structure and densification. It was determined activation of mass-transfer processes occurring in sintering.

2. Introduction

Low defect concentration and mobility as well as hindering of diffusion processes are known to be characteristic of crystal lattice of non-metal nitrides. These features cause considerable difficulties in obtaining the polycrystal sintered materials because of processes of diffusion-ductile flow which are responsible for mass transfer and for densification in solid phase sintering are hampered [1].

The formation of polycrystalline monophase materials in sintering of samples pressed from AlN (wurtzite modification), Si_3N_4 (structure like phenacite), BN (graphite-like modification) powders is determined by development of diffusion atomic mass-transfer processes [2].

In this work hydrogen-thermal treatment (HTT) for sintering activation of initial non-metal nitride powders was carried out. The treatment of powders with hydrogen leads to cleaning of its surface and verges of cracks. HTT increases activity of surface atoms before sintering [3-5].

3. Experimental details

In order to study the structure forming of compact materials in sintering the samples were made from AlN, Si_3N_4 and BN powders both untreated and treated with HTT. The sintering was carried out in Tamman furnace in dried nitrogen ambient at temperatures 1973 and 2073 K.

N. Veziroglu et al. (eds.), Hydrogen Materials Science and Chemistry of Metal Hydrides, 243–250.

4. Results and discussion

The samples of AlN are intensively densified in sintering at temperature 2073 K (Fig. 1). At the same time the samples pressed from untreated AlN powder are little densified in sintering because of their low defectiveness. It should be pointed out that samples treated with HTT are rather intensively densified in sintering already at temperature 1973 K. However the shrinkage of samples made from untreated AlN powder considerably decreased in sintering at 1973 K. The results of structure study are listed in Table I and Fig. 2, 3.

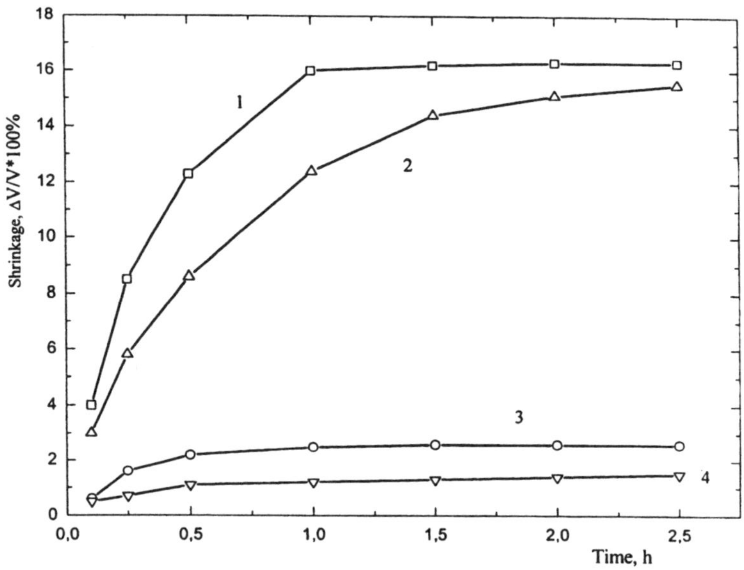

Figure 1. Dependence of volume shrinkage of AlN bricks on the sintering duration. Samples made of powders: after HTT (sintering temperature 2073 K) (1), after HTT (1973 K) (2), untreated (2073 K) (3), untreated (1973 K) (4).

It is seen from Table I that samples made of HTT treated powder are characterized by more grain mean size as well as by less content of monolithic granules. The former feature is due to more intensive grain growth and the latter is due to less concentration of granules in powder after HTT. The loosening of granules in powder occurs due to HTT resulting in increase in powder specific surface. The latter factor is also indicative of lower density of samples made from such powder as compared with samples made from untreated powder (Fig. 4). At the same time more intensive development of grain boundary diffusion processes which occur in recrystallization facilitates more densification of samples made from powder after HTT in sintering. General feature of samples micro-structure forming is that the recrystallization through gas phase takes part in this process.

The Si_3N_4 samples made of powder treated with HTT are densified in sintering better than those made of untreated powder (Fig. 5). However the shrinkage of samples

TABLE I. Grain structure of sintered samples made from AlN, Si_3N_4 and BN powders (untreated and after HTT)

Samples made from powders	Structural components, μm	Grain characteristics
Untreated AlN	0,3 - 1,0 3,0 - 5,0	Equi-axial form (80%). Granules (20%).
AlN after HTT	0,5 - 1,6 2,0 - 3,0	Equi-axial form (95%). Granules (5%).
Untreated Si_3N_4	0,2 - 0,3	Equi-axial form (90%). Anisometric (10%).
Si_3N_4 after HTT	0,1 - 0,3 1,0 - 1,5	Equi-axial form (60%). Anisometric (20%). Zone densifications, aggregates (20%).
Untreated BN	3 - 10 plane (0001)	Intercrossed twins; disoriented sections
BN after HTT	2 - 7 plane (0001)	– " " –

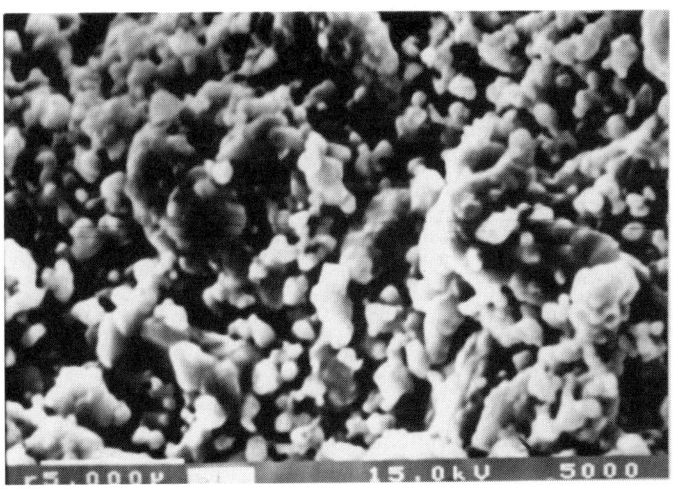

Figure 2. Typical grain structure of sintered samples made from untreated AlN powder.

made of HTT treated powder increases only in 2 times as compared with untreated powder. At the same time in the case of AlN the shrinkage increases in 7-8 times. This is caused by lower Si_3N_4 self-diffusion coefficient. The results of structure investigation of Si_3N_4 samples are shown in Table I and Fig. 6, 7.

It is evident from Table I that the samples made of HTT treated powder differ in less mean grain size, more part of anisometric grains and availability of high-dense formations from disperse grains (zone inclusions). These features are explained by structure state of powder after HTT. First of all such powder contains high-dense granules along with disperse grains (0.1-0.3 mm). Such granules do not break in

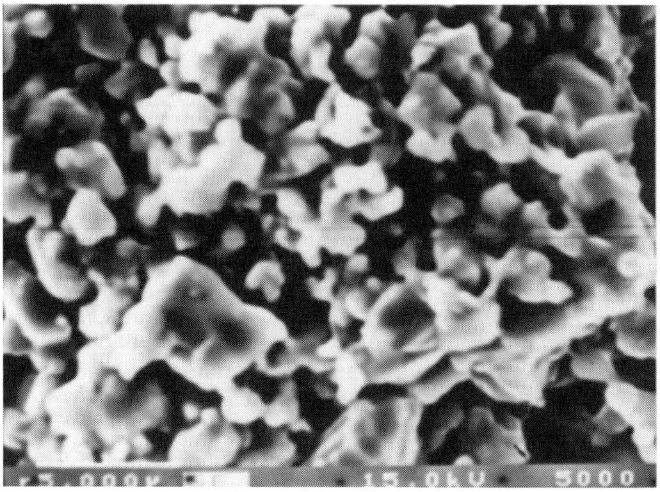

Figure 3. Typical grain structure of sintered samples made from AlN powder after HTT.

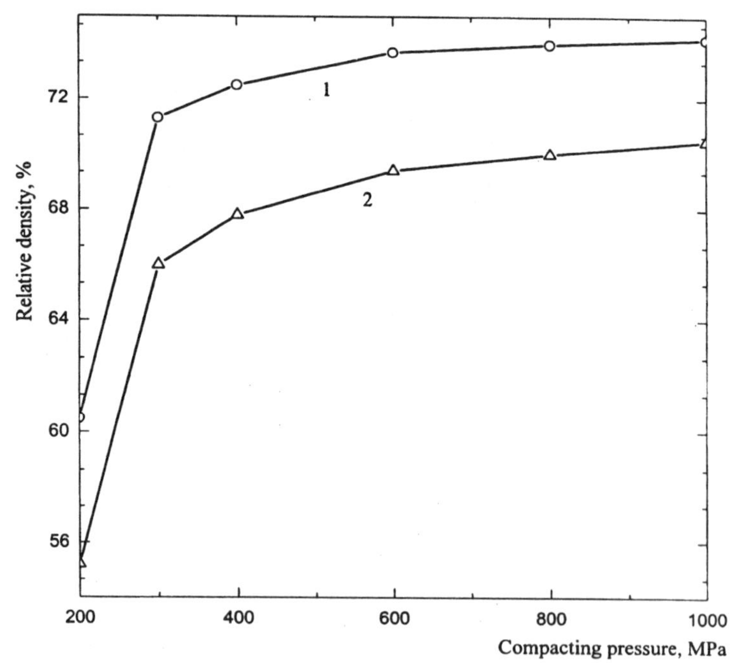

Figure 4. Relative density of pressed AlN samples (vs compacting pressure) made of powders: untreated (1), after HTT (2).

pressing and remain in polycrystalline samples after sintering. These granules determine the heterogeneity of material grain structure as well as the presence of large voids. The recrystallization through gas phase occurs in these voids resulting in forming of anisometric (as hexagonal prisms) grains. As seen from Table I the content of such grains in Si_3N_4 samples after HTT is increased.

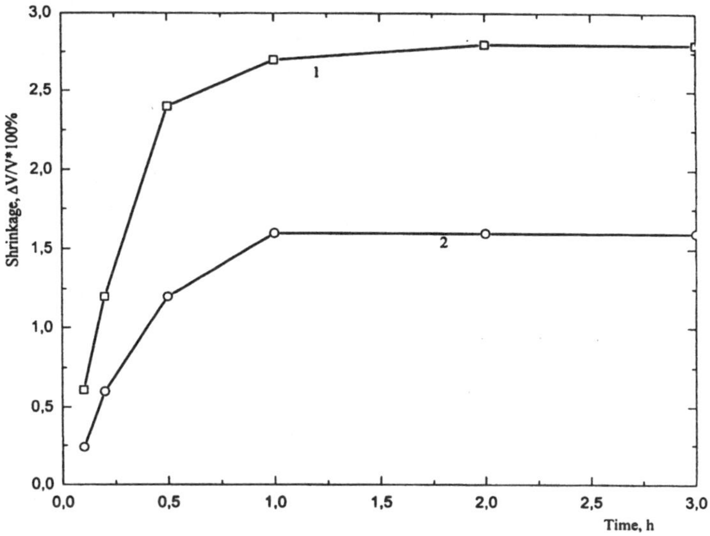

Figure 5. Dependence of volume shrinkage of Si_3N_4 bricks on the sintering duration. Samples made of powders: after HTT (sintering temperature 2073 K) (1), untreated (2073 K) (2).

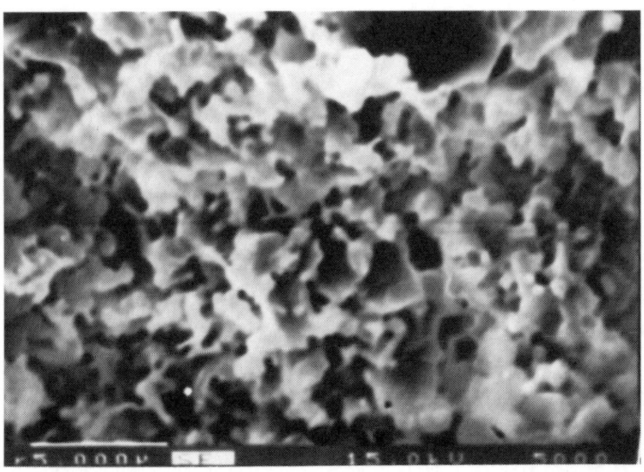

Figure 6. Typical grain structure of sintered samples made from untreated Si_3N_4 powder.

Figure 7. Typical grain structure of sintered samples made from Si_3N_4 powder after HTT.

In Si_3N_4 samples of both types the disperse grains have rounded shape. This proves that surface diffusion proceeds in sintering. To this indicates also the fact that in Si_3N_4 samples after HTT the granules (zone aggregates) surface has smoothened microrelief, i.e. separate grains are not practically revealed.

An increase in volume (negative shrinkage) occurs in sintering the BN samples (Fig. 8). This phenomenon is characteristic of materials with laminate structure and is

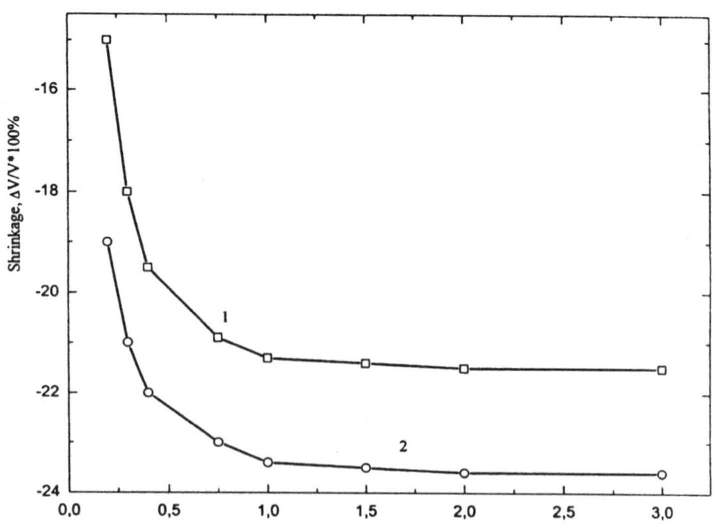

Figure 8. Dependence of volume shrinkage of BN bricks on the sintering duration. Samples made of powders: after HTT (sintering temperature 2173 K) (1), untreated (2173 K) (2).

explained by grain growth in sintering. The results of BN samples structure investigation are shown in Table I and Fig. 9, 10.

General features of micro-structure of both BN type samples are texture break; defectiveness of grains (twins, shifts by basal layers), grain accretion by prismatic and basal planes, grain growth. The analysis of BN samples micro-structure reveals that material made of HTT treated powder has the following features: less grain size is characteristic of it, grain accretion by basal planes is more intensive, texture break is less evident.

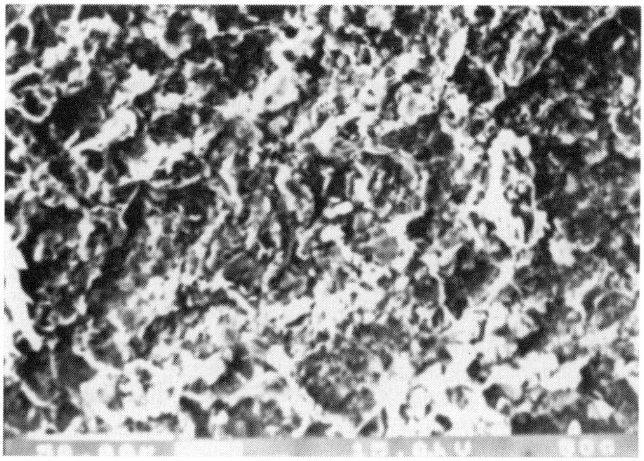

Figure 9. Typical grain structure of sintered samples made from untreated BN powder.

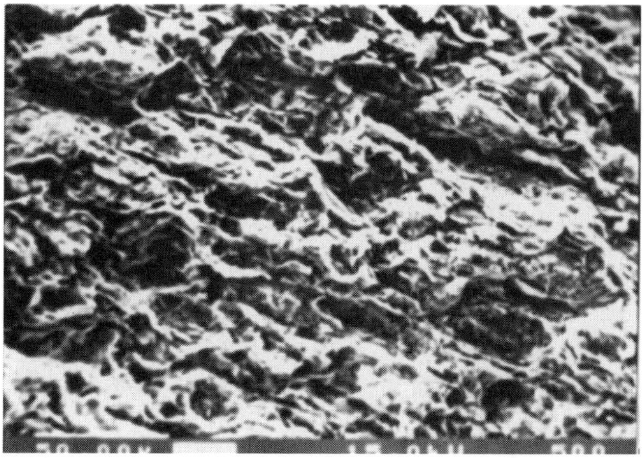

Figure 10. Typical grain structure of sintered samples made from BN powder after HTT.

The above features cause a higher density of HTT-treated BN samples in sintering. The decisive factor is the grain accretion by basal planes that is explained by enhanced BN activity due to HTT. Since the grain accretion due to recrystallization is insignificant the nature of such activity consists in the development of surface diffusion processes.

5. Conclusion

The technologies of receiving compact materials from AlN, Si_3N_4 and BN powders after hydrogen-thermal treatment were elaborated. HTT of powders influences the structure and densification processes occurring in sintering. The positive effect of HTT on AlN, Si_3N_4 and BN sintering is determined by activation of mass-transfer processes. The same processes determine also the grain structure formation.

6. References

1. Samsonov GV. Non-metal Nitrides. M: Metallurgiya. 1969. (in Russian)

2. Uskokovich DP, Samsonov GV and Ristich MM. Activated sintering. Beograd: International Institute of Sintering Science. 1974.

3. Trefilov VI, Morozov IA, Morozova RA et al. Effect of Hydrogen on improving purity of WC and AlN Powders. Int J Hydrogen Energy 1996;21(11-12):1097-1099.

4. Trefilov VI, Morozov IA, Satanin SV et al. Effect of chemicothermal treatment of AlN powders on structure and properties of hot-pressed ceramics on its basis. In: High-temperature Nitrides. Kiev: Izd IPM NANU. 1996. p. 56-65. (in Russian)

5. Morozova RA, Morozov IA, Itsenko AI et al. Study the influence of hydrogen-thermal treatment on the structure state of non-metal nitride powders. In: Abstracts of NATO Int Conf «Hydrogen Materials Science and Chemistry of Metal Hydrides». ICHMS'99. 1999. p. 254-255.

THERMAL-RADIATION PROCESSES IN THE Zr-Ni-H_2 SYSTEM AND SYNTHESIS OF HYDRIDES

H.G. HAKOBYAN, S.K. DOLUKHANYAN, V.SH. SHEKHTMAN, K.A.ABRAHAMYAN, N.G. TAMAMYAN[(1)], N.L.MNATSAKANYAN
Institute of Chemical Physics NAS of Armenia,
P. Sevak St., 5/2, Yerevan, 375044, Armenia
[(1)]Yerevan Physics Institute
Alikhanyan Br. St., 2, Yerevan, 375036, Armenia

Abstract

The processes of hydrogenation of intermetallic compounds Zr_2Ni and ZrNi under action of the accelerated electron beam were studied. It was shown that in conditions of thermal-radiation synthesis and of "cold synthesis" in hydrogen, the hydrides Zr_2NiH_{5-x} and $ZrNiH_3$ were formed. The main parameters (dispersity of an intermetallic compound, dose rate, total radiation dose, etc.), providing the radiation-stimulated hydrogenation were obtained. The probable mechanism of thermal-radiation processes in systems an intermetallic compound - hydrogen was discussed.

1. Introduction

There is a restricted quantity of works on the accelerated electrons effect on the process of intermetallic compounds hydrogenation [1,2]. In the work [1], the absorption of hydrogen by an intermetallic compound $Zr_{67}Fe_{33}$ was studied at irradiation. It was noted, in particular, that the structure of obtained after irradiation compound was similar to that of products of hydrogenation in thermal conditions. But there were essential differences in hydrogen emission. In another work [2], the catalytic properties of the synthesized under electron beam hydrides, $ZrNiH_{2.8-x}$ and $ZrCoH_{2.8-x}$, were investigated.

In the present paper, the results of systematic investigation of hydrogen interaction with intermetallic compounds on the basis of zirconium and titanium under action of accelerated electron beam are presented.

2. The Experimental Technique

For a sample irradiation by accelerated electron beam, the linear accelerator LAE-5 with energy 5 MeV and current up to 450 μA was used. The samples under investigation were pressed from a dispersed intermetallic alloy to tablets (diameter 20

N. Veziroglu et al. (eds.), Hydrogen Materials Science and Chemistry of Metal Hydrides, 251–258.

mm and width no more of 5 mm), and placed into the camera specially designed for these experiments. The volume of the camera was 3 l, it was supplied by remotely controlled flow valves for both filling and reset of gas and for connection to the vacuum system. The temperature was controlled during all processes: irradiation, radiation heating up, synthesis and cooling of the product by a platinum to platinum-rhodium thermocouple located in the sample center.

The initial intermetallic compounds ZrNi and Zr_2Ni beforehand were synthesized by a self-propagating high-temperature synthesis method (SHS) from zirconium and nickel powders, correspondingly, 99.90mass% and 99.90 masses % [3].

The hydrides, obtained during the experiments, were compared to hydrides of the same intermetallic compounds obtained earlier by SHS method in our laboratory [3]. In the last case the maximum concentration of SHS-hydride, $ZrNiH_3$, was 1.96mass% , and of hydride Zr_2NiH_5 - 2.03mass%.

The products of hydrogenation were analized on hydrogen content by burning a shot in dry oxygen flow. X-ray analysis of products was performed on the equipment DRON-0.5 on Cu K_α -radiation.

3. Results and Discussion

3.1. THE SYSTEM Zr_2Ni-H_2.

The initial intermetallic compound Zr_2Ni had the following structural characteristics: a tetragonal lattice with parameters $\underline{a}$ = 6.540 and $\underline{c}$ = 5.340 Å.

During preparation of initial sample of intermetallic compound Zr_2Ni, the special attention was paid to a dispersity of a metal matrix, as this intermetallic compound, as was shown earlier, during SHS-hydrogenation process can:

a) To create a single-phase IMC hydride by reaction

$$Zr_2Ni + H_2 \rightarrow Zr_2NiH_{5-x}$$

(at hydrogen pressure < 4 atm and particle size of 0.2 – 1.4 mm; the developed hydrogenation temperatures do not exceed 120 - 400°C);

b) To undergo to hydrogenolysis – the process of IMC matrix decomposition to several phases under effect of hydrogen:

$$Zr_2Ni + H_2 \rightarrow ZrH_2 + ZrNiH_{3-x},$$

where x < 1; in this case the hydrogenation temperature are more than 400°C.

Therefore, to prevent the hydrogenolysis during hydrogenation both by the SHS mode and under the irradiation, the optimal IMC sizes, 0.3-1.2 mm, were used. In this case, the temperature of the hydrogenation process should not exceed 400°C and the unique product of IMC Zr_2Ni interaction with hydrogen was the hydride Zr_2NiH_{5-x}.

The thermal-radiation synthesis (TRS) was carried out according the following scheme: a sample of intermetallic compound was evacuated in the camera, then the camera was filled by gaseous hydrogen and the electron beam was switched on. The effect of electron beam on the reaction IMC + H_2 was investigated in the interval of dose rates 0.025 - 0.6 Mrad/sec. The thermograms of TRS process are similar to them of other Me-H_2 systems [4-6].

In a Figure 1 a typical thermogram of TRS process at dose rate 0.2 Mrad/sec is presented. A smoothly rising of temperature in the beginning corresponds to radiation heating of sample - with increase of a radiation dose temperature of sample grows. Then, a sharp temperature jump appropriate to a course of exothermal reaction of IMC interaction with hydrogen follows. If to stop the irradiation at once, the smooth lowering of temperature would follow and an intensive cooling of sample in hydrogen would occur.

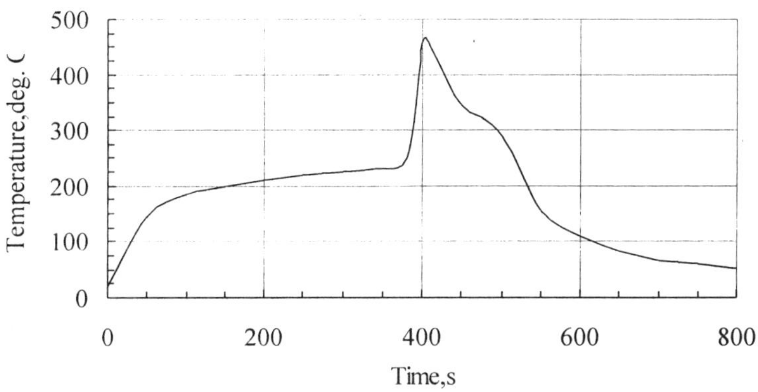

Figure 1. The thermogram of TRS in Zr_2Ni - H_2 system at dose rate 0.2 Mrad/sec.

It is important to note that the hydrogenation reaction took place at rather small radiation doses (10 Mrad) and dose rates. The lower limit of dose rate for reaction in TRS conditions was 0.025 Mrad/sec. The temperature of hydrogen interaction with IMC did not exceed 600°C. The synthesized hydride contained 1.47-1.85 mass.% of hydrogen.

The characteristics of TRS process in the system Zr_2Ni - H_2 and of the obtained products are presented in Table 1.

One can see that the high radiation doses (100 Mrad) at radiation dose rates between 0.2-0.4 Mrad/sec did not promote the synthesis of a hydrogen-rich product. On the contrary, the most hydrogen-rich hydride of an intermetallid was synthesized at rather low radiation doses (9-15 Mrad). At higher radiation doses, apparently, a baking of created dot defects and a loose of part of the active centers occurred.

In experiments in the mode of "cold synthesis" (CS), the samples of IMC Zr_2Ni were evacuated in the camera, irradiated with an electron beam in vacuum, then were cooled to low temperatures (≤100°C). Afterwards, the camera was filled by hydrogen (1 atm) and the exothermal reaction (350-640°C) started. The experiments were carried out in the interval of dose rates 0.05-0.6 Mrad/sec up to identical total dose 100 Mrad. The final product contained till 1.97 mass. % of hydrogen.

TABLE 1. The characteristics of TRS in the system Zr_2Ni - H_2

Dose rate, Mrad/sec	Dose, Mrad	$T_{begin.}$, °C	T_{max}, °C	Hydrogen concentration, *wt.*.%	Phase composition
0.6	100	255	580	1.79	$Zr_2NiH_{4.40}$
0.4	100	220	542	1.63	$Zr_2NiH_{4.01}$
0.2	100	230	462	1.47	$Zr_2NiH_{3.69}$
0.1	11	110	472	1.85	$Zr_2NiH_{4.55}$
0.05	12	85	423	1.91	$Zr_2NiH_{4.70}$
0.025	15	55	435	1.85	$Zr_2NiH_{4.55}$
0.025	9	55	330	1.80	$Zr_2NiH_{4.42}$

The characteristics of CS process in the system Zr_2Ni-H_2 are presented in Table 2. It have to note, that in CS mode, the metal matrix contacted to hydrogen after irradiation and cooling in vacuum down to 100°C and below. After hydrogen filling, the thermocouple fixed temperature jump evidencing start of chemical reaction IMC + H_2. The reaction with hydrogen was initiated by radiation defects arisen at irradiation in vacuum and saved at cooling.

TABLE 2. The characteristics of CS in the system Zr_2Ni - H_2

Dose rate, Mrad/sec	Dose, Mrad	$T_{heat.}$, °C	$T_{cool.}$ and H_2 filling, °C	$T_{max.}$ of CS T_{CS}, =, °C	Hydrogen concentration, *wt.*%	Phase composition
0.6	100	710	100	596	1.96	$Zr_2NiH_{4.82}$
0.4	100	510	100	640	1.90	$Zr_2NiH_{4.68}$
0.2	100	360	35	465	1.91	$Zr_2NiH_{4.70}$
0.1	100	365	100	623	1.91	$Zr_2NiH_{4.70}$
0.05	100	250	100		Reaction absent	

Let's note, that during cooling, the loose of radiation defects and of a matrix activity is possible. Therefore in a number of the experiments presented below, the contact to hydrogen was carried out at maximal temperatures of radiation heating. For each dose rate, this temperature was determined experimentally (Table 3). During experiments in these conditions, the unusual low threshold of radiation dose (2.9 Mrad) and of dose rate (0.007 Mrad/sec) were observed.

TABLE 3. Hydrogenation of preliminary irradiated in vacuum IMC

Dose rate, Mrad/sec	Dose, Mrad	Temperature of contact with H_2, $T_{contact}$, °C	T_{max} of reaction $T_{react.}$, °C	Hydrogen concentration, wt.%	Phase composition
0.2	54	440	540	2.03	$Zr_2NiH_{5.00}$
0.1	27	275	550	1.97	$Zr_2NiH_{4.85}$
0.05	13.5	180	580	2.02	$Zr_2NiH_{4.97}$
0.025	7.0	95	523	1.92	$Zr_2NiH_{4.72}$
0.0125	3.87	70	510	1.94	$Zr_2NiH_{4.77}$
0.007	2.9	45	340	1.95	$Zr_2NiH_{4.80}$

The data in Tables 2 and 3 it evidence, that in the experiments when a preliminary radiation processing in vacuum was performed, the temperature of reaction was higher of the optimal range 120-400°C. However in these cases, single-phase IMC hydrides $Zr_2NiH_{4.80-5.00}$ were formed, and the hydrogenolysis reaction was not observed. It is known that the interaction of defects caused by irradiation with them existing in a crystal lattice before irradiation frequently resulted in a hardening of metal crystal [7].

May be, therefore, though the reaction temperature was high, the initial alloy did not break up and the hydride of intermetallid was synthesized.

The above described experiments have allowed to conclude, that the preliminary evacuation of a sample in a combination to irradiation by electron beam promoted rise of a metal matrix reactivity, therefore a more hydrogen-rich IMC hydride was synthesised.

The finial IMC hydride was single-phase on the composition, crystal lattice of the synthesized hydride was tetragonal, $\underline{a}$=6.860; $\underline{c}$ = 5.657 Å. The hydrogen concentration in it achieves 1.95-2.03 mass.%, corresponding to the formula $Zr_2NiH_{4.80-5.00}$.

3.2. THE ZrNi-H_2 SYSTEM.

The initial intermetallic compound ZrNi had the following structural characteristics: orthorhombic lattice with parameters $\underline{a}$ = 3. 290; $\underline{b}$ = 9.998; $\underline{c}$ = 4.080 Å.

For experiments in TRS conditions, the samples of alloy ZrNi were exposed to the electron beam of dose rates between 0.1-1.00 Mrad/sec in hydrogen atmosphere, 1 Atm.

On the thermograms of all experiments, the radiation heating of a metal matrix (300-650°C) was registered without characteristic for chemical reaction of temperature jump. However, if the electron beam was turned off on a segment of temperature curve appropriate to cooling of the sample, a characteristic for given IMC plateau near 200-230°C occurred. This fact testifies the sorption of H_2 by metal matrix and the creation of a hydrogen-containing phase. The result product has the following parameters: H_2

concentration - 1.69 mass.%, orthorhombic crystal lattice with parameters $\underline{a}$ = 3.528 Å; $\underline{b}$ = 10.468 Å; $\underline{c}$ = 4.302 Å.

The behavior of IMC ZrNi at irradiation in CS conditions a little differed from that of IMC Zr_2Ni. The irradiation of ZrNi was carried on by dose rates of 0.3 - 0.8 Mrad/sec, up to total doses 50-150 Mrad. The experiments have shown, that CS took

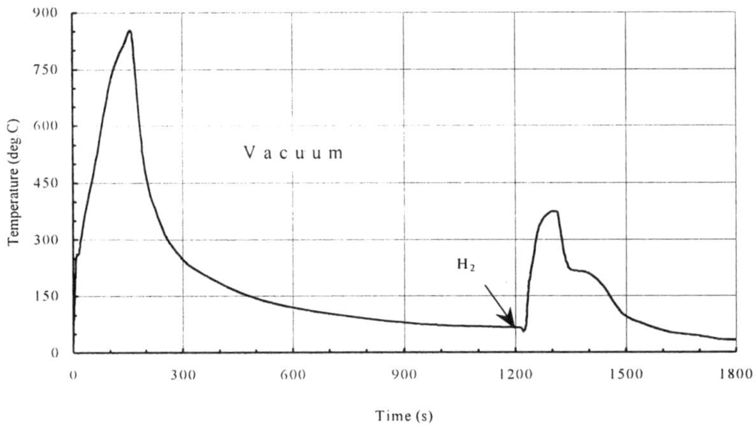

Figure 2. The thermogram of CS in the system ZrNi - H_2 at dose rate of preliminary irradiation 0.5 Mrad/sec.

place at observance of two conditions: 1) the temperatures of radiation heating in vacuum should not be lower than 600°C; 2) the camera have to be filled by H_2 at temperatures not lower than 70°C. After H_2 filling up to 1 atm, a typical for CS thermogram was registered (Fig. 2).

Similar to CS mode, the experiments were performed on exclusive thermal activation of a metal matrix in vacuum followed by hydrogen filling into the reactionary space. ZrNi sample was evacuated and heated in a furnace up to 800°C. Then the sample was cooled in vacuum down to 180°C and the hydrogen was injected into the system. After cooling, the product was analyzed. A partial hydrogenation occurred: product was two-phase - ZrNi + $ZrNiH_3$, hydrogen content - 1.43 mass. %. Thus, it was shown that in conditions of only thermal activation, hydrogenation of IMC ZrNi was incomplete.

Obtained in CS conditions hydride was single-phase, hydrogen content in it was 1.96 mass%, corresponding to the formula $ZrNiH_3$. It has the following structural characteristics: orthorhombic crystal lattice with parameters $\underline{a}$ = 3.501; $\underline{b}$ = 10.465; $\underline{c}$ = 4.30 Å. The reflections of peaks on diffractograms in comparison with TRS product are wider, apparently, evidencing nano-scale dimensions of the obtained products.

4. Discussion

The performed experiments have allowed to conclude, that under effect of radiation in systems ZrNi-H_2 and Zr_2Ni -H_2, a hydrogenation reaction occurred. However a hydrogenation of IMC in TRS conditions was worse than in other two modes, the final products contained less hydrogen. Why the concentration of hydrogen in TRS products was lower, than in CS products? What is a role of the defect nature in chemical interaction of reagents? The answers to these problems while are unclear. Apparently, for reactivity of a metal it is essential not only the quantity of the created defects, but also their configuration. One of the possible variants of the answer on these problems is the following:

Among the reactive centers, it is necessary to select 1) the defects, existing in the initial metal matrix, including the thermal and 2) the radiation defects of different nature generated by a beam (shifts and local moving of atoms, vacancies and their associations, dislocations, etc.).

If the atoms in a heteroatomic crystal differed by masses under radiation at first, light atoms would be shifted, and then - the heavy [8]. As our experiments testified, the radiation defects play a dominating role at hydrogenation under irradiation. They considerably reduced energy barriers of hydrogenation reaction and promoted more soft initiation of IMC + H_2 reaction. Let's remark, that in IMC the active centers (defects, generated by electron beam) are created already at 0.007 Mrad/sec.

Apparently, at experiments in TRS conditions, i.e. at irradiation in the presence of gaseous hydrogen, the initially generated radiation defects enter to the chemical reaction at once, not allowing the accumulation and the development of more composite and more reactive defects in metal matrix of IMC. Probably, the obtained low absorptive characteristics of TRS products might be explained by such events. In conditions, when the radiation processing was carried out in vacuum (CS), and the IMC contact with hydrogen happened later, i.e. when more composite defects were already generated, and, therefore, metal matrix became more reactive, the products of reaction were saturated with hydrogen in the greater degree. Besides these composite configurations of defects, apparently, have also greater "lifetimes", and saved their reactivity allowing to realize the interaction later after cooling down to lower temperatures.

5. Conclusions

Under effect of the accelerated electron beam, the hydrides $Zr_2NiH_{4.80\text{-}5.00}$ and $ZrNiH_3$ are synthesized. At comparing the regimes radiation processing of metal became apparent, that on a degree of active reaction of a matrix with hydrogen, the most favorable conditions were created at post - radiation hydrogenation. It means, that the preliminary activation of a hard phase by electron beam played an important role for reaction of metal matrix with hydrogen. The metal matrix of IMC sorbs hydrogen more actively at the radiation stimulation carried out beforehand in vacuum.

Acknowledgements. The work has been supported by ISTC. Project A-192.

6. References

1. Adigamov, B.Ja., Lunin, V.V. et.al. (1989) Features of behaviour of hydrides of intermetallic compounds on the basis of zirconium at an irradiation by a beam of the accelerated electrons, *Zhurnal fizicheskoi khimii*, **11,** 3119-3123.
2. Solovetsky, Yu.I., Lunin, V.V., Panteleev, I.M., and Panteleev, D.M. Formation and Decomposition of Zr-Me-Hx (Me=Ni, Co, Fe) Hydrides under the beam of accelerated electrons with the energy up 2,0MeV. Abstract Booklet of Int. Symp. on Metal Hydrogen Systems. Fundamentals and Applications. Les Diablerets, Switzerland. Aug.25-30,1996. F5:28P.
3. Hakobyan, H.G., Dolukhanyan, S.K et.al. (1983) Combustion synthesis of hydrides of intermetallic compounds in Zr-Ni system. *Izvestia AN SSSR. Seria "Neorganicheskie materiali"*, **19** (6)**,** 881-885.
4. Dolukhanyan, S.K., Shekhtman, V.Sh., Aghajanyan, N.N., Abrahamyan, K.A., Harutyunyan, Kh.S., Aleksanyan, A.G., Hakobyan, H.G., and Ter-Galstyan, O.P. (2000) Radiation-thermal synthesis of titanium hydride. *Khimicheskaya Fizika*, **19** (12), 21-25.
5. Aleksanyan, A.G., Dolukhanyan, S.K., Shekhtman, V.Sh., Abrahamyan, K.A., Hayrapetyan, V.S., Mnatsakanyan, N.L. (2001) Radiation-chemical synthesis of titanium hydridonitrides. *Khimicheskaya Fizika*, **20**(1), 32-36.
6. Aghajanyan, N.N., Dolukhanyan, S.K., Shekhtman, V.Sh., Harutyunyan, Kh.S., Abrahamyan, K.A.and Ter-Galstyan, O.P. (2001) Thermal-radiation synthesis of titanium carbohydrides. *Khimicheskaya Fizika*, **20**(3): 40-44.
7. Shalaev, A.M. (1972) *Radiation stimulated diffusion in metals,* Moscow, Atomizdat,
8. *Radiation damage of refractory compounds.* (1979) Ed. M.S. Kovalchenko, V.V. Ogorodnikov, Ju.I.Rogovoi and A.G. Kraini, Moscow, Atomizdat,.

INFLUENCE OF THE TEMPERATURE ON HYDROGEN INTERACTION WITH $LaNi_{5-X}Co_X$ ($2 \leq X \leq 3$)

E.A.BERDONOSOVA, N.A. YAKOVLEVA, K.N. SEMENENKO, N.V. BYKOVA
Department of Chemistry, Moscow State University
Leninskie gory, Moscow, 119899, Russia

Abstract

The influence of synthesis conditions on thermodynamic properties of hydrides of $LaNi_{2.5}Co_{2.4}Mn_{0.1}$ and $LaNi_2Co_3$ intermetallic compounds has been studied by means of the Tian-Calvet differential heat-conduction microcalorimetry. Two series of experiments were carried out at 308 and 353 K. The thermodynamic parameters in the $LaNi_2Co_3$ — H_2 system depend sufficiently on thermal prehistory of the intermetallic. It can be explained by possibility of different ways of transition metal ordering in their $CaCu_5$-type lattice. The characteristics in $LaNi_{2.5}Co_{2.4}Mn_{0.1}$ — H_2 system being invariable indicates only one way of transition metals ordering.

1. Introduction

Upon study and use of $LaNi_5$-based intermetallics, the requirement that the absorption characteristics are independent on time, number of hydrogenation-dehydrogenation cycles, preliminary heat treatment, etc., is crucial. The most important factors which provide such stability are the chemical nature of substituent, and the extent of its ordering in the transition metal sublattice.

The degree of order of the initial IMC is particularly dependent on synthesis conditions. Fast heating or cooling usually results in complete disorder of transition metal atoms in the lattice while long annealing at constant temperature yield ordered structures [1]. Moreover, hydrogen atoms affect transition metal sublattice upon hydrogenation and thus enhance ordering.

Due to peculiarities of the $CaCu_5$ structure type, the influence of hydride synthesis conditions on IMC absorption ability could exhibit itself most pronouncedly in the $LaNi_{5-x}M_x$—H_2 systems, where M is a transition metal with a radius similar to that of Ni, but different hydrogen affinity (Co, Cu), and x is close to 2 or 3.

2. Peculiarities of the $CaCu_5$ structure type

It is known that $LaNi_5$ lattice has two quite different transition metal positions. Upon doping $LaNi_5$ with transition elements, substitution can proceed both statistically and in ordered way. In the latter case one of 3g positions (equatorial plane) or 2c positions (basal plane) will be preferred (Fig. 1). Since the number of 3g and 2c positions have 2:3

N. Veziroglu et al. (eds.), Hydrogen Materials Science and Chemistry of Metal Hydrides, 259–266.

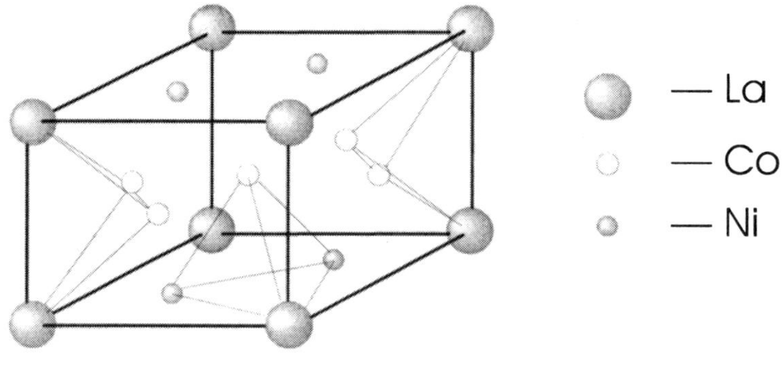

1

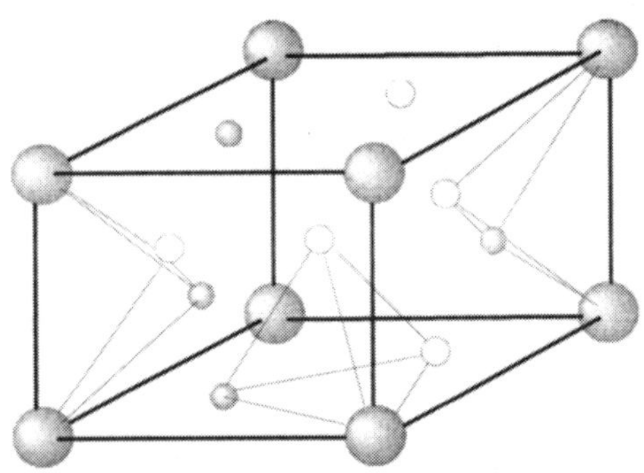

2

Fig. 1. The different versions of transition metal sublattice in $LaNi_2Co_3$ intermetallic compound: (1) — ordered arrangement, (2) — statistical arrangement.

ratio, such isomerism would most likely lead to $LaNi_2M_3$ and $LaNi_3M_2$ stoichiometry. Evidently, the ordered and disordered forms would have similar X-ray difractograms, but different physical, chemical and thermodynamic properties.

At least two structures for IMC of such composition, which can belong to one or different structure types, are described in previous works [2-6]. Whether one or other type of structure is realized depends on hydrogenation conditions and sample thermal prehistory. As it can be anticipated, the physical properties of these modifications (magnetic sensitivity, specific heat, electric resistance) as well as hydrogen absorption characteristics (equilibrium plateau, hydrogen capacity) differ drastically. In particular [7], two various sets of isotherms in $LaNi_3Cu_2$ — H_2 and $LaNi_2Cu_3$ — H_2 systems are obtained dependent on preliminary thermal and hydrogen treatment.

In spite of variety of possible substituents, only Co atoms (besides Cu) substitute Ni atoms forming solid solutions in $LaNi_5$ lattice. And intermetallic compounds, which absorb hydrogen at relatively low temperatures and pressures, are formed. The atom radii of Ni and Co are more close (R_{Ni}=1.24 Å, R_{Co}=1.25 Å), then those of Ni and Cu, but Co content in IMC $LaNi_{5-x}Co_x$ has a great influence on phase composition of obtained hydrides and hydrogen interaction thermodynamics. Nevertheless, as literature analysis [4, 6] indicates, one can expect existence, at one temperature, of at least two forms of same compound with completely different properties if Co:Ni ratio is between 2 and 3.

Hence, it seemed interesting to investigate the influence of hydride synthesis conditions in the $LaNi_{5-x}Co_x$ — H_2 systems, where $2<x<3$, on the fine structure of transition metal sublattice by methods which allow to monitor dynamics of transition metal atoms ordering or disordering.

3. Preparation of initial IMC and investigation technique

The initial intermetallic compounds $LaNi_{2.5}Co_{2.4}Mn_{0.1}$ and $LaNi_2Co_3$ was obtained by arc-melting the high purity metals (La – 99.8%, Ni – 99.96%, Co – 99.90%) under flow of purified argon. Manganese was introduced in the form of Ni—Mn ligature.

The desired composition has confirmed by chemical analysis. The X-ray phase analysis showed that the samples to be single-phase with hexagonal $CaCu_5$ structure. The lattice parameters are a=5,049 Å, c=3,998 Å ($LaNi_{2.5}Co_{2.4}Mn_{0.1}$) and a=5,056 Å, c=3,991 Å ($LaNi_2Co_3$).

The extra pure hydrogen used with a very low (< 10^{-5} vol %) concentration of impurities, was obtained by the desorption from the hydride phase $LaNi_{4.75}Al_{0.25}H_x$.

To investigate the influence of thermal and hydrogen treatment on the fine structure of transition metal sublattice in a $CaCu_5$-type intermetallide the conditions closest to isothermal are required. The combination of Tian-Calvet heat-conduction microcalorimetry with technique of titration by small hydrogen portions is most optimal. The isothermal conditions are maintained due to calorimeter design and minimization of the local non-isothermal nature of the hydrogenation process when hydrogen portions are ca. 0.0002-0.0003 mol. Moreover, this method permits to obtain calorimetric data simultaneously with the P-C-T- diagram and to monitor dynamics of transition metal ordering in the lattice. The technique applied was previously described in detail in [8].

To investigate the influence of hydride synthesis conditions on its thermodynamics two series of experiments were carried out for $LaNi_{2.5}Co_{2.4}Mn_{0.1}$ and $LaNi_2Co_3$ intermetallic compounds.

In the first series, isotherms and calorimetric data were obtained after 5-10 cycles of hydrogen absorption and desorption at 308 and 393 K. The second series were conducted after a sharp temperature shift (increase or decrease). Such treatment is known to cause disorder in the transition metal sublattice.

4. Results

4.1. THE $LaNi_{2.5}Co_{2.4}Mn_{0.1}$ — H_2 SYSTEM

For $LaNi_{2.5}Co_{2.4}Mn_{0.1}$, there is an invariable equilibrium pressure plateau and hydrogen capacity on the isotherms. The P-C-T-diagram is stable unrelated to time, number of cycles and thermal prehistory (fig. 2). The isotherms at 308 and 353 K obtained during first as well as second series of experiments are similar. It is interesting to note an only hydride forms in absorption process. That fact agrees with literature data [4, 6], but is not typical for a cobalt-rich IMC. It looks most likely that in 2.5:2.5 stoichiometry, an only one way of ordering is possible due to probability of atom entering 2c or 3g position in $CaCu_5$ crystal structure being almost the same.

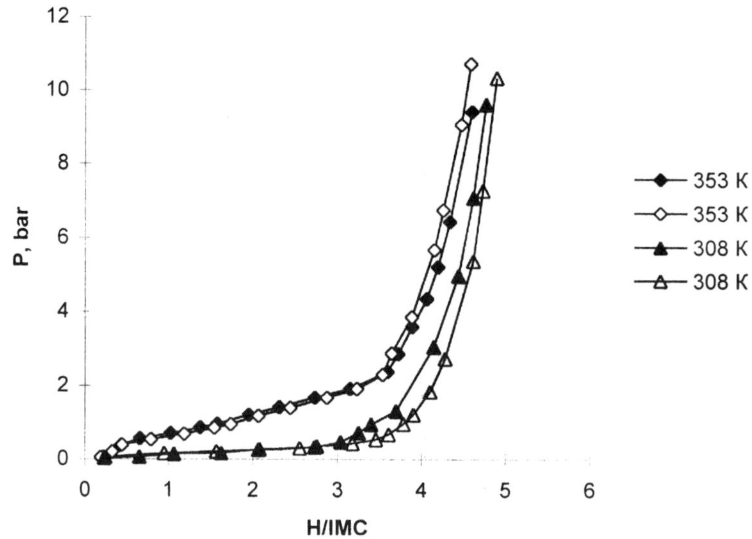

Fig. 2. The hydrogen absorption isotherms in the $LaNi_{2.5}Co_{2.4}Mn_{0.1}$—$H_2$ system, that obtained after multiple absorption-desorption cycling at 308 and 353 K (empty rhombi and triangles) and after sharp temperature shift (filled rhombi and triangles).

4.2. THE $LaNi_2Co_3$ — H_2 SYSTEM

The IMC $LaNi_2Co_3$ absorption and desorption isotherms obtained after 10 cycles of hydrogenation-dehydrogenation at 308 and 353 K are presented in fig. 3. The thermodynamic parameters and absorption characteristics in the $LaNi_2Co_3$—H_2 system obtained in the first series are reproducible and agree well with the literature data [4, 6]. On the P,C,T-diagrams there are two plateaus corresponding to formation of $LaNi_2Co_3H_{3.2}$ and $LaNi_2Co_3H_4$. The equilibrium pressure of first hydride was 0.3 bar at 308 K and 1 bar at 353 K, and for second hydride — 1.5 bar at 308 K and 9 bar at 353 K. Owing to the temperature change the second plateau equilibrium pressure was increased greater then that of first plateau. No hysteresis observed, which is typical for cobalt-substituted systems.

The second series of experiments was carried out at 308 K after the sharp temperature shift from 353 K and at 353 K — after the sharp temperature shift from 308 K. The isotherms obtained at 308 K in the first series of experiments and those obtained after temperature decreasing were similar. Two hydrides — $LaNi_2Co_3H_{3.2}$ and $LaNi_2Co_3H_4$ — were also observed. The enthalpies of hydrogenation and dehydrogenation at 308 K obtained in course of first as well as second series were equal within experimental limits. One can mention but a slight increase of equilibrium pressure of second plateau (on 0.3 bar) and slowing down of the hydrogenation-dehydrogenation processes after high temperature treatment.

A different situation was observed at 353 K. Fig. 4 represents two isotherms for the same $LaNi_2Co_3$ sample: (1) — after multiple absorption-desorption cycling at 353 K,

(2) — after sharp temperature shift from 308 to 353 K. The main difference of second type isotherm is that the second plateau disappears and hydrogen capacity increases.

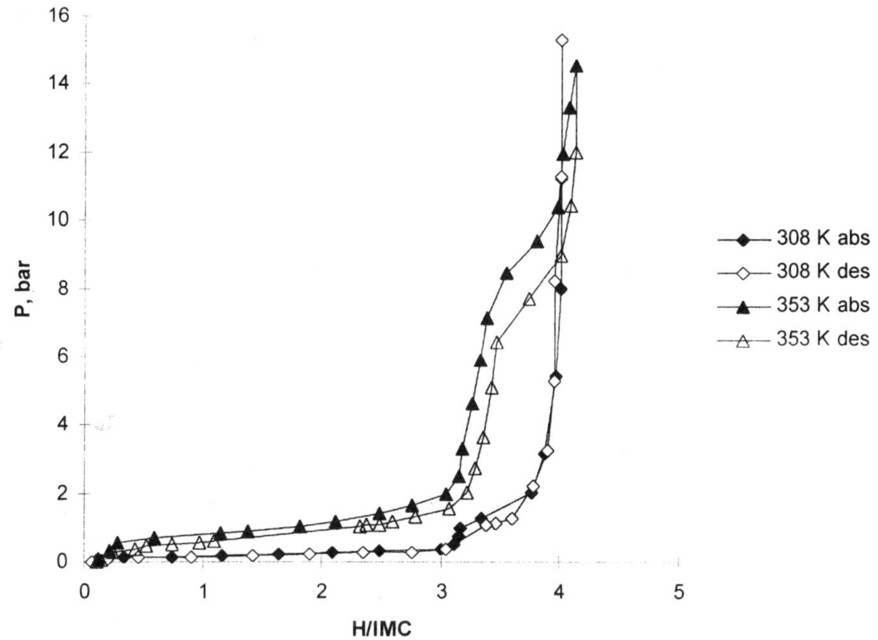

Fig. 3. The hydrogen absorption and desorption isotherms in the $LaNi_2Co_3$—H_2 system at 308 and 353 K.

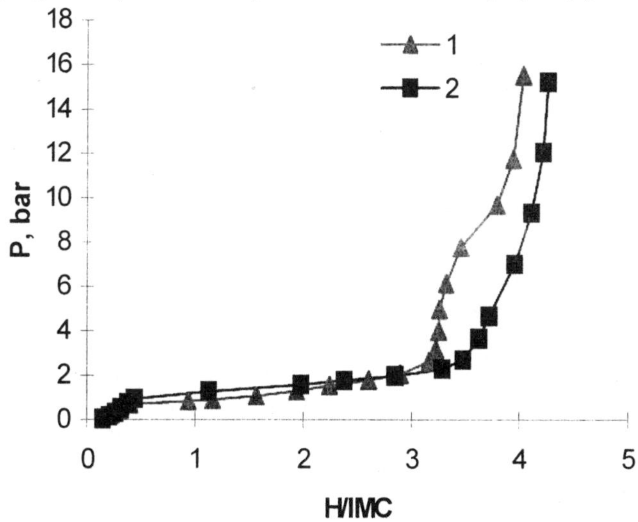

Fig. 4. The hydrogen absorption isotherms in the $LaNi_2Co_3$—H_2 system, that obtained:
1 — after multiple absorption-desorption cycling at 353 K,
2 — after sharp temperature shift from 308 to 353 K.

Such a result agree well with assumption that a sharp temperature shift influences on thermodynamic properties and composition of IMC hydrides. The curve (2) in course of following hydrogenation at 353 K is converted into the curve (1). And the curve (2) can be reproduced only after the same thermal treatment — sharp temperature shift from 308 to 353 K.

5. Discussion

The phase diagram for intermetallic compound $LaNi_2Co_3$ at 353 K is sufficiently dependent on hydrogenation conditions, thermal prehistory and initial hydrogen treatment at different temperatures.

We suppose the reason to be transition metal disorder in dehydrided sample caused by temperature shift described above. The Co and Ni atoms are statistically arranged on 2c and 3g positions of metal sublattice without any preferences. Then, hydrogen treatment upon each hydrogenation-dehydrogenation cycle enhances ordering with Co atoms occupying mostly 3g positions in the $LaNi_2Co_3$ and the first type of structure (1, "ordered") forms. Evidently the IMC structure changes influence on thermodynamic properties of these compounds. And it was demonstrated experimentally in course of cycles while hydrogen-induced ordering takes place. The second plateau appeared on P,C,T-diagrams and hydrogen capasity was decreased from values that described for modification (2) to values of modification (1).

The conversion of solid solution into ordered form of intermetallide can be monitored by means of a plot reflecting the dependence of differential molar enthalpy of hydrogen absorption or desorption vs. hydride composition. As fig. 5 indicates, the first desorption after sharp temperature shift from 308 to 353 K is more non-equilibrial, than previous absorption. It was deduced due to appearance of the second plateau and the considerable dispersion of molar enthalpy of hydrogen desorption in the two-phase region.

The difference of hydrogen absorption properties between ordered and disordered $LaNi_2Co_3$ samples could be easily explained in terms of differences in their crystal structures. It is likely that hydrogen atoms in such compounds occupy the sites corresponding to 6m and 12n in $LaNi_5$. In the ordered form, the Co atoms are preferably arranged in 3g positions in equatorial plane of the cell. Therefore, the 6m sites have constant surroundation La_2Co_2, and 12n sites — $LaCoNi_2$ (fig. 1). In case of solid solution, the number of suitable sites for hydrogen occupation surrounded by maximum of Ni atoms, as well as hydrogen absorption capacity are greater then in ordered form.

The existence of two plateaus on the phase diagram could be observed for the ordered structure rather then for the solid solution, because the suitable sites for hydrogen occupation in disordered state are more equivalent in energy. It was indicated by existence of an only one plateau on the phase diagram of $LaNi_2Co_3$ obtained after sharp temperature shift and on the P,C,T-diagrams of $LaNi_{2.5}Co_{2.4}Mn_{0.1}$.

As our experiments testify, the ordering/disordering transition proceeds very easily, with low activation energy. The ordered structure is more stable in the presence of hydrogen atoms. To obtain one or other modification the suitable thermal prehistory and hydrogenation conditions are required.

The results obtained in this work for $LaNi_2Co_3$ and in previous works [7] for $LaNi_2Cu_3$ and $LaNi_3Cu_2$ are similar. It is possible that other intermetallic compounds of $CaCu_5$-structure with transition metals ratio 2:3 could demonstrate such properties.

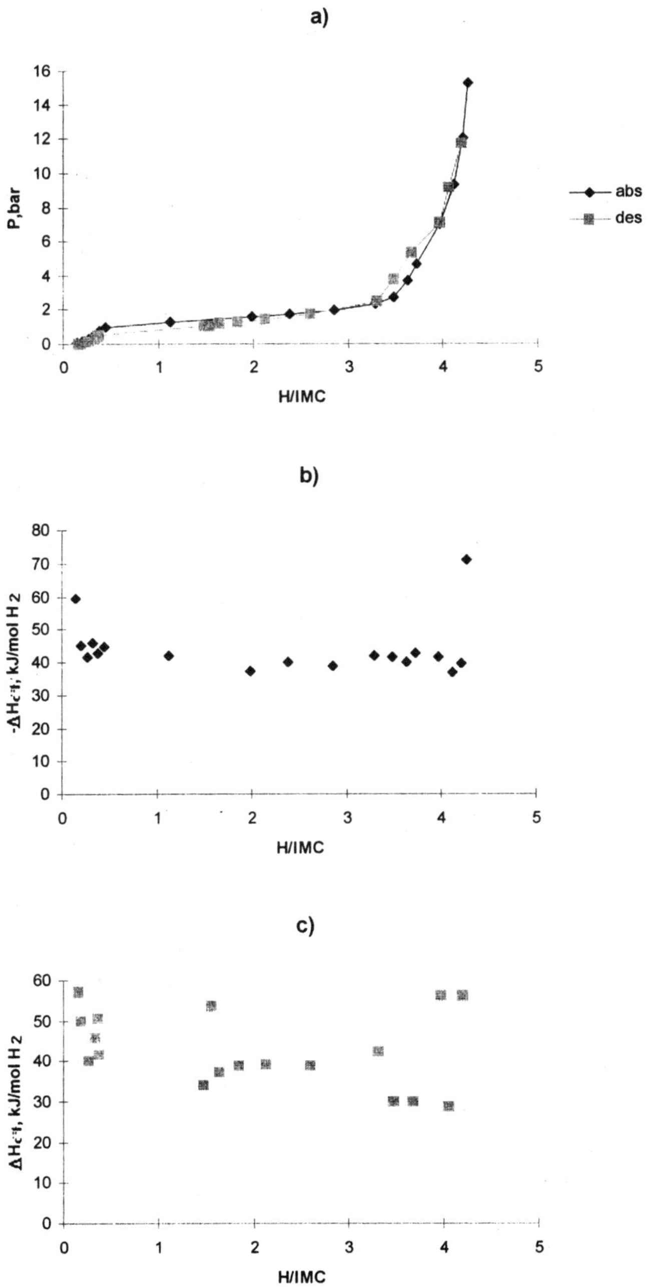

Fig. 5. Thermodynamic parameters of the $LaNi_2Co_3$ hydrogenation that obtained after sharp temperature shift from 308 to 353 K: a) hydrogen absorption and desorption isotherms, b) dependence of molar enthalpy of hydrogen absorption vs. hydride composition, c) dependence of molar enthalpy of hydrogen desorption vs. hydride composition.

6. Acknowledgements

This work was supported by Russian Foundation for Basic Research, grants # 99-03-32626, 00-15-97457, 01-03-06236 and Federal Program «Integration», grant # A-2.1.0113.

7. References

1. Naray-Sabo, I. (1969) Inorganic crystallochemistry, *Hungary Ac. of Sc. Publ.*
2. Flandorfer, H., Rogl, P., Hiebl, K., Bauer, E., Lindbaum, A., Gratz, A., Godart, C., Gignoux, D. and Schmitt, D. (1994) Physical properties and structural chemistry of $Ce(Ni_{1-x}Ga_x)_5$ alloys, *Physical review B* **50**(21), 15527-15540.
3. Girodin, D., Allibert, C.H., Givord, F. and Lemaire, R. (1985) Phase equilibria in the $CeCo_5$-$CeCu_5$ system and structural characterization of the $Ce(Co_{1-x}Cu_x)_5$ phases, *J. Less-Common Metals* **110**, 149-158.
4. Kuijpers, F.A. (1973) RCo_5—H and related systems, *Philips Res. Repts Suppl.* **2**, 1-102.
5. Moze, O. and Buschow, K.H.J. (1995) Neutron diffraction investigation of preferred site occupation in several $CaCu_5$-type compounds of cerium, *J. Magn. Magn. Mater.* **146**, 111-116.
6. Van Mal, H.H., Buschow, K.H.J. and Kuijpers, F.A. (1973) Hydrogen absorption and magnetic properties of $LaCo_{5x}Ni_{5-5x}$ compounds, *J. Less-Common Metals* **32**, 289-296.
7. Ganich, E.A., Yakovleva, N.A. and Semenenko, K.N. (1999) Investigation of hydrogen interaction with intermetallic compounds $LaNi_{5-x}Cu_x$, where x=2 and 3, *Zh. Obshch. Khim.* **69**, 1288-1295.
8. Yakovleva, N.A., Ganich, E.A., Rumyantseva, T.N. and Semenenko, K.N. (1996) Interaction mechanism of hydrogen with the $CaCu_5$-type crystal structure intermetallic compounds, *J. Alloys and Compounds* **241**, 112-115.

SYSTEM OF $CECO_2NI-H_2$ UNDER PRESSURE OF HYDROGEN UP TO 2000 AT.

S.A. LUSHNIKOV, S.N. KLYAMKIN, V.N. VERBETSKY
Department of Chemistry, Moscow Lomonosov State University, Moscow, 119992 Russia

Abstract

Interaction in $CeCo_2Ni-H_2$ system under high pressure of hydrogen of up to 2000 at. has been investigated. Obtained hydride $CeCo_2NiH_{5,6}$ and initial intermetallic compound (IMC) both have $PuNi_3$ structure type. Absorption isotherms of $CeCo_2Ni-H_2$ system are quite irreversibly. X-ray analysis has shown, that additional absorption of hydrogen by hydride leads to increasing his volume lattice at 2,9%.

1. Introduction

Intermetallic compounds RT_3 (R-earth rare elements, T-Ni,Co) are interesting for investigation as a metallic matrix for obtaining hydride phases rich with hydrogen. IMC RT_3 has $PuNi_3$ or $CeNi_3$ structure type.
Interaction IMC RT_3 with hydrogen leads to anisotropic expansion of their lattice at 30-35%. This expansion of lattice leads to forming new octahedral and tetrahedral vacations and these vacations apparently may be filled with hydrogen under high pressure.
X-ray analysis has shown that maximum content of hydrogen in RT_3 hydrides is equally composition RT_3H_6 [1]. In work [2] authors suppose more detailed calculation, which is equally composition $RT_3H_{5,6}$ and based on neutron diffraction. This determination of RT_3 hydrides composition was confirmed in [3], where authors obtained hydride phases under high pressure in $PrCo_3-H_2$ and $ErCo_3-H_2$ systems. For these systems small plateau was noticed under pressure of hydrogen 1000 at. and maximum composition is equally $RCo_3H_{5-5,5}$.
In the work [4] interaction IMC $CeCo_3$, having $PuNi_3$ structure type, with hydrogen under high pressure has been studied. Obtained hydride had composition $CeCo_3H_{6,1}$ and kept structure type of investigated IMC.
At present work has been studied interaction hydrogen with pseudobinary IMC $CeCo_2Ni$ under pressure 2000 at.

2. Experimental Details

$CeCo_2Ni$ has been obtained of melting in electrical furnace pure components under argon atmosphere. Obtained sample was quenched during 240 hours at 900^0C in quartz ampoule with argon.
Interaction of IMC with hydrogen was conducted on installation, described in [5] under pressure of hydrogen up to 2000 at. and temperatures interval from (-195^0C) to 300^0C.

N. Veziroglu et al. (eds.), Hydrogen Materials Science and Chemistry of Metal Hydrides, 267–271.

Calculation amount of hydrogen, absorbed under high pressure, was made with modified Van der Vaals equation [6]. During synthesis of hydride hydrogen was given with small amounts, consequently until his saturation. This allowed avoiding hydride decomposition.
After obtaining hydride under high pressure (to 2000 at.), the sample was cooling to low temperature (-195^{0}C) and was kept at open air during short time (5 minutes).
X-ray analyses were conducted on installation DRON-2 (Co, CuKα, Ni filter).
Amount of hydrogen was checked with thermodesorption method.
X-ray analysis has shown that the sample has single phase. Substitution one Co atom for Ni atom in IMC $CeCo_3$ hasn't been changed his $PuNi_3$ structure type. Volume of IMC $CeCo_2Ni$ lattice is a little less, than ones of IMC $CeCo_3$, because of atom Ni radius is less than atom Co radius.

3. Results and Discussions

$CeCo_2Ni$ starts to absorb of hydrogen after small induction period (about 5 minutes) under pressure about one atmosphere. Hydride composition under such pressure is equally $CeCo_2NiH_{4,0}$, and being a little different of results in [7], where hydride with composition $CeCo_2NiH_{3,0}$ was obtained.
X-ray data of IMC $CeCo_2Ni$, hydrides $CeCo_2NiH_{4,0}$ and $CeCo_2NiH_{5,6}$ in comparison with IMC $CeCo_3$, hydrides $CeCo_3H_{4,0}$ and $CeCo_3H_{6,1}$ [1] presented in the table 1.

TABLE 1. X-ray date of IMC and their hydrides

Composition	Str. type	a (Å)	Δa/a (%)	c (Å)	Δc/c (%)	V(Å^3)	ΔV/V (%)
$CeCo_3$	$PuNi_3$	4,952±0,002	-	24,80±0,02	-	529,0	-
$CeCo_3H_{4,4}$	$PuNi_3$	4,961±0,001	0,2	32,69±0,01	31,8	697,0	32,1
$CeCo_3H_{6,1}$	$PuNi_3$	5,030±0,003	1,6	32,98±0,02	33,0	723,0	36,8
$CeCo_2Ni$	$PuNi_3$	4,960±0,003	-	24,65±0,02	-	525,3	-
$CeCo_2NiH_{4,0}$	$PuNi_3$	4,950±0,002	-0,2	32,52±0,03	31,9	688,8	31,1
$CeCo_2NiH_{5,1}$	$PuNi_3$	4,970±0,004	0,2	32,88±0,01	33,4	703,8	34,0

Obtained diffractograms of IMC $CeCo_2Ni$, hydrides $CeCo_2NiH_{4,0}$ and $CeCo_2NiH_{5,6}$ are shown on figures 1.
Obtained hydride under low pressure $CeCo_2NiH_{4,0}$ has anisotropic expansion of the lattice due to increasing of «c» parameter («Δc/c» 31,9%). At the same time, parameter «a» is less than in IMC $CeCo_2Ni$ («Δa/a» -0,2%).
Hydride phase under low pressure of $CeCo_3H_{4,4}$ also has anisotropic lattice expansion due to increasing of «c» parameter («Δc/c» 31,8%), but «a» parameter slightly increases («Δa/a» 0,2%).
Behavior of $CeCo_2Ni$-H_2 system and studied early [4] $CeCo_3$-H_2 system under high pressure of hydrogen are similar.

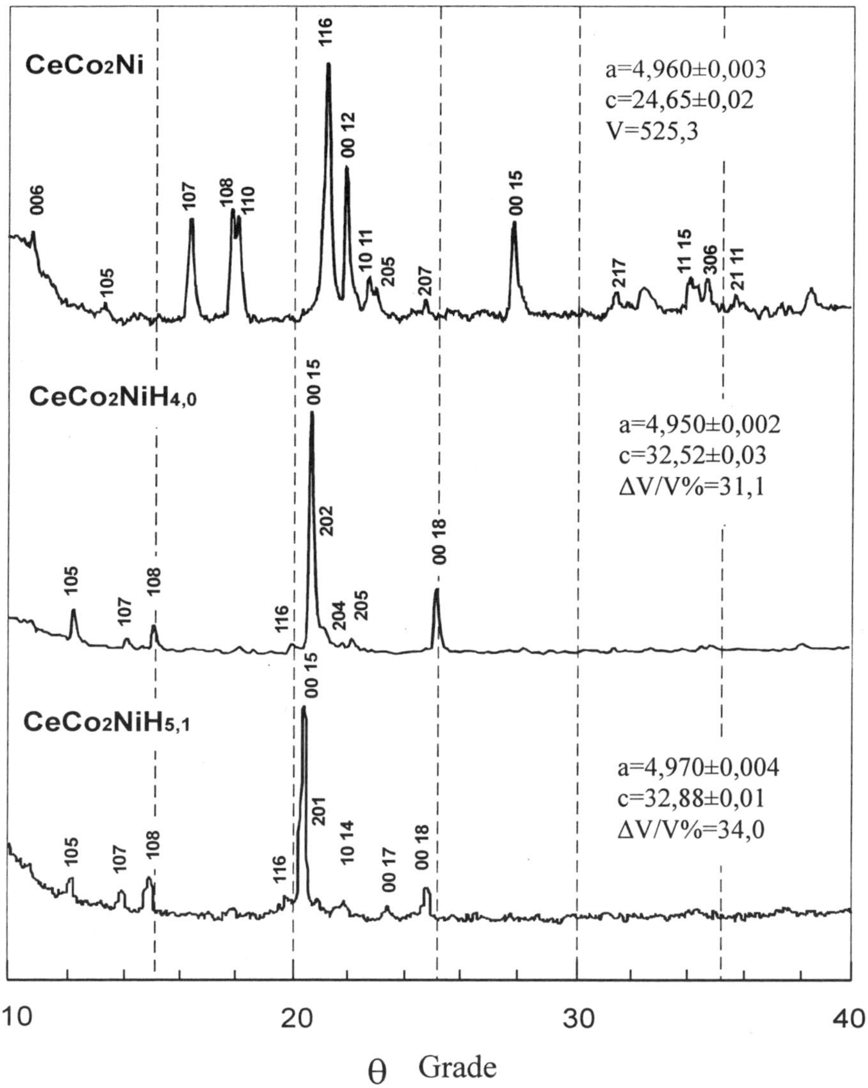

Figure1. Diffractograms of $CeCo_2Ni$, $CeCo_2NiH_{4,0}$ and $CeCo_2NiH_{5,1}$.

Analysis of absorption-desorption isotherms of $CeCo_2Ni$-H_2 system under high pressure (fig.2) has shown, that into whole interval of pressures it was equable absorption of hydrogen at room temperature.

There was change of composition of hydride phase from $CeCo_2NiH_{4,0}$ to $CeCo_2NiH_{5,1}$ with increasing pressure from 18 at. to 1820 at. Decreasing of hydrogen pressure to atmosphere leads to phase composition, which is equal $CeCo_2NiH_{4,8}$. This composition is considerable higher than composition of $CeCo_2NiH_{4,0}$ hydride phase, obtained under low pressure of hydrogen.

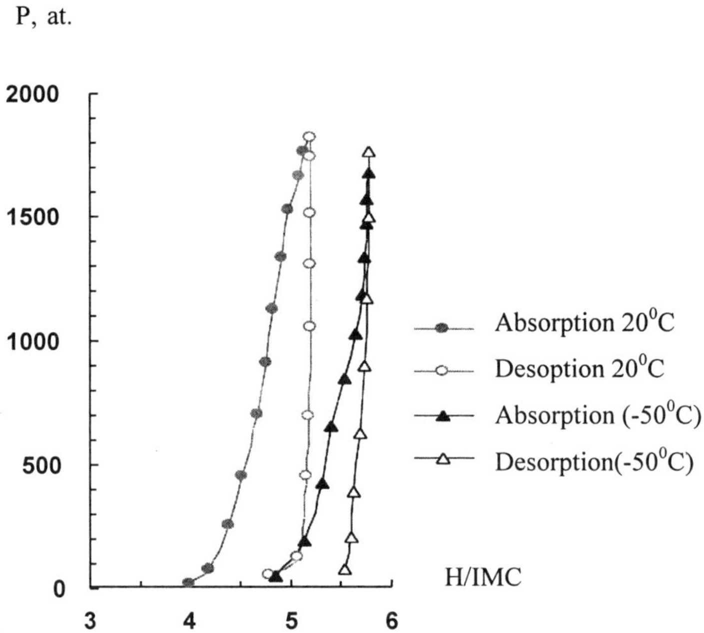

Figure 2. Absorption-Desorption isotherms in $CeCo_2Ni$-H_2 system.

Absorption of hydrogen at low temperature (-50^{0}C) and pressure of hydrogen 1767 at. leads to forming hydride phase with composition $CeCo_2NiH_{5,8}$. Decreasing pressure to atmosphere at the same temperature practically hasn't changed of obtained hydride composition ($CeCo_2NiH_{5,5}$).

So, with additional absorption of hydrogen, the increasing of stable of hydride phases at room and low temperature is observed.

For $CeCo_3$-H_2 system also increasing of stable of hydride phases under high pressure was observed [4]. Absorption of hydrogen at room temperature leads to composition of hydride phase equal $CeCo_3H_{6,1}$ at pressure 1730 at., and after decreasing pressure to atmospheric the hydride phase composition becomes $CeCo_3H_{4,9}$.

Absorption of hydrogen at low temperature (–70^{0}C) and pressure 1730 at. leads to forming hydride with composition $CeCo_3H_{6,8}$, and decreasing of pressure to atmosphere leads to composition $CeCo_3H_{5,7}$.

X-ray analysis of hydride phases under high pressure in $CeCo_2Ni$-H_2 and $CeCo_3$-H_2 systems has shown, that additional absorption of hydrogen leads to futher increasing of the volume lattice their hydrides. For $CeCo_2NiH_{5,1}$ this increasing is 2,9% and for $CeCo_3H_{6,1}$ it is 4,7%.

4. Conclusions

Substitution in IMC $CeCo_3$ one Co atom for Ni atom hasn't changed his structure type.

For $CeCo_2Ni$-H_2 system under high pressure was obtained hydride phase with composition $CeCo_2NiH_{5,1}$ at room temperature and pressure 1820 at. and $CeCo_2NiH_{5,8}$ at pressure 1767 at. and temperature (-50^{0}C).

For $CeCo_3$-H_2 system hydride phase at 1730 at. and room temperature is equally composition $CeCo_3H_{6,1}$ and at the same pressure of hydrogen 1730 at. and temperature (–70^{0}C) is equally composition $CeCo_3H_{6,8}$ [1].
Absorption-desorption isotherms have shown irreversibility of hydrogen desorption, and consequently, more high stability of hydride phase under high pressure.
Completed X-ray analysis has shown increasing of volume lattice at 2,9% hydride phase $CeCo_2NiH_{5,1}$ under high pressure in comparison with hydride phase $CeCo_2NiH_{4,0}$ under low pressure.
This work was supported by Russian Foundation for Basic Research, grant #00-15-97457

5. References

1. Yartis, V.A. (1992) New aspects of IMC hydrides chemistry: isotropic and anysotropic structures, *J. Koordin. Chemistry* **18**, 401-408.
2. Takeshita, T., Wallace, W.E., and Craig, R.S. (1974) Solubility of hydrogen in RCo_3 compounds, *J. Inorg. Chemistry* **13**, 2283-2284.
3. Lakner, J.F., Uribe, F.S., and Steward S.A. (1980) Hydrogen and deuterium sorption by selected rare earth intermetallic compounds at pressures up to 1500 at., *J. Less-Common Metals* **72**, 87-105.
4. Lushnikov, S.A., Klaymkin, S.N., and Verbetsky, V.N. (2001) Interaction of RT_3 intermetallic compounds with hydrogen under high pressure, *J.Inorgan. Mater.* **37**, 457-460.
5. Klaymkin, S.N., and Verbetsky, V.N.(1993) Interaction of intermetallic compounds with hydrogen at pressures up to 250 Mpa: the $LaCo_{5-x}Mn_x$-H_2 and $CeNi_5$-H_2 systems, *J. Alloys and Compounds* **194**, 41-45.
6. Hemmes, H., Driessen, A., and Griessen, R. (1986) Thermodynamic properties of hydrogen at pressures up to 1 Mbar and temperatures between 100 and 1000 K, *J. Phys.C: Solid State Phys.* **19**, 3571-3585.
7. Petrii, O.A., Semenenko, K.N., Korobov, I.I., and Vasina, S.Ya. (1987) Investigation of materials based on intermetallic compounds of the $CeNi_3$-$CeCo_3$ system and a polymeric binding material, *J. Less-Common Metals* **136**, 121-134.
8. Burnasheva, V.V., Klimeshin, V.V., and Semenenko, K.N. (1979) Study of equilibrium in the systems RCo_3-H_2 (R- Ce,Pr,Tb,Dy,Er), *J. Inorgan. Mater.* **15**, 251-255.

INFLUENCE OF HYDROGENATION ON MAGNETIC ANISOTROPY OF R_2Fe_{17} SINGLE CRYSTALS

I.S. TERESHINA[1,2], S.A. NIKITIN[1,2], N.Yu. PANKRATOV[1,2], N.V. TRISTAN[2], T. PALEWSKI[2], V.N. VERBETSKI[3], A.A. SALAMOVA[3]

[1] *Faculty of Physics, M.V. Lomonosov Moscow State University, Vorobevy Gory, Moscow, 119992 Russia.*
[2] *International Laboratory of High Magnetic Fields and Low Temperatures, 95 Gajowicka str, 53-421 Wroclaw, Poland.*
[3] *Faculty of Chemistry, M.V. Lomonosov Moscow State University, Vorobevy Gory, Moscow, 119992 Russia.*

The magnetocrystalline anisotropy of R_2Fe_{17} (R = Gd, Tb, Er) and their hydrides are studied by analyzing magnetization curves of single crystal like samples in the temperature range 4.2-300 K in magnetic fields up to 140 kOe. No noticeable influence of hydrogenation on magnetic anisotropy in the Gd_2Fe_{17} has been found. An easy-plane to easy-cone transition was detected for $Tb_2Fe_{17}H_3$ single crystal. The hydrogenation leads to the disappearance of the transition of FOMP type in $Er_2Fe_{17}H_3$ and to the increase of the field of the magnetic anisotropy. The obtained results are discussed on the basis of a model taking into account the electrostatic interaction of the quadrupolar moment of the asymmetric 4f shell with the electric field gradient. It is established that the orientation of the quadrupolar moment of the asymmetric 4f shell of the rare earth ion with respect to the direction of the resulting magnetic moment of 4f electrons plays an important role.

1. Introduction

The purpose of the this work was to investigate the influence of hydrogen atoms on magnetocrystalline anisotropy (MCA) of the R_2Fe_{17} compounds and to establish physical mechanisms responsible for change of MCA caused by the influence of insertion atoms. As was already established [1,2], the hydrides retain crystal structure of the parent samples (Th_2Ni_{17}-type) however their atomic volume increases significantly what leads to Curie temperature increase by approximately 35-40 K upon insertion of one hydrogen atom per formula unit.

N. Veziroglu et al. (eds.), Hydrogen Materials Science and Chemistry of Metal Hydrides, 273–280.

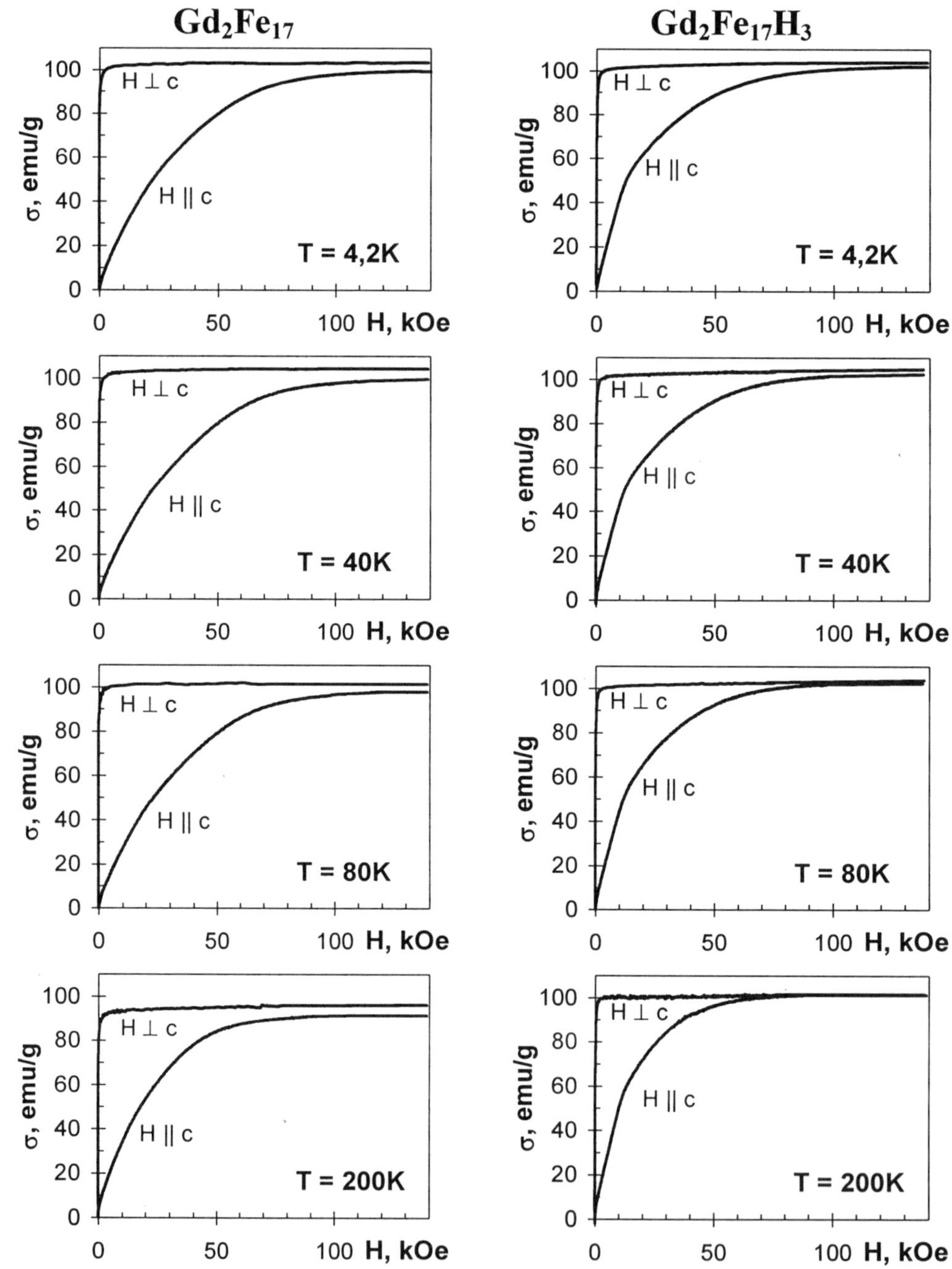

Figure 1. The isothermal magnetization curves at T = 4.2, 40, 80 and 200 K measured on single crystals of the Gd_2Fe_{17} and $Gd_2Fe_{17}H_3$ compounds for magnetic field applied parallel and perpendicular to the c - axis.

The change of the magnitude of the intra- and intersublattice exchange interaction constants upon hydrogenation was calculated on the basis of the molecular field theory using magnetic ordering temperatures [3]. The dependence of the saturation magnetization on concentration of the absorbed hydrogen has a complex character [4]. Neutron diffraction investigations [5] showed that rare earth and iron atoms could occupy three and six nonequivalent positions, respectively, in the hydrogenated compounds as well as in the parent ones.

In contrast to previous works where investigations of MCA were carried out on magnetically aligned powders of the hydrogenated samples of the R_2Fe_{17} compounds [1], we have investigated single crystals of the hydrides what allowed us to determine with high accuracy temperature ranges of the existence of the magnetic phases having certain directions of magnetization vector.

2. Results and Discussions

The detailed description of the preparation methods of the single crystals $R_2Fe_{17}H_x$ ($x = 0; 3$) used in the work was given elsewhere [3].Magnetic measurements were carried out on the pendulum and torque magnetometers in the temperature range 77-700 K in magnetic fields up to 13 kOe and on the capacity magnetometer in the temperature range 4.2-300 K in magnetic fields up to 140 kOe.

The ranges of the existence of the different magnetic phases depending on temperature and magnetic field were found for original compounds as well as for their hydrides. The hydrogenation was found to strongly influence the temperature and field ranges where the magnetic phases with easy magnetization axis directed perpendicular or along the hexagonal *c* axis exist.

The influence of the hydrogenation on magnetic properties of the iron sublattice was earlier investigated by the present authors using samples with R = Y [6] or Lu [7]. The insertion of three hydrogen atoms does not lead to any noticeable change of neither magnetization nor magnetic anisotropy. The Curie temperature, however, was found to increase by approximately 100 K. The neutron diffraction investigations [8] showed that in presence of small concentration of hydrogen ($x \leq 3$), the insertion atoms settle in octahedral sites and, therefore, do not influence appreciably magnetization and magnetic anisotropy of the iron sublattice.

In this work the influence of hydrogen atoms, inserted into octahedral positions, on the magnetization and magnetic anisotropy of the R_2Fe_{17} compounds, where R is Gd, Tb and Er, was investigated by measurements magnetization curves. The magnetization curves were measured along main crystallographic directions of the parent and hydrogenated single crystals in the magnetic fields up to 140 kOe.

The magnetization curves measured on Gd_2Fe_{17} and $Gd_2Fe_{17}H_3$ along hexagonal axis and in basis plane at T =4.2 K are practically the same (see Fig. 1). Therefore, there is no noticeable influence of hydrogenation on magnetic anisotropy in the compounds as well as in Lu_2Fe_{17}, Y_2Fe_{17} and their hydrides [6,7]. It is worth to mention that trivalent ions of Lu and Y do not have paramagnetic moment since 4f electron shell of the elements is filled. The trivalent Gd ion has a considerable

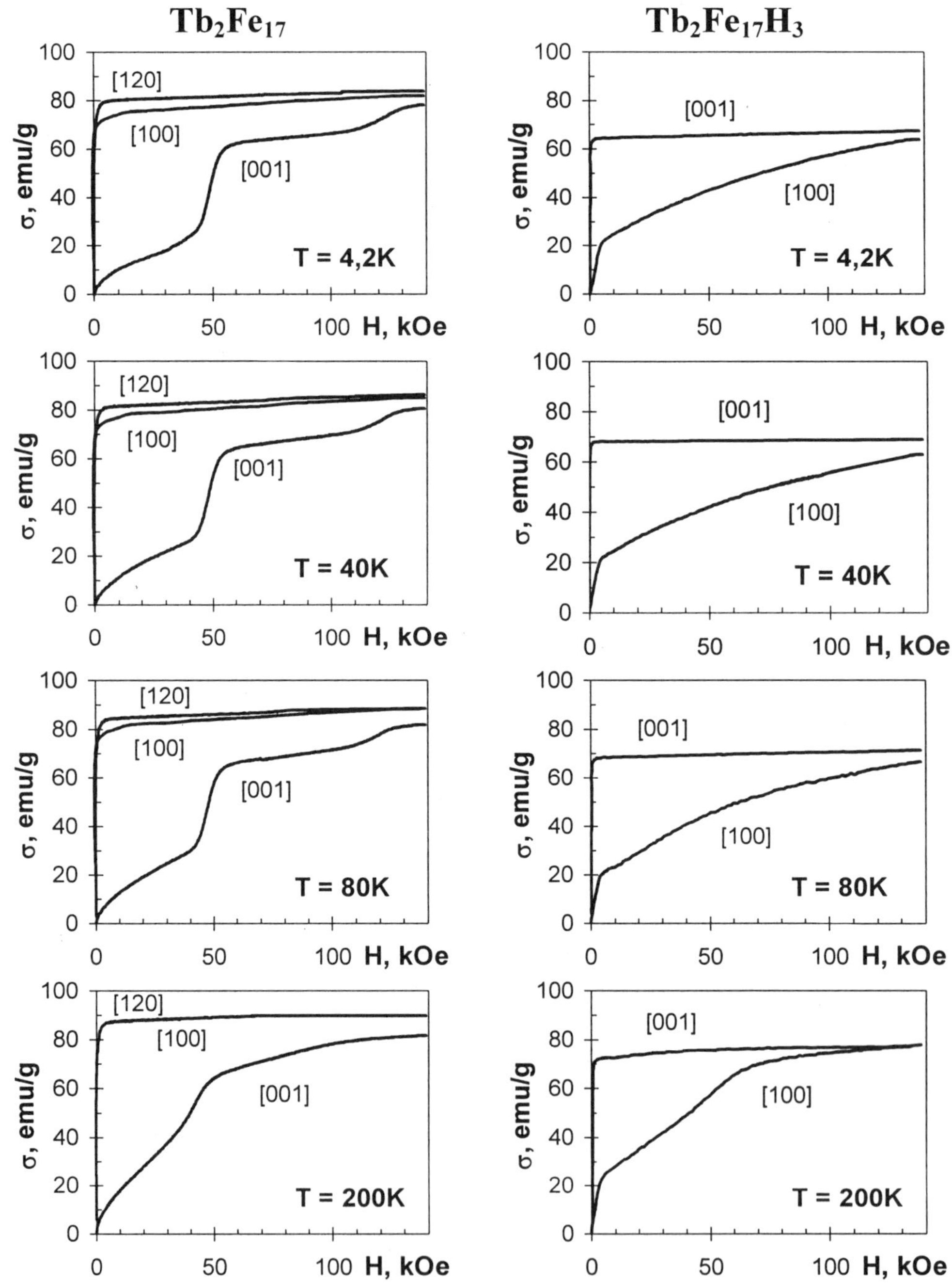

Figure 2. The isothermal magnetization curves at T = 4.2, 40, 80 and 200 K measured on single crystals of the Tb_2Fe_{17} and $Tb_2Fe_{17}H_3$ compounds for magnetic field applied along the [001], [100], [120] directions.

localized magnetic moment in the S state, the orbital moment of Gd is equal zero and 4f electron shell is a spherical one.

Because there is no effect of hydrogenation on magnetic anisotropy of the R_2Fe_{17} compounds where R = Y, Lu, Gd, it is interesting to understand how the insertion of hydrogen atoms into crystalline lattice would influence the magnetic anisotropy in those R_2Fe_{17} compounds where rare earth ion has nonzero orbital moment. The R_2Fe_{17} compounds, where R = Tb and Er, were chosen for the investigation because the rare-earth moments have nonzero orbital moments but at the same time the orientation of quadruple and magnetic moments of the 4f electron shell is different for Tb and Er.

One can see from Fig. 2 that insertion of hydrogen into crystalline lattice of the $Tb_2Fe_{17}H_3$ gives rise to the appearance of the magnetic phase with easy magnetization axis directed along hexagonal c or [001] axis. At the same time in parent Tb_2Fe_{17}, the [001] axis is the axis of hard magnetization.

Fig. 2 shows that there is a jump of magnetization in the original compound Tb_2Fe_{17} at T = 4.2 K in the magnetic field H = 50 kOe oriented along the hexagonal axis [001] which is the hard direction of magnetization in this case. This jump is caused by the change of magnetization vector direction from [100] axis to a direction of local minimum appearing under the influence of the internal magnetic field and located at an angle to [001] axis. Such transitions were earlier observed in Tb_2Fe_{17} compounds [9] and were named transitions of the FOMP (First Order Magnetization Process) type [10]. Thus, our data give the evidence that the hydrogenation not only suppresses this transition in a $Tb_2Fe_{17}H_3$ but also transforms the hexagonal axis [001] into easy axis of magnetization. At the same time the [100] axis, placed in basal plane and perpendicular to [001] axis, becomes the hard axis of magnetization. On the contrary, as our measurements showed, the insertion of hydrogen in the crystalline lattice of $Er_2Fe_{17}H_3$ does not change the direction of the easy axis magnetization and [100] axis is still placed in basal plane perpendicular to the hexagonal c axis (see Fig. 3.).

The transition of FOMP type is clearly observed along the [001] axis in the parent Er_2Fe_{17}. However, the hydrogenation leads to the disappearance of the transition in $Er_2Fe_{17}H_3$ and to the increasing of the field of the magnetic anisotropy.

It is worth to notice that Tb ion has negative value of Stevens factor α_J, while Er ion has positive one and that the sign of Stevens factor determines the orientation of electron density distribution of 4f electrons relative to rare earth ion magnetic moment direction: if $\alpha_J > 0$ then quadruple moment is parallel to the magnetic moment and if $\alpha_J < 0$ then these two moments are perpendicular to each other.

Fig.4 shows the schematic model of the interaction of the non-spherical distribution of 4f-electron charge density with surrounding hydrogen atoms, placed into octahedral positions of crystalline lattice. It is worth to notice that the distribution of hydrogen atoms into octahedral vacations is anisotropic regarding to the c axis. One can see that hydrogen atoms are situated in the basal plane and surround the 4f-electron shell whose long axis is also situated in the basal plane.

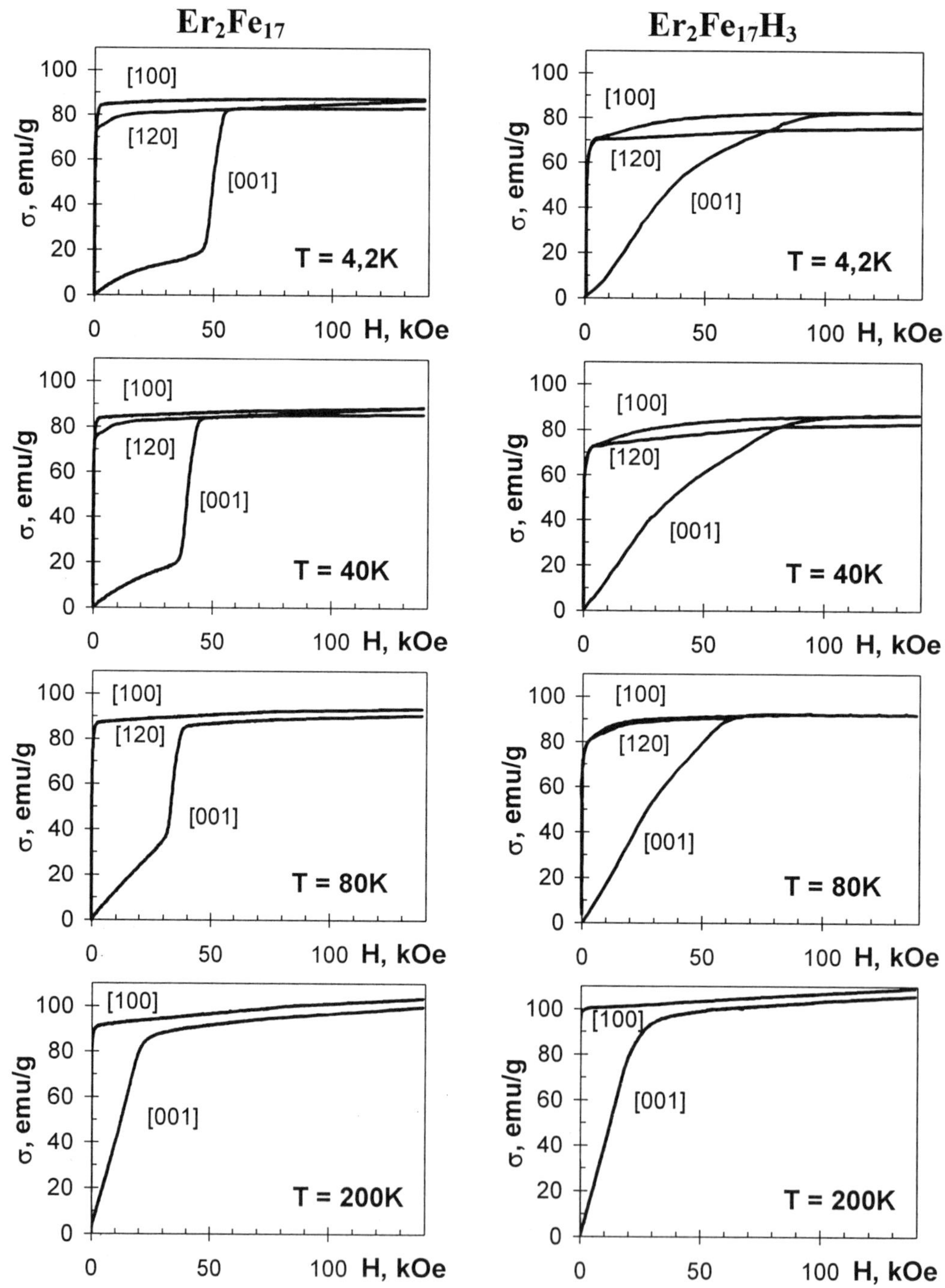

Figure 3. The isothermal magnetization curves at T = 4.2, 40, 80 and 200 K measured on single crystals of the Er_2Fe_{17} and $Er_2Fe_{17}H_3$ compounds for magnetic field applied along the [001], [100], [120] directions.

Therefore, the inserted hydrogen atoms influence the 4f-electron shell the way that its quadruple moment lays in the basal plane. In the case of ions with negative Stevens factor, for example Tb ion, this leads to fixation of magnetic moment of the 4f-shell along the hexagonal axis ***c***. In $Tb_2Fe_{17}H_3$ this develops into the situation when [001] axis becomes an easy axis of magnetization and perpendicular to it [100] axis becomes a hard axis of magnetization in contrast to situation in parent Tb_2Fe_{17}. For the ions with $\alpha_J > 0$, for example Er ion, the opposite situation is observed: the fixation of magnetic moment of the 4f-shell and increasing of magnetic anisotropy takes place in basal plane.

The explanation of experimentally obtained behavior is thus possible on the basis of the model which takes into account the electrostatic interaction between the quadruple moment of the 4f electron subshell with electrical field gradient. In the site of rare earth ion the gradient is caused by electrostatic interaction with neighbor ions and, probably, with a heterogeneous distribution of the valence and conductivity electron density [11,12]. The hybridization of 5d-electrons of rare earth ions with s and p electrons of surrounding ions plays also an important role. The insertion of hydrogen atoms into crystalline lattice modifies all contributions into electric field gradient. The experimental results are in agreement with our model, pointing to an important role of the rare earth ion 4f sublattice quadruple moment orientation relative to crystalline fields.

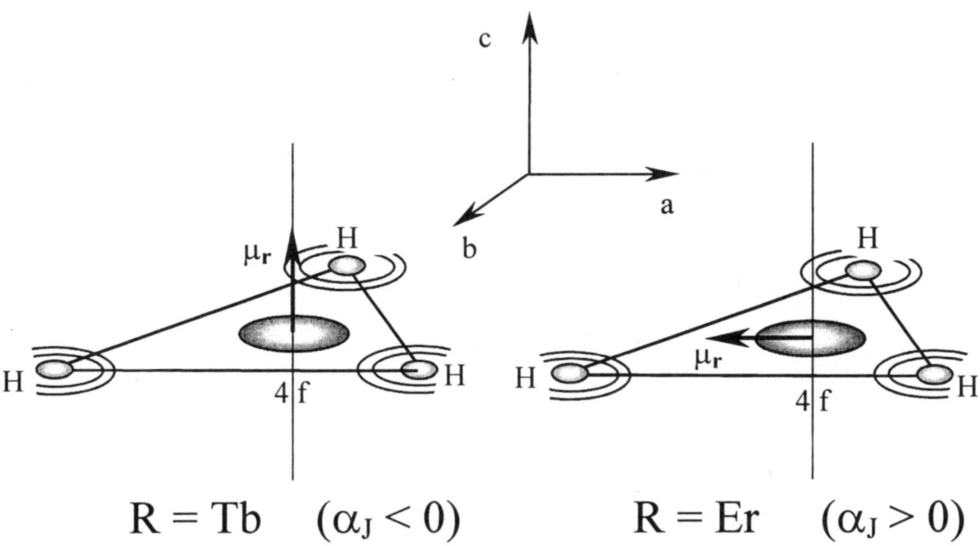

Figure 4. The scheme of the interaction of the quadrupole and magnetic moments of the 4f electron shell of the rare earth ion with interstitial hydrogen atoms.

3. Conclusion

The insertion of hydrogen into crystal lattice of R_2Fe_{17} compounds influences magnetic anisotropy and, hence, results in a change of the magnetic anisotropy of the compounds based on rare earth elements according with sign of Stevens factor of the rare earth ion.

4. Acknowledgments

This work was supported by Federal Program on Support of Leading Scientific Schools N 00-15-96695 and RFBR Grants N 02-02-16523. The part of the work performed by N.Yu. Pankratov was supported by INTAS fellowship grant YSF 2001/2-92 and RFBR Grant N 02-02-06264.

5. References

1. O.Isnard, S.Miraglia, J.L.Soubeyroux, D.Fruchart, P.l'Heritier. (1994) A structural analysis and some magnetic properties of the $R_2Fe_{17}H_x$ series. *J.Magn.Magn.Mater.* **137**, 151-156.
2. S.A.Nikitin, E.A.Ovtchenkov, A.A.Salamova, A.Yu.Sokolov, V.N.Verbetsky. (1997) Effect of interstitial hydrogen and nitrogen on the magnetocrystalline anisotropy of R_2Fe_{17} (R = Tb, Dy, Ho, Er). *J.Alloys Comp.* **261**, 15-18.
3. I.S.Tereshina, S.A.Nikitin, V.N.Verbetsky, A.A.Salamova. (in press) Effect of interstitial elements on the magnetic anisotropy and magnetostriction of intermetallic compounds based on 4f and 3d transition metals *Fiz.Met.Metallov.*
4. D.Fruchart, S.Miraglia. (1991) Hydrogenation hard magnetic alloys from fundamental to applications (invited). *J.Appl.Phys.* **69(8)**, 5578-5583.
5. D.Givord, R.Lemaire, J.M.Moreau, E.Roudaut. (1972) X-ray neutron determination of a so-called Th_2Ni_{17}-type structure in the lutetium-iron system. *J.Less.Comm.Met.* **29**, 361-369.
6. S.A.Nikitin, E.A.Ovtchenkov, A.A.Salamova, V.N.Verbetsky. (1997) Effect of interstitial hydrogen and nitrogen on the magnetocrystalline anisotropy of Y_2Fe_{17}. *J.Alloys Comp.* **260**, 5-6.
7. I.S.Tereshina, S.A.Nikitin, J.Stepien-Damm, L.D.Gulay, N.Yu.Pankratov, A.A.Salamova, V.N.Verbetsky, W.Suski. (2001) Structural and magnetic properties of $Lu_2Fe_{17}H_x$ (x = 0; 3) single crystals. *J.Alloys Comp.* **329**, 31-36.
8. O.Isnard, S.Miraglia, C.Colbeck, E.Tomey, J.L.Soubeyroux, D.Fruchart. (1992) Neutron diffraction study of the structural and magnetic properties of the $R_2Fe_{17}H_x(D_x)$ ternary compounds (R=Ce, Nd and Ho).*J.Alloys Comp.***178**,15-22.
9. R.Verhoef, P.H.Quang, R.J.Radwanski, C.Marquina and J.J.M.Franse. (1992) Magnetisation curves of single-crystalline Tb_2Fe_{17}. *J.Magn.Magn.Mater.* **104-107**, 1473-1474.
10. G.Asti, F.Bolzoni. (1980) Theory of first order magnetization processes: uniaxial anisotropy. *J.Magn.Magn.Mater.* **20**, 29-43.
11. R.Coehoorn, K.H.J.Buschow, M.W.Dirken, R.C.Thiel. (1990) Valence-electron contributions to the electric-field gradient in the *hcp* metals and at Gd nuclei in intermetallic compounds with the $ThCr_2Si_2$ structure.*Phys.Rev.B* **42(7)**,4645-4654.
12. V.Yu.Irkhin, Yu.P.Irkhin. (1998) Charge screening and magnetic anisotropy in metallic rare-earth systems. *Phys.Rev.B* **53**, 2697-2700.

KINETICS OF HYDRIDE PHASE GROWTH IN THE AREA OF STRUCTURAL DEFECTS

N.M.VLASOV, I.I.FEDIK
Scientific Research Institute
Scientific Industrial Association "Luch",
24, Zheleznodorozhnaya St., Podolsk, Moscow region, 142100, Russia

Abstract

The comparative analysis of the stress fields of the known structural defects has been carried out. The most powerful defect is wedge disclination (a model of threefold junctions of grain boundaries). Then a microcrack tip and edge dislocation follow. The kinetics of diffusion growth of hydride phase in the wedge disclination area has been investigated. The dilatation field of this defect depends logarithmically on a radial coordinate. It allows obtaining an exact analytical solution of a diffusion equation in the force field. The analytical expressions describing kinetics of impurity segregation and hydride phase growth in the wedge disclination area are given. The results of theoretical analysis are attracted for explanation of hydrogen embrittlement of metals when forming and growing hydride phase.

1. Introduction

Metals and alloys have different structural defects. The stress field in their area increases diffusion mobility of hydrogen atoms. Impurity segregations are formed near the structural defects [1]. Segregation kinetics is defined by a non-stationary diffusion equation in the force field. If hydrogen concentration exceeds the solubility limit at given temperature, hydrides are formed in some metals (for example, in zirconium). Their formation process is a phase transition of the second kind [2]. It runs free of appearing new phase nucleus. It is correct for a metal having no structural defects. Near the latter ones a phase transformation takes place and hydrides are formed. They can be considered as the nucleus of hydride phases within volume of a metal. Such nucleus grow at the expense of hydride atom diffusion. Their growth kinetics is subject to the diffusion equation for a mobile interphase boundary. The analytical solution of this problem is possible just for the structural defects simulated by the wedge disclinations. The stress disclination field depends logarithmically on a radial coordinate. Such dependence allows obtaining an exact analytical solution of kinetics of hydride phase growth [3].

The comparative analysis of the stress fields shows that the wedge disclinations are the most powerful structural defects.Then a microcrack tip and edge dislocation follow. Therefore, hydride phases are formed mainly near the wedge disclinations. They simulated different structural material imperfections. Among them threefold junctions of grain boundaries play an important role. In the field of external loads (power or temperature) these linear defects are the sources of internal stresses. Their appearance is

N. Veziroglu et al. (eds.), Hydrogen Materials Science and Chemistry of Metal Hydrides, 281–288.

caused by orientated difference of elastic and thermophysical characteristics of separate polycrystal grains. The stress fields of the opposite signs mutually compensate each other at the distance equal to a half of typical grain size. The stress field of threefold junctions of grain boundaries is simulated by the wedge disclination. Therefore, we will consider namely this structural defect.

We consider zirconium as an example of formation and growth of hydride phase. This metal and alloys on its base are used in structural elements of nuclear engineering. Using zirconium is caused by successful combination of physico-mechanical properties and a low capture cross-section of thermal neutrons. Hydride formation depends on hydrogen solubility in different zirconium phases. It follows from a structural diagram of Zr-H: 1) β-Zr phase dissolves hydrogen in large quantity; 2) Hydrogen solubility in α-Zr at 550°C reaches 6,1% (at.), and at the room temperature it is 0,0015% (at.). Therefore, hydride phases in zirconium are formed at low temperatures (< 200°C). In dependence on temperature and hydrogen concentration there are the following types of hydrides:γ-ZrH, δ-ZrH_2, ε -ZrH_2. Parameters of a hydride crystal lattice are changed in comparison with zirconium. Stresses and microcracks occur on the interphase boundary. Their growth determines hydrogen embrittlement of zirconium. An incubative period of this process is caused by the diffusion growth of hydride phases in the are of the wedge disclination (a model of threefold junctions of grain boundaries).

The objective of the work is studying kinetics of hydride phase growth in the area of the wedge disclination. The results of theoretical analysis are attracted for explanation of hydrogen embrittlement of zirconium and its alloys.

2. Hydrogen segregation in the area of wedge disclination

Impurity segregations from hydrogen atoms are formed in the area of structural defects of metals. Kinetics of the process depends on the stress field of a defect. The most powerful defect is wedge disclination. The first invariant of stress tensor logarithmically depends on a radial coordinate (plane deformation) [4].

$$\sigma_{ll} = \frac{\mu\omega(1+\nu)}{2\pi(1-\nu)}\left\{1 + 2\ell n\frac{r}{R} + \frac{2\left(\frac{r_0}{R}\right)^2}{1-\left(\frac{r_0}{R}\right)^2}\ell n\frac{r_0}{R}\right\}, \tag{1}$$

where μ- shear modulus, ν - Poison's ratio, ω -modulus of disclination rotation vector, r_0 and R – internal and external radiuses of disclination. For disclination with a positive dilatation ω<0 is accepted. It corresponds to a positive value of σ_{ll} (tension stress). For disclination with a negative dilatation ω>0 is accepted. It corresponds to a negative value of σ_{ll} (compression stress). The values of σ_{ll} for the microcrack tip and edge dislocation can be taken from the expressions

$$\sigma_{ll} = \frac{\sqrt{2a}\sigma_0(1+v)}{\sqrt{r}}\cos\frac{\theta}{2} \quad \text{(microcrack tip)} \tag{2}$$

$$\sigma_{ll} = \frac{\mu b(1+v)}{\pi(1-v)r}\sin\theta \quad \text{(edge dislocation)}$$

where 2a – microcrack length, σ_0–external stress, b-modulus of Burgers vector, r_0 and θ- polar coordinates ($0 \le 0 \le 2\pi$). The rest designations correspond to the accepted before. The comparative analysis of relations (1) and (2) shows that the wedge disclination is the most powerful structural defect. The dilatation field in its area is changed according to a logarithmic law. It leads to a higher rate of impurity segregation in comparison with the other structural defects. The next power structural defect is a microcrack tip. The dilatation field in the area of its tip is changed according to the law $\sim \frac{1}{\sqrt{r}}$. The dilatation field of the edge dislocation is changed according to the law $\sim \frac{1}{r}$. Therefore, a rate of impurity segregation in its area compares unfavorably with the one for disclination and microcrack tip.

Coordinate dependences of σ_{ll} determine the rate of impurity segregations for different structural defects. Kinetics of segregation depends on interaction potential of hydrogen atoms and a structural defect [5].

$$V = -\frac{\sigma_{\ell\ell}}{3}\delta\upsilon, \tag{3}$$

where δυ - crystal volume change at hydrogen atoms placement. The value of δυ for all metals remains unchanchable and equal to $3 \cdot 10^{-30}$m^3. A negative value of potential V corresponds to the attraction of hydrogen atoms to the structural defect. For the wedge disclination with a positive dilatation expression (3) takes the form

$$V = -A\,\ell n\frac{r}{R}, \qquad A = \frac{\mu\omega(1+v)}{3\pi(1-v)}\delta\upsilon. \tag{4}$$

Relationship (4) is written accurate up to the constants. They do not influence hydrogen atom diffusion. Diffusion flow is proportional to the gradient of potential V. In this case ΔV=0 as V- a harmonic function. Expression (4) takes into account just a dimensional effect in binding energy between a hydrogen atom and wedge disclination. The other types of interaction (module, electrostatic, chemical) are not under consideration. Their contribution into interaction energy is taken into account with changing constants in relationship (4). Diffusion migration of hydrogen atoms in the stress field of the structural defects is defined by interaction potential V. For the wedge disclination with a logarithmic dependence V from a radial coordinate it is possible to obtain an exact analytical solution of diffusion equation in the field of forces. For the other values of potential V obtaining the analytical solution causes considerable mathematical difficulties. Therefore, some approximate methods are used. One of them takes into account in the full flow of hydrogen atoms only dependence on the interaction potential gradient. It is correct just when describing hydrogen diffusion near a structural defect.

The stress field of a defect gives the main contribution into the process of hydrogen diffusion.

Kinetics of hydrogen atom segregation in the area of the wedge disclination is defined by a non-stationary diffusion equation in potential of the interaction [6]

$$\frac{1}{D}\frac{\partial C}{\partial t} = \Delta C + \frac{\nabla(C\nabla V)}{KT}, \qquad r_0 < r < R, \tag{5}$$

$$C(r,0) = C_0, \quad C(R,t) = C_0, \quad C(r_0,t) = C_p,$$

where D- coefficient of hydrogen atom diffusion, k- Boltzmann's constant, T- absolute temperature, C_p- equilibrium concentration of hydrogen atoms on the boundary of the internal disclination radius. For the threefold junctions of grain boundaries the internal radius is accepted equal to thickness of the grain boundary, and external one to a half of the grain size. Relationship $\frac{R}{r_0}$ is rather large. When solvingsome problems on diffusion kinetics, such defect is considered in an unlimited matrix. The physical sense of the initial and boundary conditions for equation (5) is obvious. At initial moment the concentration of hydrogen atoms is equal to the average concentration. Such concentration remains also on the external disclination boundary. When migrating hydrogen atoms to the disclination with a positive dilatation, hydrogen runs to boundary r=R from the disclination with an opposite sign. The boundary condition at $r=r_0$ means that the concentration of hydrogen atoms becomes equilibrium on this boundary in a moment according to the interaction potential. Taking into account relationship (4) problem (5) is written as follows:

$$\frac{1}{D}\frac{\partial C}{\partial t} = \frac{\partial^2 C}{\partial r^2} + \frac{1+\alpha}{r}\frac{\partial C}{\partial r}, \quad r_0 < r < R,$$

$$C(r,0) = C_0, \; C(R,t) = C_0, \; C(r_0, t) = C_p, \tag{6}$$

Dimensionless parameter α

$$\alpha = \frac{\mu\omega(1+v)\delta\upsilon}{3\pi(1-v)kT}$$

defines the relationship of binding energy of hydrogen atoms with the wedge disclination to the thermal motion energy. If $|\alpha| \ll 1$, the stress field is small disturbance of the main diffusion flow of hydrogen atoms. For $|\alpha| \gg 1$ the disclination stress field plays the main role in the diffusion process. At $|\alpha| \approx 1$ diffusion hydrogen flows at the expense of $\frac{\partial C}{\partial r}$ and $\frac{\partial V}{\partial r}$ are compared. Let's define the value of $|\alpha|$ for Zr-H system. For $kT = 10^{-20}$ J, $\mu = 4\cdot 10^{10}$ Pa, $v=0{,}3$, $|\omega| = 0{,}3$ rad., $\delta\upsilon = 3\cdot 10^{-30}$ m^3 we obtain $|\alpha| \approx 1$. The wedge disclination with a positive dilatation corresponds $\alpha=-1$. The diffusion equation takes a simple form

$$\frac{1}{D}\frac{\partial C}{\partial t} = \frac{\partial^2 C}{\partial r^2}, \qquad r_0 < r < R,$$

$$C(r,0) = C_0, \; C(R,t) = C_0, \; C(r_0, t) = C_p. \tag{7}$$

The equilibrium concentration of hydrogen atoms C_p depends on potential V and at $\alpha=-1$ has a form

$$C_p = C_0 \exp\left(-\frac{V}{kT}\right)_{r=r_0} = C_0 \exp\left(\ell n\frac{R}{r}\right)_{r=r_0} = C_0\left(\frac{R}{r_0}\right).$$

The solution of equation (7) is known

$$\frac{C-C_0}{C_p-C_0} = \frac{R-r}{R-r_0} - \frac{2}{\pi}\sum_{n=1}^{\infty}\frac{\sin\frac{\pi n(r-r_0)}{(R-r_0)}}{n}\exp\left[-\frac{\pi^2 n^2 Dt}{(R-r_0)^2}\right]. \tag{8}$$

The stress field of the wedge disclination changes symmetry of the diffusion equation. The cylindrically symmetric impurity segregation is formed according to the law of plane symmetry. It increases the rate of hydrogen segregation. Such increase of hydrogen segregation rate follows from the diffusion equation. Under the same conditions for $\frac{\partial C}{\partial r} < 0$ the segregation rate $\frac{\partial C}{\partial t}$ is larger in equation (7) in comparison with equation (6).

3. Kinetics of hydride phase growth

If the concentration of hydrogen atoms exceeds the solubility limit at the given temperature, hydride phases are formed. The phase transformation occurs in the area of the structural defects. The most powerful defect is wedge disclination. The hydride phase is formed near this defect. The further hydride growth comes about at the expense of hydrogen atom diffusion. On the moving interphase boundary hydrogen concentration changes intermittently : $C=C_1$ for a hydride phase and $C=C_2$ in the surrounding metal ($C_1>C_2$, $C_2>C_0$, where C_0 – average concentration of hydrogen atoms). The boundary of hydride phase immediately catches hydrogen atoms from a solid solution and delivers them to the composition with a higher concentration. Kinetics of diffusion growth of hydride phase is defined from the task solution.

$$\frac{1}{D}\frac{\partial C}{\partial t} = \frac{\partial^2 C}{\partial r^2} + \frac{1+\alpha}{r}\frac{\partial C}{\partial r},$$

$$C(R_1,t) = C_2, \quad C(r,0) = C_0\ (r \geq R_o),\ C(\infty, t) = C_0, \tag{9}$$

$$(C_1 - C_2)\frac{dR_1}{dt} = D\left\{\left|\frac{dC}{dr}\right| + \left|\frac{C\alpha}{r}\right|\right\}_{r=R_1},$$

where R_0-nucleus radius of hydride phase, R_1-current hydride radius. The rest designations correspond to accepted before. The distance between disclinations considerably exceeds the hydride phase size. It allows considering its growth in unlimited medium. Supposing it we can obtain the exact analytical task solution. The growth of hydride phase is limited by diffusion of hydrogen atoms. Changing hydride size is subject to the law $R_1(t) = \beta\sqrt{Dt}$, where β- dimensionless parameters. Its value can be found from the equation of mass balance on the interphase boundary.

We consider the solution of task (9) for $\alpha=-1$ and $\alpha=0$. The first case describes kinetics of hydride phase growth taking into account the disclination stress field. The second case describes the process kinetics without taking into account the stress field of the structural defect. Comparison of the two processes under the same conditions defines the contribution of disclination stress field. For $\alpha=-1$ kinetics of hydride phase growth can be found from the task solution

$$\frac{1}{D}\frac{\partial C}{\partial t} = \frac{\partial^2 C}{\partial r^2},$$
$$C(R_1,t) = C_1, \quad C(r,0) = C_0\ (r \geq R_o),\ C(\infty, t) = C_o, \tag{10}$$
$$(C_1 - C_2)\frac{dR_1}{dt} = D\left(\frac{dC}{dr} + \frac{C}{r}\right)_{r=R_1}.$$

In approximation to "a stationary interphase boundary" we obtain a quadratic equation for determining parameter β

$$\beta^2 - \frac{2\beta}{\sqrt{\pi}}\left|\frac{C_2 - C_0}{C_1 - C_2}\right| - \left|\frac{2C_2}{C_1 - C_2}\right| = 0. \tag{11}$$

The other approximations (for example, using a new variable $\frac{r}{2\sqrt{Dt}}$) do not change a qualitative process picture but just complicate the equation for determining parameter β [7]. Diffusion growth of hydride phase is a model approximation. Kinetics of hydride growth depends on competition of two processes: diffusion of hydrogen atoms and their absorption on the interphase boundary. Kinetics of hydride phase growth is changed from $\sim\sqrt{t}$ up to~t. Diffusion hydride growth supposes momentary absorption of hydrogen atom on the interphase boundary. Kinetics of hydride phase growth for $\alpha=0$ is defined from the task solution

$$\frac{1}{D}\frac{\partial C}{\partial t} = \frac{\partial^2 C}{\partial r^2} + \frac{1}{r}\frac{\partial C}{dr},$$
$$C(R_1,t) = C_1, \quad C(r,0) = C_0\ (r \geq R_o),\ C(\infty, t) = C_o, \tag{12}$$
$$(C_1 - C_2)\frac{dR_1}{dt} = D\left(\frac{dC}{dr}\right)_{r=R_1}.$$

In this approximation hydride phase growth occurs just at the expense of the gradient of hydrogen atoms concentration. Dimensionless parameter β_1 in relationship $R_1(t) = \beta_1\sqrt{Dt}$ we can find from the equation of mass balance on the interphase boundary. Parameter β_1 is a root of transcendental equation

$$\beta_1 = \frac{2}{\sqrt{\pi}} \left| \frac{C_2 - C_0}{C_1 - C_2} \right| \frac{K_1\left(\beta_1 \frac{\sqrt{\pi}}{2}\right)}{K_0\left(\beta_1 \frac{\sqrt{\pi}}{2}\right)}, \tag{13}$$

where K_0 (x) and K_1 (x)-modified cylindrical functions. The solution of equations (11) and (13) defines the contribution of the stress field of the wedge disclination into the kinetics of hydride phase growth. For $C_0=2\cdot10^{-4}$ (at.), $C_2=10^{-4}$ (at.) and $C_1=3\cdot10^{-4}$ (at.) we obtain

$$\beta^2 - \frac{\beta}{\sqrt{\pi}} - 1 = 0, \quad \beta_1 = \frac{1}{\sqrt{\pi}} \frac{K_1\left(\beta_1 \frac{\sqrt{\pi}}{2}\right)}{K_0\left(\beta_1 \frac{\sqrt{\pi}}{2}\right)}. \tag{14}$$

At $\frac{K_1(x)}{K_0(x)} > 1$, $\beta_1 > \frac{1}{\sqrt{\pi}}$. From the first equation (14) we also obtain $\beta > \frac{1}{\sqrt{\pi}}$. It is impossible to obtain inequality of $\beta > \beta_1$ without a numerical analysis. The solution of these equations gives $\beta=1.3$ and $\beta_1= 0.8$. The stress field of the wedge disclination increases the rate of hydride phase growth. Using the other values of limiting concentrations of hydrogen atoms will lead to changing numerical values of parameters β and β_1.

Hydride phase formation is accompanied by volumetrical changes. For zirconium and its alloys these changes are from 9 up to 12%. It exceeds the size change at the expense of different values of a coefficient of linear expansion of hydride and metal. Some stresses occur on the interphase boundary. They change kinetics of diffusion process. Oп the early stages of hydride phase growth such changes do not take into account. Formation of hydride phases in the form of long cylindrical needles can be observed experimentally. It takes place in the area of the threefold junctions of grain boundaries. The wedge disclination simulates the stress field of this structural defect. The threefold junctions come in contact with the grain boundaries. When increasing the size of a cylindrical hydride, its growth in the form of plates occurs along the grain boundaries. Hydride plates are the stress concentrators that lead to embrittlement of zirconium and alloys on its base[8].

4. Conclusions

Hydride phase formation occurs in the area of structural defects. They increase rate of diffusion migration of hydrogen atoms. The most powerful defect is wedge disclination. It models stress fields of different structural imperfections. Threefold junctions of grain boundaries are main ones at hydrogen embrittlement. In their area impurity segregations and hydride phases are formed in the first turn. The dilatation field of the wedge disclination logarithmically depends on a radial coordinate. Such dependence allows obtaining an exact analytical solution for changing the size of hydride phase. Kinetics of hydride growth occurs at the expense of hydrogen atom diffusion in the stress field of the wedge disclination. Hydride phases in the form of long cylindrical needles are formed along the lines of threefold junctions of grain boundaries. Then hydride plates in radial direction occur along the grain boundaries. Hydride plates cause embrittlement of zirconium and alloy on its base.

5. References

1. Hirth J.P., Carnahan B. (1978) Hydrogen absorption at dislocations and cracks in Fe. *Acta Metallurgica* **26**(12), 1795-1803.
2. Khachaturyan A.G. (1974) *Phase transformations and solid solution structures.* Nauka, Moscow.
3. Vlasov N.M., Zaznoba V.A. (1998) Phase transformations in the area threefold junctions of special grain boundaries. *Dokl. Acad. Nauk (Physics)* **363**(4), 472-474.
4. Vlasov N.M., Fedik I.I. (2000) Diffusional relaxation of the residual stresses. *Dokl. Acad. Nauk (Mechanics)* **375**(3), 334-337.
5. Teodosiu C. (1985) *Elastic models of crystal defects.* Mir, Moscow.
6. Vlasov N.M., Zaznoba V.A. (1999) Diffusion processes in the area of threefold junctions of special grain boundaries. *Fiz. Tverd. Tela* **41**(1), 64-67.
7. Lubov B.Ja. (1981) *Diffusion processes in heterogeneous solid media.* Nauka, Moscow.
8. Dutton R. (1984) *Materials degradation problems in hydrogen energy systems.* Int. J. Hydrogen Energy **9**(1/2), 147-155.

APPLICATION OF THE ELECTROCHEMICAL EXTRACTION METHOD TO THE STUDY OF A METAL HYDRIDE ELECTRODE - HYDROGEN SYSTEM

L.G. SHCHERBAKOVA
Frantsevich Institute for Problems of Materials Science, NAS of Ukraine
03142, 3, Krzhizhanovskii St., Kyiv, Ukraine

Abstract

Information about using the electrochemical extraction method for studying the processes of hydrogen transport in hydrogen-absorbing alloys is absent. By using this method, the kinetics of extraction of hydrogen from electrodes made of the alloy $MmNi_{3.5}(CoMnAl)_{1.5}$ was investigated. Time dependencies of the extraction current for different degrees of saturation of the electrode by hydrogen have the appearance typical of diffusion-controlled processes. By integrating the experimental extraction curves, the quantity of electricity spent on the ionization of hydrogen absorbed was evaluated, and then the total amount of hydrogen extracted was calculated. The ionization process is controlled by hydrogen diffusion in the solid phase; the kinetic parameters of the process were calculated. The change of the type of the time dependence of the extraction current with increasing degree of saturation of the electrode by hydrogen is explained by both a change in the character of hydrogen distribution in the sample and the possibility of transformation of the α phase into the $\alpha+\beta$ phase as the atomic ratio [H]/[alloy] rises.

1. Introduction

Due to the ability to absorb significant volumes of hydrogen and the high reversibility of processes of hydrogen absorption/desorption, materials of the AB_5 type are extensively used as negative electrodes in alkaline accumulators. During cycling, processes of hydrogen transport determine to a high extent the operating characteristics of the current source. The electrochemical properties of metals that absorb hydrogen are determined by the characteristics of the electrochemical process on the metal/electrolyte interface and hydrogen diffusion processes in the bulk of the material.

In the determination of the diffusion coefficient and the rate of penetration of electrolytic hydrogen into metals, the bipolar membrane method, based on the registration of a hydrogen flow through a thin membrane, is extensively used [1,2]. Among material disadvantages of this method are sophisticated apparatus and the complexity of a theoretical interpretation of results, taking into account the possibility of realizing hydrogen exchange reactions on two interfaces. With using this method, the kinetics of penetration of hydrogen through $LaNi_{3.94}Si_{0.54}$ alloy films applied to the metal base has been investigated [3].

N. Veziroglu et al. (eds.), Hydrogen Materials Science and Chemistry of Metal Hydrides, 289–294.

The method of electrochemical extraction is free from disadvantages characteristic of procedures using membranes, since it enables one to perform saturation followed by hydrogen extraction in a conventional electrochemical cell under conditions of the controlled hydrogen content in a sample. The major merit of the method is that a curve hydrogen ionization current-time can be recorded at any initial distribution of hydrogen across the section of a sample, i.e., the requirement concerning the saturation of a tested electrode by hydrogen across the whole thickness is absent. The theoretical principles of the method of electrochemical extraction have been presented in work [4] under the assumption that hydrogen transfer is controlled by stages of diffusion in the material and by an interface penetration reaction.

Information about using the extraction method for studying the processes of hydrogen transport in hydrogen-absorbing alloys is absent. In the present work, an attempt to extend this method for studying the process of hydrogen ionization during discharging of an AB_5 type alloy was made.

2. Experimental details

An alloy of composition $MmNi_{3.5}(CoMnAl)_{1.5}$ (Mm is a mesh metal enriched to 52 mas.% in cerium) was made by argon-arc melting followed by the mechanical crushing of the ingot and grinding of the powder by cycling in hydrogen. The alloy powder with a particle size -60 mesh (0.2 g) was mixed with polyvinyl alcohol (3 mass %), used as a binder, and pressed under a pressure of 250 kg/cm^2 in a pellet 1.0 cm in diameter and 0.05-0.06 cm in thickness. An electrode was placed in a nickel cup-shaped holder (served as a current cap), mounted into a polytetrafluorethylene bottom of a clamping glass cell. A three-electrode cell with an anodic and a cathodic space separated from one another, a platinum plate, which served as a sintered electrode, and a Hg/HgO reference electrode were used. A 6M KOH working solution was not deaerated before experiments. All investigations were performed with using a P-5848 potentiostat and a KSP-4 recorder.

The investigated electrode was electrolytically saturated with hydrogen by the method of galvanostatic polarization (regimes are given in Table 1). Then the electrode was subjected to potentiostatting at a potential providing the ionization of hydrogen absorbed, and a time dependence of the ionization current was taken.

TABLE 1. Results of data treatment for extraction curves

Curve No.	Charging current, mA	Charging time, Min	C_H, g/cm^3	H/alloy ratio	D, cm^2/s
1	5.25	5	$3.7\cdot10^{-4}$	0.03	$4.20\cdot10^{-7}$
2	5.25	15	$1.12\cdot10^{-3}$	0.10	$4.40\cdot10^{-7}$
3	5.25	30	$2.24\cdot10^{-3}$	o.19	$1.73\cdot10^{-7}$
4	10.0	30	$4.28\cdot10^{-3}$	0.37	$2.02\cdot10^{-7}$
5	20.0	18	$4.90\cdot10^{-3}$	0.43	$1.80\cdot10^{-7}$

3. Results and discussion

During galvanostatic polarization of the electrode from the alloy $MmNi_{3.5}(CoMnAl)_{1.5}$ by cathode current, the electrode potential shifted in the direction of negative values ranging from -0.98 to -1.02 V, and a dependence of the electrode potential on the charging time looked like a typical charging curve. After switching-off, the electrode potential settled rapidly in the range from -0.96 to -0.94 V and remained practically unchanged, which suggests that the rate of the process of spontaneous desorption of hydrogen from the electrode is small.

The ionization potential of hydrogen was chosen such that potentiostatting provides zero surface concentration of hydrogen. This can be fulfilled by the polarization of the electrode at potentials of the current-potential curve when hydrogen is ionized at a limit current

$$MH_x + xOH^- \rightarrow M + xH_2O + xe^-.$$

Peaks of the anode current in voltammograms ($\partial E/\partial \tau$ = 1mV/s) obtained on electrodes with different amount of hydrogen absorbed were observed at potentials from -0.80 to -0.75 V. The value of the current of the anode peak increased proportionally to the concentration of hydrogen absorbed. The extraction of hydrogen was performed at a more positive potential (E_{ext} = -0.45 V). The anodic dissolution currents of the initial alloy in the potential range -0.70 V < E < -0.40 V did not exceed 0.25 mA/cm^2, which is two orders of magnitude smaller than the values of the limit currents of hydrogen ionization. That is why, in treatment of results, the correction for the anodic dissolution of the alloy was not made. The potentiostatting of the electrode was continued until the extraction current settled at a stationary value of the extraction current ($i_{ext.}$). The value of the extraction current and the time required to attain it depended on the concentration of hydrogen absorbed.

Time dependences of the extraction current for different degrees of saturation of the electrode by hydrogen have the appearance typical of diffusion-controlled processes (see Figure 1). By integrating the experimental extraction curves, the quantity of electricity spent on the ionization of hydrogen absorbed was evaluated, and then the total amount of hydrogen extracted (c_H) was calculated. Depending on regime of saturation, they were $(0.33\text{-}4.5)\cdot 10^{-3}$ g/cm^3 (see Table 1). Good agreement (within 5%) between the concentrations of ionized hydrogen calculated from current-potential curves and those calculated by the extraction method was observed.

The treatment of the data obtained was performed with using relations presented in [4]. According to Eq.(22) [4], extraction curves plotted on the semilogarithmic coordinates $\ln i_{ext.}$ - t are linear if the process is limited by the stage of hydrogen diffusion in the solid phase. However, only for the electrodes with an initial hydrogen concentration of $3.75\cdot 10^{-4}$ g/cm^3 and $1.12\cdot 10^{-3}$ g/cm^3, the extraction curves plotted on the coordinates $\lg i_{ext}$- t had the linear section in the time interval 10 s < t < 300 s (see Figure 2,a). Values of the calculated diffusion coefficients (D_H) for both degrees of saturation by hydrogen are practically identical and range from $4.2.10^{-7}$ to $4.4\cdot 10^{-7}$ cm^2/s. The calculated depth of penetration of hydrogen into the electrode with taking into account the extraction time does not exceed 0.02 cm, i.e., hydrogen is localized in the upper part of the electrode.

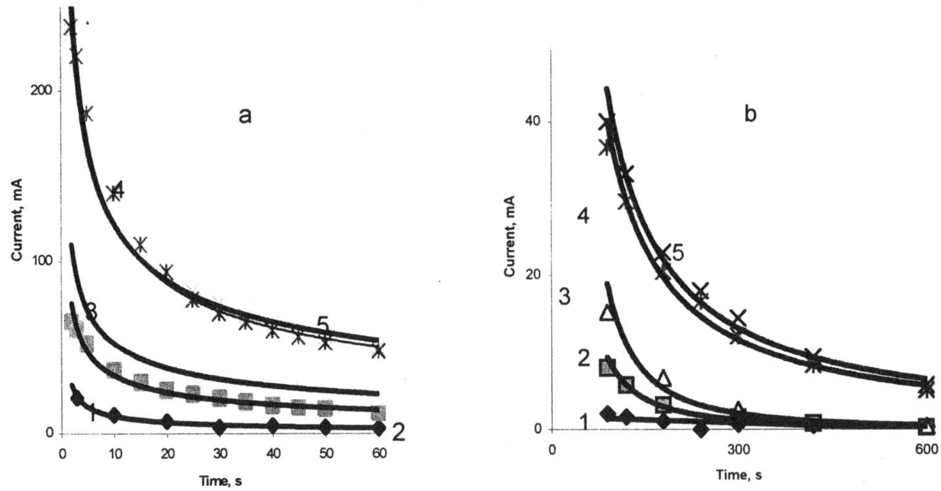

Figure 1 Time dependences of hydrogen exstraction currents (a,b), a - initial section (The saturation conditions are given in Table 1).

For electrodes with higher degrees of saturation by hydrogen , experimental values are approximated with a high degree of accuracy by the dependence $i = at^{-x}$ at $x \approx 0.5$. Chronoamperograms plotted on the coordinates $i - t^{-0.5}$ in the time range $3 < t < 600$s are linear (see Figure 2,b). The linearity of the $i - t^{-05}$ relationship is a criterion of processes controlled by nonstationary volume diffusion in selective metal dissolution [5] and reduction of a depolarizer on electrodes [6], and is also realized in the process of diffusion transport of hydrogen through the coating-base system [7]. The diffusion coefficients of hydrogen calculated by the Cottrell equation [6] have values of (1.7-2.0) 10^{-7} cm^2/s. With taking into account the diffusion rate and extraction time, it can be supposed that hydrogen is distributed across the whole thickness of the electrode.

The pressed porous electrode made of the alloy $MmNi_{3,5}(CoMnAl)_{1,5}$ can be considered as a system with parameters distributed twice, namely across the thickness of the electrode and along the particle radius. That is why the rate of the process of hydrogen ionization during discharging can be determined by both the hydrogen diffusion in alloy particles and volume diffusion. Consequently, a change of the type of the time dependence of the extraction current can be connected with either difference in the character of hydrogen distribution across the thickness of the sample or the transformation of the α phase of hydrogen dissolved in the alloy into the $\alpha+\beta$ phase as the atomic ratio [H]/[alloy] increases, or the combination of both factors.

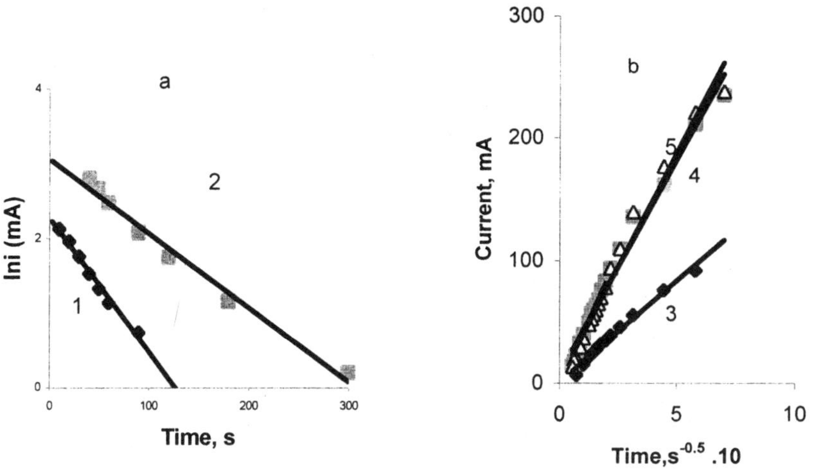

Figure 2 .Time dependence of hydrogen extraction currents for degrees of saturation by hydrogen $3{,}75.10^{-4}$ g/cm^3 (1), $1{,}12.10^{-3}$ g/cm^3 (2), $2{,}24.10^{-3}$ g/cm^3 (3) , $4{,}28\ 10^{-3}$ g/cm^3 (4) and $4{,}90.10^{-3}$ g/cm^3 (5).

4. Conclusions

By using the method of electrochemical extraction, the kinetics of extraction of hydrogen from electrodes made of the alloy $MmNi_{3.5}(CoMnAl)_{1.5}$ was investigated. The ionization process is controlled by hydrogen diffusion in the solid phase; the kinetic parameters of the process were calculated. The change of the type of the time dependence of the extraction current with increasing degree of saturation of the electrode by hydrogen is explained by both a change in the character of hydrogen distribution in the sample and the possibility of transformation of the α phase into the $\alpha+\beta$ phase as the atomic ratio [H]/[alloy] rises.

5. References

1. Devanathan, M. and Stachurski, Z. (1962) The adsorption and diffusion of electrolytic hydrogen in palladium, *Proc.Roy.Soc.* **A270**, 90-96.
2. Chernenko, V.I., Krapivnyi, N.G., and Zverev, N.N. (1979) On realization of limit conditions in the study of hydrogen diffusion through metallic membranes, *Problems of Chemistry and Chemical technology*, Kharkov, Issue 54, pp. 28-35.
3. Zhu,Z., Yuang, J., Yao, W., Zhang, Y. (1995) Diffusion studies of microcrystalline and crystalline $LaNi_{3,94}Sn_{0.54}$ films, *J.Alloys Comp.***231**, 297-301.
4. Krapivnyi, N.G (1982) Application of electrochemical extraction to the study of hydrogenation of metals , *Electrohimiya* **18**; N9; 1174-1178.
5. Marshakov, I.K., Vvedenskii, A.V., Kondrashin, V.Yu. and Bokov, G.A.(1988) *Anodic selective corrosion of alloys,* Voronezh, Russian, p.294.

6. Galyus, Z. (1974) *Theoretical foundations of electrochemical analysis*, World Publishers, Moscow.
7. Krapivnyi, N.G.(1981) Regularities of the diffusion of electrolytic hydrogen in a base during metal electrodeposition, *Electrokhimiya*, **17**, 678-685.

CONSTRUCTION PRINCIPLES OF METAL-HYDRIDE GAS SUPPLY SYSTEMS FOR VACUUM-PLASMA INSTALLATIONS

YU.F.SHMAL'KO, YE.V.KLOCHKO, M.V.LOTOTSKY,
V.N. BORISKO
A.N.Podgorny Institute of Mechanical Engineering Problems of National Academy of Sciences
of the Ukraine, 2/10 Pozharsky Str., Kharkov, 61046 Ukraine
(smalko@ipmach.kharkov.ua)
V.N.Karasin Kharkov National University, 4 Svobody Sq., Kharkov, 61077 Ukraine

Abstract

It is shown, that the most perspective way of metal hydrides usage in gas supply systems of hydrogen-utilizing vacuum-plasma installations is the creation of multifunctional metal-hydride elements on the basis of hydride-forming materials – reversible low-pressure hydrogen sorbents. These elements are the component parts of a working zone of a vacuum-chamber in consumer installation (e.g. the electrodes of a plasma-forming step). In so doing, apart from complex solution of auxiliary problems of hydrogen isotopes compact storage, their purification and controlled leak-in into the working space of an installation vacuum chamber, the installation autonomy as to its evacuation and gas supply is provided, as well as the working efficiency of the installation as a whole is increased. On the basis of the analysis of the features of metal hydride systems for hydrogen isotopes "internal" leak-in, the general principles of building-up such the systems are formulated.

1. Introduction

The development and exploitation of various vacuum-plasma installations anticipates the usage of low-pressure hydrogen isotopes as a working medium. One of effective ways to create the appropriate gas supply systems is the usage of reversible hydrides of metal, alloys and intermetallic compounds. The main advantages of these materials are high hydrogen packing density in the metallic matrix; wide pressure ranges of hydrogen uptake and release depending on the kind of a metal hydride sorbent, with the help of thermal action for effective control of hydrogen supply pressure; possibility to decrease power inputs for hydrogen excitation, dissociation and ionization at the expense of hydrogen thermo-sorption activation effect; etc.

The analysis of existing possibilities of metal hydrides usage in hydrogen isotopes supply systems for vacuum-plasma installations of technological and power purposes [1,2] has shown that the unique combination of hydrogen-sorption and thermo-emission

N. Veziroglu et al. (eds.), Hydrogen Materials Science and Chemistry of Metal Hydrides, 295–301.

properties of hydrides – reversible hydrogen getters – allows one to realize the principally new schemes of plasma-forming gas "internal" leak-in. Such the schemes are provided by the introducing a metal-hydride element on the basis of hydrogen-saturated getter material into construction units in consumer devices. It allows one to eliminate leak device, input gas line, storage and pre-treatment systems for higher-pressure hydrogen. Schematically such a solution is shown in Fig. 1.

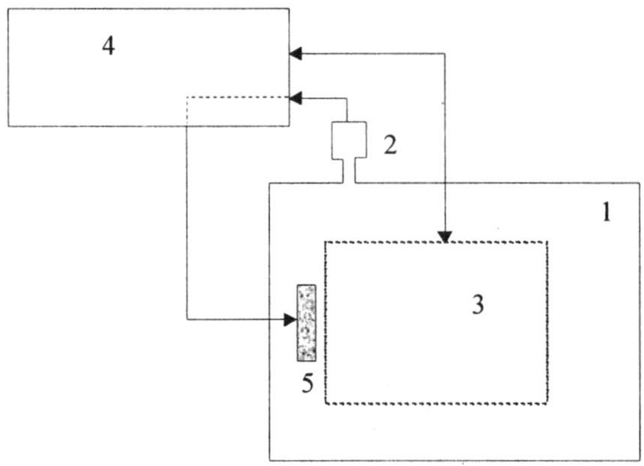

Figure 1. The scheme of "internal" hydrogen leak-in into a vacuum-plasma installation.

The metal-hydride element (5) is placed directly in the vacuum chamber (1) of an installation. The latter contains a vacuum sensor (2), which is the feedback one in pressure controlling circuit as a component part of the control block (4) in consumer installation. Pressure increase in the working space (3) of the vacuum chamber (1), or the decrease of the element (5) working temperature results in hydrogen absorption by metal hydride. Using such a feature, it is possible to relieve or fully eliminate an evacuation system of the vacuum chamber (1). At the same time, one or several elements (5) combine the functions of hydrogen source and getter facility for its evacuation.

We have shown [2] that such schemes have a number of advantages as compared to conventional gas supply systems. However, the direct realization of such an approach requires the active thermal action onto metal hydride element to provide its required operation mode. It is connected with an essential complication of the system for leak-in control.

This work deals with the new approaches to the creation of metal-hydride systems of hydrogen isotopes "internal" leak-in into vacuum-plasma installations. The approaches use a possibility of direct interaction of metal hydride surface with hydrogen plasma. It is an important factor promoting the enhancement of technical characteristics both of a metal-hydride hydrogen injector and of an installation as a whole.

2. The features of metal-hydride systems for “internal” hydrogen leak-in into vacuum-plasma installations

The possibility to change hydrogen sorption – desorption dynamics in course of the interaction of a metal hydride element with hydrogen plasma opens the perspectives for creation of new gas supply systems which do not require the complicated external control blocks [3]. The essence of such an approach is in using metal hydrides on the basis of reversible hydrogen getters as the materials of electrodes in a vacuum-plasma device. The role of controlled action onto the metal-hydride elements is played, in such a case, by the element bombarding with high-energy plasma particles. It, in turn, stipulates the desorption of the plasma-forming gas from the electrode material.

Fig. 2 illustrates schematically the application possibilities of hydrogen-gettering materials in gas-discharge energy-technological devices using hydrogen isotopes. The ranges of the plasma-forming gas working pressures are plotted along the ordinate axis, and the abscise axis represents the available temperature, up to which the constructive elements of gas-discharge devices can be heated.

The working pressure and temperature regions represented in the diagram (Fig. 2) are in good agreement with the temperatures and pressures of dissociation of the typical hydrogen-gettering materials.
It allows one to combine the main functions of the construction elements (e.g. electrodes) of gas-discharge devices with controlled plasma-forming gas supply into discharge region.

The introducing a metal hydride with moderately high thermal stability into the composition of gas-discharge cell cathode can, in a number of cases, result in an interesting effect of formation of pressure-auto stabilizing mode of the gas discharge which burns in a closed volume in hydrogen medium [3,4]. The intensive ion bombardment of the cathode stipulates hydrogen desorption from metal hydride into discharge burning region. Under these conditions the stable discharge burning is observed, and the pressure in a gas-discharge cell is unambiguously defined by a single external factor, namely discharge current (Fig. 3,4). It opens the perspectives to create the installations, which are completely autonomous as to their evacuation and gas leak-in [3].

The advantage of the proposed approach to building-up metal hydride gas supply systems is a possibility of plasma-forming gas (hydrogen isotopes) supply directly into the region of its usage. At the same time the hydrogen being supplied is in the excited state [5] that allows one to increase the efficiency of processes of negative and positive ions formation.

The example illustrating the realization of such the statement is the developed and studied by the author’s plasma source of hydrogen ions on the basis of the reflective discharge with metal hydride cathode [6].

In the Fig. 5 the dependency of current density of H^- ions extracted from such an ion source on the pressure of the bleeding hydrogen (curve 1) is shown. In the same Figure the similar dependence for the plasma source of the same geometry but having the cathode made of stainless steel (curve 2) is also presented.

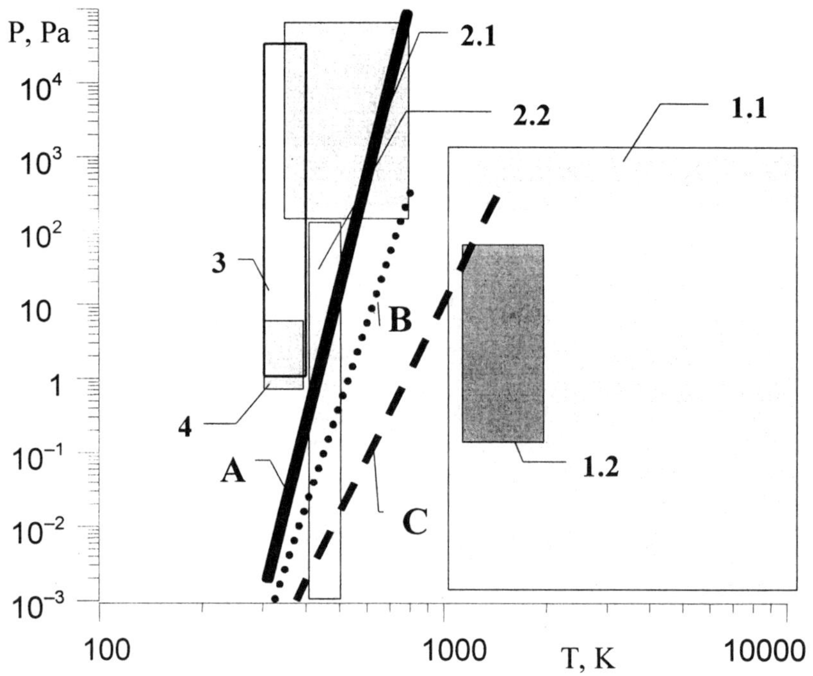

Figure 2. The regions of operating parameters of vacuum-plasma devices and the possibilities of hydrogen getters applications for their gas supply.

The estimated "pressure – temperature" dependencies for the materials of compositions:
A – ZrNi; **B** – Zr–V–B_2O_3; **C** – Zr–Al.

Discharge types in vacuum-plasma installations:

1.1 – Low-pressure arc with cathode spots (ion sources; plasma accelerators; low-thrust arc jet engines; neutron generators; generators of the intensive optical, UV- and sort X-irradiation; hydrogen lasers; plasma current switch-boxes, dischargers; technological systems of diamond films deposition).

1.2 – Low-pressure arc with incandesced cathode (ion sources; injectors of neutral atom beams into controlled fusion installations; plasma technological sources of powerful electron beams).

2.1 – Glow discharge without magnetic field (technological systems of diamond-like films synthesis; plasma-chemical systems; plasma current switch-boxes).

2.2 – Glow discharge in magnetic field (ion sources; technological systems of diamond-like films synthesis; plasma accelerators; solar wind action simulators; plasma-optical systems).

3 – Re-charging targets, protective gas targets, gas recuperators of the energy of powerful charged particle beams.

4 – HF-discharges (ion sources; technological sources of monatomic hydrogen; hydrogen frequency standards, hydrogen masers; plasma-chemical reactors; technological systems for amorphous silicon films synthesis).

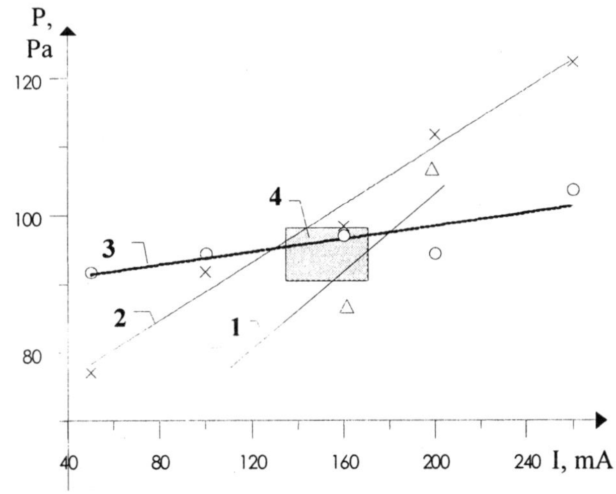

Figure 3. The dependence of hydrogen pressure (P) in vacuum chamber on the current of magnetron discharge (I) burning in the closed volume in hydrogen medium. 1 – the cathode made of hydrogen-saturated getter alloy of Zr–V–Fe + B_2O_3; 2 – the cathode made of hydrogen-saturated getter alloy of Zr–V–Fe; 3 – steel cathode (external hydrogen supply); 4 – discharge ignition region.

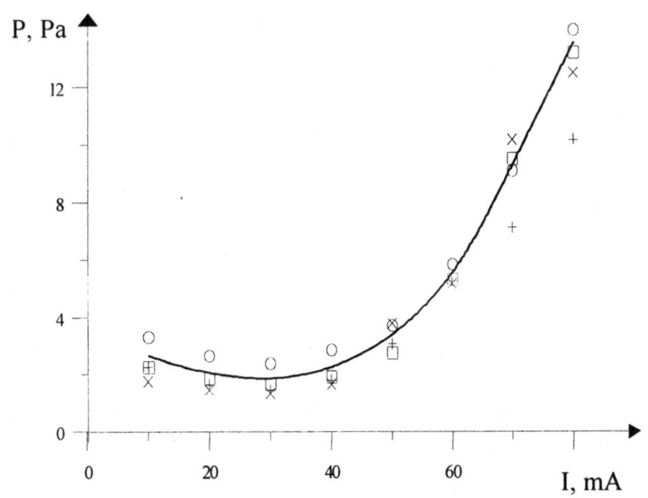

Figure 4. The dependence of hydrogen pressure (P) in vacuum chamber on the current (I) of reflective discharge burning in the closed volume in hydrogen medium (the cathode made of hydrogen-saturated getter alloy of Zr–V–Fe + B_2O_3) at the values of the applied magnetic field strength: H=400 Oe (×); H=600 Oe (+); H=800 Oe (□); H=1000 Oe (○)

The data analysis (Fig. 5) denotes the existence of the critical pressure for the source with metal hydride cathode whose exceeding results in significant (approximately by the order of magnitude) increase of H^- beam current. Thus the current efficiency of such ion source achieves to 40–50%. It essentially exceeds the similar characteristic for the constructions of such devices, which are known to the authors.

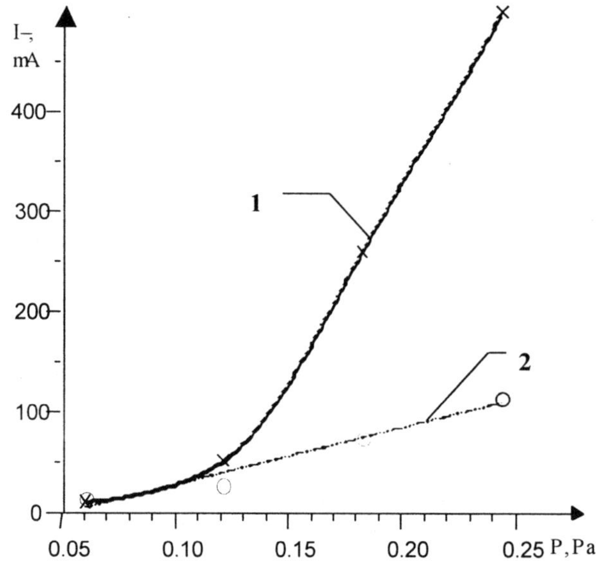

Figure 5. The dependency of the current I^- of extracted negative ions H^- beam on hydrogen pressure P in gas discharge chamber of the ion source:
1 – metal hydride cathode; 2 – stainless steel cathode.

The high purity and non-equilibrium state of hydrogen supplied from metal hydride elements into a working zone of a vacuum-plasma technological installation allows one to influence effectively on the elementary processes in the plasma volume and on the "plasma – solid" interface. It opens the new possibilities to improve plasma technologies. For example, the application of metal hydride gas supply systems in the installations for the synthesis of diamond- and diamond-like films allows us to improve essentially their quality.

Apart from providing the plasma-forming gas supply into vacuum-plasma devices, metal hydride "internal" leak-in systems can be used for creation of gas targets, which protect the construction elements of gas discharge devices against sputtering [7]. The creation of re-charging gas targets is also available.

3. Conclusions

Starting from the mentioned above, it can be formulated the following principles of building-up the "internal" schemes of gas supply for vacuum-plasma installations which are based on getter hydride-forming materials.

1. Introducing the hydrogen-saturated hydride-forming material into the structure of construction elements in a vacuum-plasma installation, which are placed close to the plasma-forming region.
2. The selection of hydride-forming material component composition providing its hydride phase dissociation pressure on the level of working

pressure in the gas-discharge cell, at the temperature closed to the one, up to which metal hydride element bombarded by plasma particles is heated.

3. Providing leak-in stability due to internal feedback between hydrogen desorption intensity from metal hydride element and gas discharge parameters.

This work is carried out partially in the framework of the Project # 959 of the Science and Technology Center in Ukraine (STCU).

4. References

1. Shmal'ko, Yu.F., Solovey, V.V. and Lototsky, M.V. (1994) Use of metal hydrides in systems for supplying vacuum physical-energy installations, in D.L.Block, T.N.Veziroglu (eds.), *Hydrogen Energy Progress X. Proc. 10-th Word Hydrogen Energy Conf. (Cocoa Beach, Florida, U.S.A., 20 – 24 June, 1994,* **2**, 1311–1319.
2. Klochko, Ye.V., Lototsky, M.V., Solovey, V.V. and Shmal'ko, Yu.F. (1999) Metal hydride systems for processing hydrogen isotopes in physical-technical installations, in *ICHMS'99: VI Int. Conf. "Hydrogen Materials Science & Chemistry of Metal Hydrides". Katsively, Yalta, Ukraine, Sept. 2–8, 1999,* 232–233 (in Russian).
3. Borisko, V. N., Klochko, Ye. V., Lototsky, M.V., Petrushenya A.A. and Shmal'ko Yu.F (2000) The auto stabilized on pressure systems of a gas feed for vacuum-plasma installations with use of a convertible hydrogen getters, *Vacuum Engineering and Technology* **10**, No. 1, 15–19 (in Russian).
4. Klochko, Ye.V., Lototsky, M.V. Popov,V.V., Shmal'ko, Yu.F. and Borisko, V.N. (1997) Sorption and electro transfer characteristics of hydrogen-gettering material in contact with a hydrogen plasma, *J. Alloys and Compounds* **261**, 259–262.
5. Borisko, V.N., Shmal'ko, Yu.F., Lototsky, M.V., Klochko, Ye.V. and Solovey, V.V.(2000) About oscillatory excitation of hydrogen molecules, desorbed from metal hydrides *Dopovidi NAN Ukraine (Reports of National Acad. Sci. of the Ukraine)* **11**, 91–95 (in Ukrainian).
6. Borisko, V.N., Klochko, Ye.V., Lototsky, M.V. and Shmal'ko, Yu.F. (1998) Technological plasma sources of negative ions *VANT (Problems in Nuclear Science and Engineering, Series: Physics of Radiation Damages and Radiation Material Science)* **3(69), 4(70)**, 179–182 (in Russian).
7. Borisko, V.N., Bobkov, V.V., Chebotarev, V.V., Garkusha, I.E., Lototsky, M.V., Poltavtsev, N.S., Rudaya, I.A., Shmal'ko, Yu.F., Starovojtov, R.I., Tereshin, V.I. and Byrka, O.V. (2000) Influence of plasma treatment on erosion characteristics and structure of reversible hydrogen getters *VANT (Problems of Atomic Science and Technology. Series: Plasma Physic)* **6**, 166–168 (in Russian).

ROLE OF METALS HYDRIDES AT PRODUCING OF MAGNETIC MATERIALS BASED ON Nd-Fe-B SYSTEM

P.A. MANORIK, A.A. PAVLYUKOV[(1)], V.Y. TSIVILITSIN[(1)], O.S. OPANASENKO[(1)], I.B. BONDAR[(1)], V.K. BUKHTIYAROV, S.V. DEMESHKO, M.A. PHEDORENKO, I.V. TROTSUK
L.V.Pisarzhevsky Institute of Physical Chemistry NASU, Pr. Nauki 31, Kiev, 03039 Ukraine
[(1)]Institute for Problems of Materials Science NASU Krgiganovcky str. 3, Kiev, 03142 Ukraine

1. Introduction

At present time the greatest level of energy product the permanent magnets based on Nd-Fe-B alloys have. Its are produced with methods of powder metallurgy according to the following technological scheme: alloys melting, its fine mechanical crushing, powder pressing in a magnetic field with the subsequent operations of sintering and heat treatment of products. In residual induction and power product they surpass the nearest class of Sm-Co magnets in 1,5 times. Operating temperature of Nd-Fe-B magnets in which the part of neodymium is replaced with dysprosium or terbium, now reach 200°C. Cost of these magnets in 3-4 times is less, than magnets based on Sm-Co that opens wide opportunities of their industrial usage.

The lack of the magnetic alloys based on Nd-Fe-B is their rather low corrosion resistance both in as-cast and in sintered alloys, that negatively influences on all magnetic properties of permanent magnets. The protection of magnetic materials toward corrosion is a rather urgent task, as it allows increasing of values and stability of the magnetic characteristics of the permanent magnets.

2. Results and discussion

The initial alloys were melted in the electroarc furnace with tungsten electrode in an argon atmosphere, alloying neodymium, ARMCO iron and iron - boron ligature (10-12 weight percents of boron). After melting the as-cast alloy was grinded in jaw crusher in argon environment at first. Then powder was grinded in a ball mill during various intervals of time in protective isopropanol environment at cooling in the refrigerating chamber for reception of powders with an average diameter of particles 4-10 μm. Further a suspension of powder in alcohol was pressed in a magnetic field directed perpendicularly to a direction of effort of pressing. Sintering of received moulds was carried out in the vacuum furnace with vacuum not less than 10^{-5} millimetres of mercury at temperature 1040-1100°C.

N. Veziroglu et al. (eds.), Hydrogen Materials Science and Chemistry of Metal Hydrides, 303–308.

X-ray phase analysis was carried out with the diffractometer DRON 3M with monochromatic X-ray radiation of CoK_α and CuK_α. Chemical structure of phases which are included in structure of magnetic materials, was defined with the Auger spectrometer JAMP-10S (JEOL), having the deep resolution in some nuclear layers and spatial resolution 50 nm. The researched surfaces were received by means of samples chipping in conditions of high vacuum inside spectrometer.

According to [1] the hard magnetic sintered alloys $Nd_{15}Fe_{77}B_8$ consists of three phases: the main phase $Nd_2Fe_{14}B$ and two additional, one of which is enriched with neodymium - more than 80 % (weights), and another - with boron $Nd_2Fe_7B_6$. The additional phases are located, as a rule, along borders of grains of the main phase.

Last years the reports on participation of hydrogen in corrosion and destruction of magnets based on Nd-Fe-B alloys have appeared. This effect, so termed «the Negative Harris Effect» [2], consists in adsorption with magnets of hydrogen from environment with hydrides formation, for example, $Nd_2Fe_{14}B$-H and NdH_x, which at the further interaction with water vapours, give neodymium hydroxides $Nd(OH)_3$. This phase has much greater atomic volume in comparison with the initial alloy. Its formation results in destruction of the as-cast alloy and magnets on based on Nd-Fe-B system.

Possible formation of oxygen-bearing phases at corrosion of an alloy Nd-Fe-B was reported by us earlier [3, 4]. The additional investigation of oxidation process of Nd-Fe-B alloy Auger-spectroscopy had confirmed also the possibility of direct oxidation by oxygen, contained in air, both as the main as the additional phases. Auger-electron spectra of all phases of cast alloy recorded after its fracture inside the JAMP are presented in figure 1. Content of Nd, Fe and B in phases is in a good agreement with data [1]. Auger peaks of oxygen are present on spectra of any phase constituting the alloy. It can be easily seen that there is correlation between content of neodymium and oxygen. Increasing of oxygen content follows enlargement of neodymium one. It is reasonable to assume that at saving of $Nd_{15}Fe_{77}B_8$ alloy ingots under usual conditions the Nd-rich phase will be oxidised first because the phase has the greatest affinity with oxygen. The investigation of the powder X-ray spectrum (figure 2a) shown that it could correspond to hexagonal lattice (space group $C6_3/m$) with parameters a=0,6426 nm, c=0,3729 nm. Neodymium hydroxide ($Nd_2O_3 \cdot 3H_2O$) had the same spectrum [5]. This conclusion was confirmed by comparison of X-ray spectra of light grey powder annealed (dehydrated) at 1270 K and commercial Nd_2O_3. In both cases spectra corresponded to La_2O_3-type hexagonal structure (space group 3m) with parameters a=0,3826 nm, c=0,5996 nm and belonged to A-modification of Nd_2O_3. It was impossible to identify weak additional peaks that were present on figure 2b.

Nd-rich phase in $Nd_{15}Fe_{77}B_8$ cast alloy is situated along grain boundary of main phase in the form of sufficiently large oblong inclusions, outcropping of ingot (figure 3a). Morphology of sintered alloy microstructure (figure 3b) differs from the one of cast alloy. Microstructure of alloy to be produced from powder with 7.2 and 14.4 ks grinding time and sintered at 1330 K are presented in figure 3b and 3c respectively. In this case additional phases are distributed into sample volume in form of isolated inclusions of 1-10 μm. Oxidation of such relatively small inclusions disposed on sample surface lead to appearance of little quantity of powder only. However, if such inclusions interconnect each other, it can lead to destruction of final product. It is remarkable that big inclusions of Nd-rich phase are absent in microstructure at last case.

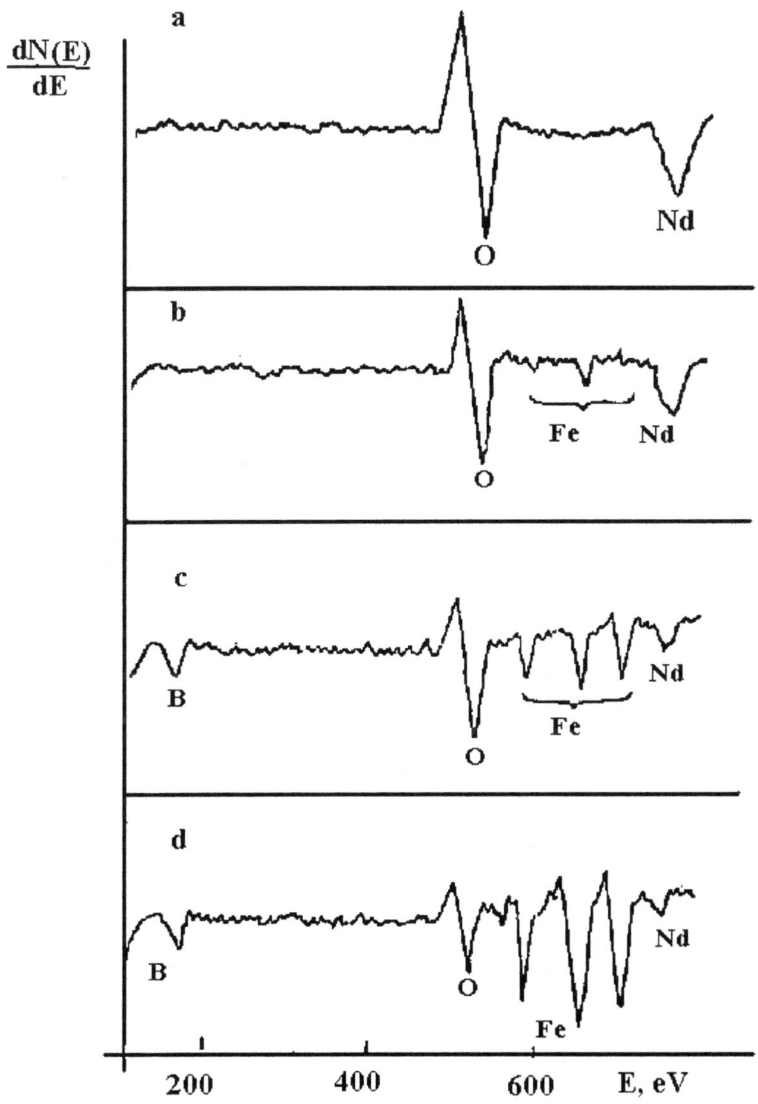

Figure 1. Auger-electron spectra: Oxide film (a), Nd-rich phase (b), B-rich phase (c) and hard magnetic main phase (d).

Just the oxidation of Nd-rich phase which is responsible for liquid phase sintering and high coercivity, results in corrosion of magnets. One of directions of reception of corrosion resistant magnets may be doping of intergranular Nd-rich phase, or protection of this phase toward oxidation. It was shown that use of hydridoborates (for example, $NaBH_4$) or aluminium hydrides (for example, $LiAlH_4$), both as at the stage of producing of the as-cast alloy as powders sintering, was the convenient method of doping of

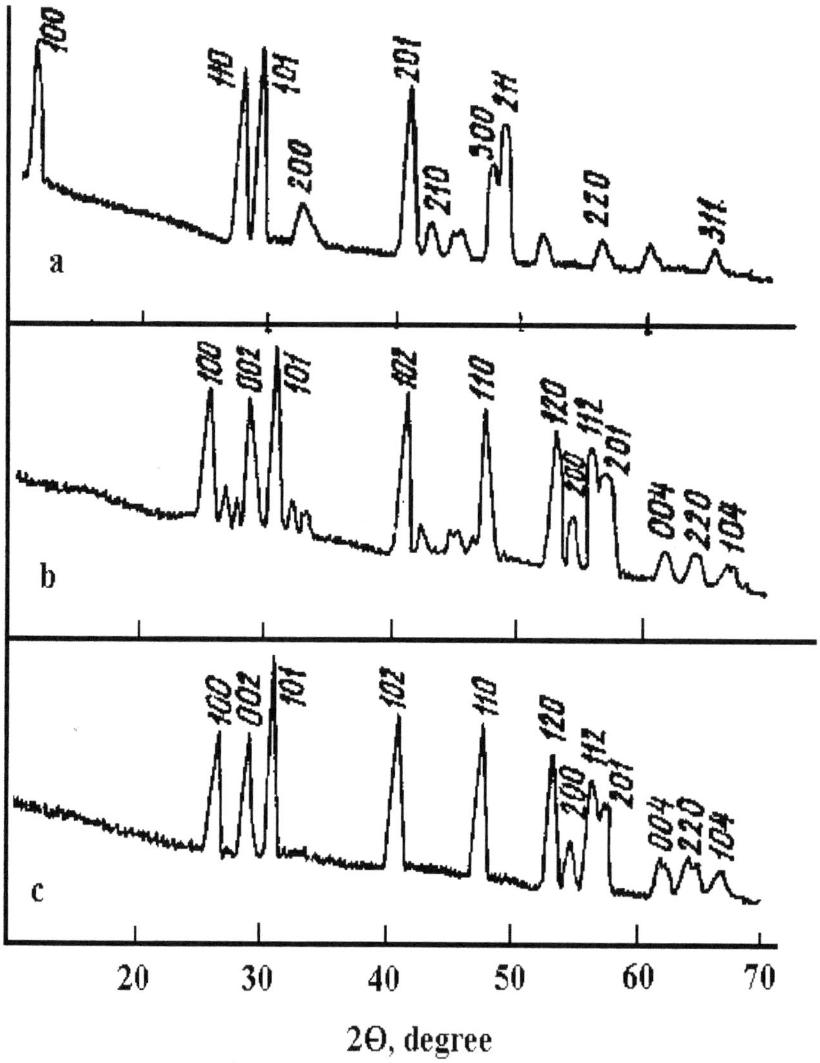

Figure 2. X-ray spectra: neodymium hydroxide $Nd_2O_3 \cdot 3H_2O$ (a), annealed (dehydrated) $Nd_2O_3 \cdot 3H_2O$ at 1270 K(b) and commercial Nd_2O_3(c).

Nd-Fe-B alloys with lithium, sodium, aluminium. Doping by the mentioned elements of intergranular layer rendering essential influence on the operational characteristics of magnets. E.g. the introducing up to 0,1% (wt.) of $NaBH_4$ into the powder of Nd-Fe-B alloy before its sintering increases intrinsic coercive force ($_iH_c$) of obtained magnets on 10 - 20 % comparing with control samples.

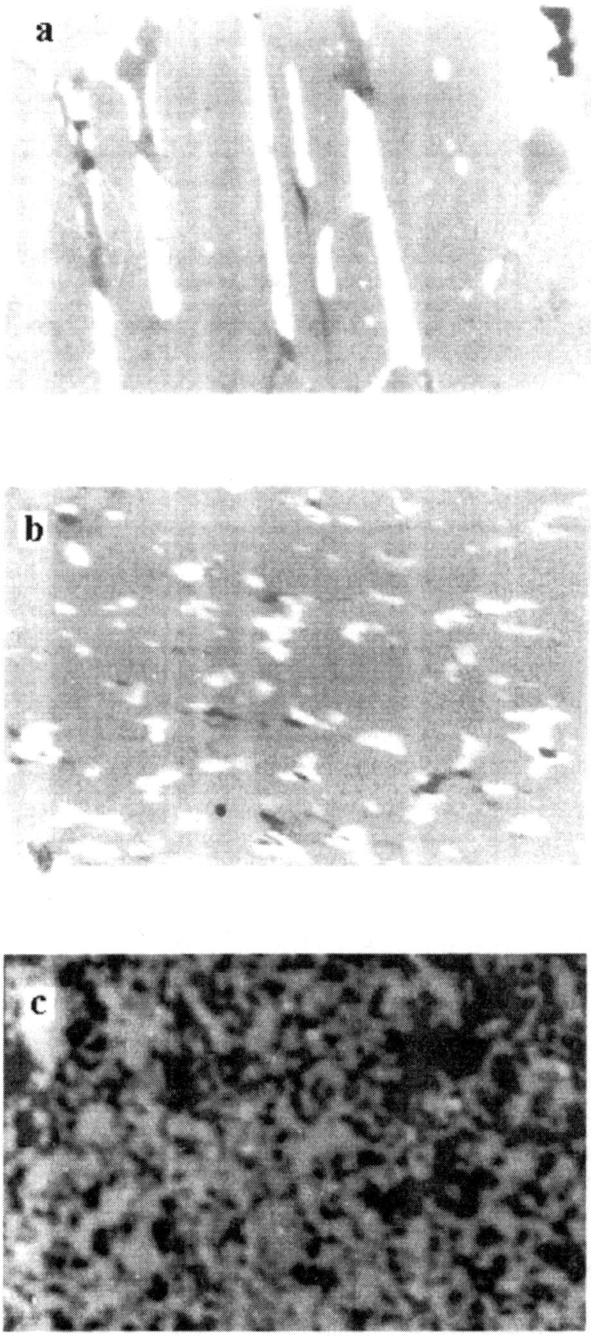

Figure 3. Microstructure of cast (a) and sintered (b, c) alloy in reflected electrons (x1000): b - grinding 7.2 ks, sintering at 1330 K; c - grinding 14.4 ks, sintering at 1330 K.

Thermostability of permanent sintered magnets and alloys on the basis of this system [6] was investigated. It is revealed that the magnets with greater density are more corrosion resistant than that which has smaller density. The covering of magnets made from alloys on basis of Nd-Fe-B with thin protective nickel, cobalt or zinc films allows protecting them toward corrosion. The application of Nd-B coverages, which are obtained during reduction of Ni(II) complexes with hydridoborates of metals is perspective. E.g. the application of solutions, containing 0,13 M $NiCl_2 \cdot 6H_2O$, 1 M ethylendiamine, 1 M NaOH and 0,03 M $NaBH_4$ the corrosion resistant Ni coverages (containing 5-7 % of boron) were obtained.

"Two alloys" technology of magnets producing, which we developed, consists in that the hard magnetic phase $Nd_2Fe_{14}B$ and corrosion resistant fusible alloy dopant to provide fluid phase sintering of magnetic powders. In the present work the perspective chemical approach with use of metals hydrides is used for magnets producing with higher corrosion resistance. The essence of it is that intermetallide $Nd_2Fe_{14}B$ can be produced with traditional method of melting or chemical one [7]. Fusible alloy dopants, which are doped with various metals and ensures fluid phase sintering, it is possible to receive with a chemical means - reduction of metals and alloys from solutions with hydrides. From solutions, containing coordination compounds of cobalt (II), iron (II), neodymium (III), copper (II) etc., by means of reduction by sodium hydridoborate are received amorphous powders, which contain the appropriate metals or their compounds with boron, or alloys, and also hydrogen. These powders, after the appropriate heat treatment, can be perspective for manufacturing of magnets of system Nd-Fe-B. In many cases these dopants turn out as amorphous powders which corrosion resistance is much higher than crystal powders have. More over, by means of chemical reduction it is possible to receive powders of plastic alloys which mechanical crushing is strongly complicated or impossible. For example Co - containing powder of composition 55 - 65 at. % Co, 25 - 45 at. % B and 8 - 12 at. % H with particles size 40-80 nm were obtained by means of reduction of cobalt salts and complexes by sodium hydridoborate.

Acknowledgement

The financial support of Science and Technology Centre in Ukraine in framework of Agreement N1230 is greatly acknowledged.

3. References

1. Sagawa, M., Fujimura, S., Yamamoto, H. et al. (1984) Permanent magnet materials based on the rare earth-iron-boron tetragonal compounds (invited), *IEEE Trans. Magn.* **20**, 1584-1589.
2. Yan, G.L., Williams, A.J., Farr, J.P.G., Harris, I.R. (1999) The effect of density on the corrosion of Nd-Fe-B magnets, *J. Alloys and Compounds* **292**, 266-274.
3. Pavlyukov A.A. (1990) The reason of low structural and magnetic stability of Fe-Nd-B alloys, *Fizika metallov i metallovedenie* **11**, 39 - 44.
4. Pavlyukov, A.A., Tsivilitsin, V.Y., Opanasenko, O.S. (1991) Structural transformations in alloys of the Fe-Nd-B system giving rise to their disintegration. *Inorganic Materials* **27**, 600-603.
5. Roy, R., Me Kinstry, H.A. (1953) Consuming the so-called $Y(OH)_3$-type structure and the structure of $La(OH)_3$, *Acta Cryst.* **6**, 365-366.
6. Pavlyukov, A.A., Tsivilitsin, V.Y. (1997) Thermal stability of sintered permanent magnets based on Fe-Nd-B alloys, *Powder metallurgy and metal ceramics* **36**, 554-557.
7. Dragieva, I., Stoeva, S., Stoimenov, P., Pavlikianov, E., Klabunde, K. (1999) Complex formation in solutions for chemical synthesis of nanoscaled particles prepared by borohydride reduction process, *Nanostructured materials* **12**, 267-270.

THE ACTION OF INTENSIVE PLASMA FLOWS ON THE METAL HYDRIDE MATERIAL

V.V.BOBKOV, V.N.BORISKO, V.V.CHEBOTAREV [(1)], I.YE. GARKUSHA [(1)], G.P. GLAZUNOV [(1)], YE.V. KLOCHKO [(2)], M.V. LOTOTSKY [(2)], N.S. POLTAVTSEV [(1)], YU.F. SHMAL'KO [(2)], R.I. STAROVOITOV, V.I. TERESHIN [(1)] AND I.YU. ZAVALIY [(3)],

V.V. Karasin Kharkov National University, 4 Svobody Sq., Kharkov, 61077 Ukraine

[(1)] National Scientific Center "Kharkov Institute of Physics & Technology",

1 Academicheskaya Str., Kharkov, 61108 Ukraine

[(2)] A.N.Podgorny Institute of Mechanical Engineering Problems of National Academy of Sciences of Ukraine,

2/10 Pozharsky Str., Kharkov, 61046 Ukraine

[(3)] G.V.Karpenko Physical-Mechanical Institute of National Academy of Sciences of Ukraine, 5 Naukova Str., Lviv, 79601 Ukraine

Abstract

In course of bombardment of metal hydrides by high-energy particles in vacuum, the particles collisions with the surface will mainly result in knockout of more movable hydrogen atoms from interstitial of metal matrix, rather than metal atoms. This method of sputtering protection of vacuum-plasma installations has been experimentally tested using hydrogen-saturated Zr-V getters.

One of the most effective ways to solve the problem of sputtering vacuum-plasma installations construction elements, subjected to intensive bombardment by high-energy particles in vacuum, would be in these elements making of metal hydrides. In this case, the particles collisions with the surface of bombarded metal hydride elements, apparently, will mainly result in knockout of more movable hydrogen atoms from interstitial of metal matrix, rather than the atoms of the latter. The endothermic character of hydrogen desorption combined with the possibility of convective heat removal with desorbed gas makes it possible to cool, in addition, the metal hydride element. The layer of desorbed hydrogen forming near bombarded surface can, in turn, create the protective

N. Veziroglu et al. (eds.), Hydrogen Materials Science and Chemistry of Metal Hydrides, 309–312.

gas target, which provides extra dissipation of the energy of a flow of bombarding particles.

In this work the action of plasma flows on the metal hydride material on the basis of hydrogen-saturated getter alloy Zr–V–Fe + B_2O_3 [1] has been investigated. The powdered metal hydride was compacted at the room temperature with the binder (copper powder) into cylindrical sample 20 mm in diameter and 5 mm in thickness.

Studying the temperature mode of the ion-bombarded sample was carried out under conditions of burning the reflective discharge in hydrogen environment. In course of it, the sample was as one of the cathodes in the gas-discharge cell; the second (reference) cathode was made of stainless steel, copper or molybdenum. The experiment equipment and technique are described in [2]. The temperature of the both cathodes was measured with introduced into them chromel alumel thermocouples that were included into measuring circuit either according to standard scheme or to differential one.

The analysis of energy spectra of ions bombarding the cathode has shown that the fraction of their kinetic energy makes of 65–70% with respect to electric power supplied into the discharge. Under such conditions, the ion-bombarded metal hydride cathode is heated up to a less extent than the reference one which is inert with respect to hydrogen. The corresponding temperature difference, already in 20 minutes, is steadied on the level of 80 K and remains constant (and practically the same for various materials of the reference cathode) during the whole experiment (up to 1.5 hours). Such a behavior is caused by endothermic heat effect of metal hydride decomposition, as well as by convective heat removal with desorbed hydrogen. The analysis of the heat balance of the metal hydride cathode shows that up to 70% of heat power evaluated herein is spent to these processes.

The metal hydride cathode sputtering was practically absent in course of the experiments. At the same time, it took place for a reference cathode, independently of its material. It was observed as appearance of metallic deposit on the surface of quartz plate near the reference cathode.

In the next series of experiments the sample was bombarded by single pulses of N^+ ions having the energy of 2 keV and the density of $2 \cdot 10^{14}$ cm^{-3} (the "Prosvet" coaxial plasma accelerator was used for this purpose). The specific flow power in the impulse 3–5 μs in duration was up to 10 MW/cm^2. It is comparable, as respect to a heat load, with operation conditions of the most energy-loaded construction materials of a fusion reactor.

In the Table the phase characteristics of non-irradiated and irradiated (after bombardment with 7 impulses) sides of the material sample are presented. The phase characteristics have been determined on the sample x-ray difractograms indexing results. The indexing was carried out using the special software, which realizes the algorithm of *R* divergence factor minimization [3].

The difractograms were indexed with the usage of the previously obtained data [1] on phase-structure and hydrogen sorption characteristics of hydride-forming materials of the similar composition.

The sample irradiation causes the noticeable (15% on average) lines broadening in the difractogram, at simultaneous decrease of their absolute intensities. It testifies that the irradiated side of the sample which, according to SEM data, has been subjected to melting, in conditions of desorbed hydrogen bubbling through the melt (i.e. it was fastly cooled after bombarding impulse), has a fine-crystalline or, likely, partially amorphous-like structure.

TABLE 1. Face content of the samples

Characteristic			Value	
			Non-irradiated side	Irradiated side
Divergence factor R			0.458	0.528
Average peak halfwidth, Δ, °			0.30	0.34
Phase (space group)	Cu (F m 3m)	Lattice periods, nm	a=0.36187	a=0.36215
		$\frac{W_i}{\sum W_i}$, %	76.21	64.21
	λ_2–Zr(V,Fe)$_2$ H$_X$ (F d 3 m)	Lattice periods, nm	a=0.77496	a=0.77110
		$\frac{W_i}{\sum W_i}$, %	15.18	17.12
	ZrCu$_5$(H$_X$) (F 4 3 m)	Lattice periods, nm	–	a=0.70549
		$\frac{W_i}{\sum W_i}$, %	–	9.95
	ε–ZrH$_X$ (I 4)	Lattice periods, nm	a=0.3621 c=0.4469	a=0.3626 c=0.4552
		$\frac{W_i}{\sum W_i}$, %	6.96	4.28
	η–Zr$_3$(V,Fe)$_3$ OH$_X$ (F d 3 m)	Lattice periods, nm	a=1.23991	a=1.23152
		$\frac{W_i}{\sum W_i}$, %	4.57	4.93

Notes:

1. The peak half width is recognized as the distance from the top where the intensity decreases in e times;
2. W_i is the weight factor, which is proportional (when distortions are absent) to mass fraction of the itch phase in a mixture.

The copper lattice periods, both on non-irradiated side and on irradiated one, are somewhat higher than the reference data (0.3615 nm), and the period of the irradiated side is higher. It is caused, likely, by nitrogen implantation from ion beam into copper crystal lattice, and further diffusion of its atoms through the sample.

The lattice periods of hydride-forming intermetallic phases – Laves phase λ_2–Zr(V,Fe)$_2$H$_X$ and η–Zr$_3$(V,Fe)$_3$OH$_X$ phase – are higher than for starting intermetallides (0.7396 and 1.2156 nm respectively), but lower than for completely saturated hydrides (0.7886 and 1.2656 nm). Passing from non-irradiated side to irradiated one, the periods of these phases are noticeably decreased. The comparison of the appropriate data with the reference ones [1] allows us to evaluate hydrogen saturation fraction for Laves phase as 72% before and 64% after irradiation. The corresponding values for the η–phase are 49% and 30%. It testifies that under bombardment the intensive hydrogen desorption from the mentioned hydride phases take place.

During bombardment, the weight intensity factors for Laves phase and η-phase (as the ratio to the sum of weight intensity factors for all phases) are low changed. At the same time, the similar parameter for copper decreases more than by 10%, and in 1.5 times for zirconium dehydride. At that, in the irradiated sample there arises the phase of the $ZrCu_5$ intermetallide having some increased as compared to the reference data (0,687 nm) lattice period (seemingly, due to hydrogen or nitrogen atoms inclusion). The most probable explanation of this fact is the interaction, during ion bombardment impulse, of zirconium hydride with melt of copper which has more low melting temperature than the other components of the composite. As a result, in a course of melt crystallization $ZrCu_5$ – the most rich in copper intermetallide in Zr – Cu system – is formed.

As a whole, the significant changes in the sample phase composition are not observed, and after bombardment the sample weight loss is of 0.1% (corresponding to 0.5 mg/cm^2 per an impulse). It testifies to low enough sputtering susceptibility by intensive plasma flows for the sample based on hydrides zirconium alloy. Therefore, the proposed method of sputtering protection for construction elements of vacuum-plasma installations is effective one.

This work is partially supported by Science and Technology Center in Ukraine (STCU), Project No. 959.

References

1. Yartys', V.A., Zavaliy, I.Yu. and Lototsky, M.V. (1992) Low-pressure hydrogen absorbents on the base of Zr-V and Zr-V-Fe alloys modified by oxide additives, *Koord.· Khimiya (Co-ordinatory Chemistry)* **18**, No.4, 409-423 (in Russian).
2. Borisko, V. N., Klochko, Ye. V., Lototsky, M.V., Petrushenya A.A. and Shmal'ko Yu.F (2000) The auto stabilized on pressure systems of a gas feed for vacuum-plasma installations with use of a convertible hydrogen getters, *Vacuum Engineering and Technology* **10**, No. 1, 15–19 (in Russian).
3. Milburhn, G. (1975) *X-Ray Crystallography,* "Mir" - Publishers, Moscow, (Translated from English into Russian).

USE OF HYDRIDE DEVICES IN A VEHICLE

YU.I. SHANIN
Research Institute of Scientific and Industrial Association Luch
24 Zheleznodorozhnaya, Podolsk, Moscow Region, 142105, Russia

Keywords: hydride, hydrogen, vehicle, hydrogen storage, hydride heat pump.

1. Introduction

The existing stock of cars absorbs about half of petroleum extracted in the world (organic fuel stock runs low very quickly and it will be enough according to the forecasts for 30-50 years) and releases 14 % of carbonic gas. The fraction of vehicles exhaust in air pollution of cities achieves 70 %. A lot of efforts is spent to make a vehicle ecologically safe (or even less harmful). The power and environment problems acquire at present paramount importance. These problems solution can be promoted by the following measures: creation of more advanced power plants, improvement of working process of internal-combustion engines, application of systems of neutralization of exhaust gas, use of new types of fuel.
Combustion engine is the heat engine which is lack finish. The maximum efficiency of modern car combustion engine in a mode of the greatest operating torque does not exceed 30%. The efficiency falls up to 10-12% at lower engine revolutions and during urban operation [1]. The other fraction of energy in the various forms of heat (exhaust gases, coolant) dissipates in an environment. Temperature of exhaust gases of combustion engine vs. number of engine revolutions can change from 100 to 450°C, temperature of cooling liquid after engine heating up does not exceed 100-110°C.
In the motor-car construction good prospects has the development of electrical power installations based on electrochemical sources of energy - storage batteries and electrochemical generators. But practical application of electric vehicles is problematic due to the following reasons:

1) absence of reliable current sources;
2) environmental pollution comes from sphere of vehicles to a sphere of thermal power stations producing energy for electric vehicles.

N. Veziroglu et al. (eds.), Hydrogen Materials Science and Chemistry of Metal Hydrides, 313–320.

2. Use of hydrogen in motor car construction

Hydrogen is one of fuel types with good prospects in the future. Hydrogen is ecologically clean, has good motor properties and its resources (in water of ocean) are practically unlimited. The essential problem, which is not overcome up to this day, is low efficiency of its large-scale production. Setting apart economic aspect, we shall note that use of hydrogen in combustion engine allows to solve its main problems: to increase efficiency of fuel and to reduce toxicity of exhaust gas. Hydrogen can be applied directly as the basic fuel and as a small additive (5-15 % of weight of total fuel consumption) to petrol. The fastest and feasible for today solution is to use hydrogen for engine heating-up, when a catalyst does not work to its full capacity and exhaust is the most harmful. When an engine begins to work in a stable operation mode, the basic fuel - petrol - is used.

Beside combustion engine hydrogen can be applied in "fuel cells", where it combines with oxygen of air producing water and electric current. The latter feed electric motors and charges the storage battery of electric vehicles.

Hydrogen can be reserved onboard a vehicle, or is produced during movement. In the first case the problems of safe storage and decrease of volume and weight of a tank with hydrogen are urgent.

Production of hydrogen directly onboard a car is carried out by Opel in model HydroGen 1 [2], in which reforming is utilized to transform usual fuel into hydrogen in several catalytic reactors. In this case toxicity of exhaust is considerably reduced. Alongside with petrol, diesel fuel, natural gas and methanol can be utilized for reforming.

V.V. Studennikov and G.I. Kudimov had proposed in 1997 [3] the simple device for hydrogen manufacture - electrohydrogen generator. Gravitational electrolysis of electrolyte with production of electric current, gaseous hydrogen and oxygen is supposed to carry out in electrohydrogen generator. Process is accompanied with artificial separation of electrolyte ions by its rotation in a special reactor with angular speed 1500-12500 rev/min. As authors states, this direction in a case of positive practical results can bring down the price of hydrogen to 0.0038 $\$/m^3$.

3. Use of hydrides in vehicles

The stable tendency of hydrogen use in vehicles and large heat losses of their engines can promote a wide expansion of hydride technologies in motor industry. It is possible to try to apply almost all known and realized on today based on hydrides devices as the following vehicle devices:

1. Hydrogen storage with use of reversible metal hydrides;
2. Hydride compressor increasing hydrogen pressure (if necessary);
3. Hydride heat pump for cold generation or air-conditioning with use of exhaust gas or cooling liquid heat;
4. Starting device for preliminary heating up of cold engine;

5. Storage batteries with hydride electrode;
6. Various sensors and regulating devices based on hydrides ability to absorb hydrogen in a certain temperature interval.

Some aspects of these devices application in vehicles are considered below.

3.1. HYDROGEN STORAGE

Hydrogen onboard a vehicle can be transported in gaseous, liquid and bonded state (for example, in reversible metal hydrides, fullerenes, carbon nanotubes and fibres). The basic requirements to tanks storing hydrogen are: 1) safety, 2) dimensions of storing tank should be comparable with a petrol tank size and provide mileage about 500 km). The required compactness can be achieved at the ratio of hydrogen weight to weight of system (gravimetric density) equal to 6.5-10 % and volumetric density of 60 kg H_2/m^3.

Hydrogen content at pressure 10-50 MPa is 0.7-1.3% wt. of various bottle. Increasing of working pressure to 60-75 MPa with use of high-strength structural materials of steel, titanium and aluminium alloys, combined and composite materials make it possible to increase relative mass fraction of hydrogen in such tanks to 0.04-0.05 [1]. But at the same time, the problem of safe operation of such bottles on vehicles is far from end.

The liquid hydrogen at normal pressure has temperature 20.4 K and density of 0.071 kg/l (i.e, density of liquid hydrogen is equal to density of gaseous hydrogen compressed up to 85 MPa) that causes application of cryogenic tanks with very high thermal insulating ability. One more problem is the lowering of hydrogen evaporation. Already today in hydrogen automobile model HydroGen 1 [2] loss of hydrogen by evaporation are lowered up to 1-3 % of volume for day time consumption due to application of special fiber glass heat insulation. With use of special rocket cryogenic tanks with multilayer isolation 10 mm in thickness it is possible to achieve ratio of stored hydrogen weight to weight of structural unit about 0.4-0.7 mass fraction and to lower evaporation to 0.8-1.0 % per day. For prevention of hydrogen loss caused by sharp increase of pressure in hydrogen tank above operating level and for reduction of hydrogen losses due to tank heating during long-time vehicle parking, it is possible to use hydride absorbing material. Cryosorption of hydrogen by carbon-based materials with low density can give the further improvement of this way of storage. Gravimetric density can reach 10 weight %, and volumetric density of 32 kg H_2/m^3 (at pressure in 5 MPa and temperature 77 K) in activated carbon AX-21 having specific surface of 3000 m^2/kg. In spite of the fact that the engineering problems of cryogenic hydrogen tank service are enough labour-consuming, but can be solved in principle, the application of cryogenic tanks will depend on energy consumption for hydrogen release and liquation, which is still very high.

The reversible sorption of hydrogen is possible by such materials as hydrides, fullerenes, nanotubes and graphite nanofibres.

Hydrogenated fullerenes have reversible sorption capacity up to 7.7 % wt. [4]. But their use followed by large engineering difficulties and requires high pressure of activation (50-85 MPa) and high operation temperatures (~ 500°C).

In experiments [5] samples of soot containing single-layer carbon tubes absorbed 5-10 % wt. of hydrogen at room temperature and pressure 0.04 MPa. The results of other experimenters give values from 1.23 to 8.25 wt. % at different ways of tubes preparation. Besides, it is impossible to deny presence of high capacity of hydrogen. The research of graphite nanofibres began from sensation, when the U.S. experts have stated that prepared by them fibres absorbed hydrogen up to 67 wt. %. This result is disputable, since it was not reproduced by other researchers. The best result achieved in Russia is 0.6 wt. %. Theoretical estimations of hydrogen sorption of nanotubes and fibres give lower value (1 wt. %) in comparison with various experimental results. Actually, researchers have up to now at their disposal only about 100 grams amounts of these unique materials, and this is a distant prospect.

Engineering feasibility of hydrogen storage by reversible metal hydrides is more probable direction. The choice of hydride for system of hydrogen storage should provide minimum weight of the system at its satisfactory operational characteristics. The following requirements should be met by such systems [1]:

1. High sorption ability;
2. Small heat of hydride formation;
3. High density for minimization of volume;
4. Easy of hydrogenation;
5. Appropriate pressure- temperature-composition (P-T-C) relationships in the temperatures range of 20 to 200°C;
6. High stability, fire- and explosion safety;
7. Low cost.

The system $FeTi\text{-}H_2$ has the most suitable characteristics as hydrogen storage material [1]. Its sorption ability is about 2 wt. %, equilibrium pressure at temperature 0 °C is 0.15 MPa, and at 50 °C - 1.0 MPa. There is a possibility to change equilibrium pressure of this alloy by partial substitution of iron by other transition metal (Mn, Ni). Copper-magnesium alloys have even more sorption capacity (up to 3 wt. % for $Mg_2Cu\text{-}H_2$ system), but sorption takes place at temperature 250 °C and pressure 1.5 MPa.

The major term of hydride application in the vehicle power supply system is the release of necessary quantity of hydrogen from hydride for engine power conditions. Therefore selecting hydride it is necessary previously to know available temperatures of exhaust gases and cooling liquid (for average vehicle driving cycle). One hydride or their combination (having large capacity on hydrogen) need to be chosen according to the temperature value.

The comparison of ways of hydrogen storage with a petrol tank is given in Table 1 [1].

It can be seen from Table 1 that the hydrogen storage system based on hydride has quite acceptable volumetric characteristics. At the same time it is obviously necessary to decrease essentially (some times) the weight characteristics of tanks using more light hydrides.

To increase hydrogen pressure up to the necessary level with the purpose of its further application as a fuel, the hydride compressor can be used. Sources of energy with two different temperatures (for example, air and liquid cooling the engine) are required for

TABLE 1. Comparison of relative parameters of vehicle tanks

Kind of capacity for storage of fuel	Weight	Volume
Petrol tank	1	1
Cryogenic tank for liquid hydrogen	2	4-5
Hydrogen storage system based on $FeTiH_x$	16	1.7
High pressure bottles (40 MPa)	32	7
Tank for hydrogen under normal conditions		2500

hydride compressor. The possibility to control a pressure level with use of hydride compressor will allow to increase hydrogen pressure and respectively to expand a list of hydrides acceptable to use as hydrogen storage material. Hydride compressor can be considered as the additional device of the hydrogen storage system, in which hydride amount is 10-20 times less than in the latter.

3.2. HYDRIDE HEAT PUMPS

Surplus waste heat of combustion engine can be used for cold generation. If we have two levels of temperature - high (T_h=100-200°C) and middle (T_m=20-40°C corresponding to environment temperature in the summer period), then appropriate hydrides can be selected allowing to achieve a level of negative temperatures (T_l) in a range from -5 to –50°C [6]. Choosing hydride for a refrigerator, it is necessary to select ones with maximum enthalpy of hydride formation that is opposed to choice for hydrogen storage. It is supposed to utilize vehicle waste heat as energy source of a hydride heat pump (HHP) - devices of periodic action without moving parts. Useful engine power is not spent for cold generation, therefore cold can be considered as a free-of-charge auxiliary product and, hence, fuel efficiency is improved. Apropos authors of [3] for electrohydrogen generator operation also proposed to utilize engine waste heat for maintenance of endothermic chemical reactions for dissolution of anodic products in water.

The details of method to select hydrides for heat pumps cycles were described in the author's work. The results of cheapest hydrides choice are presented here as illustration. It is supposed that the source of heat can have temperature (T_h) at two levels: 1) 200°C, and 2) 110 °C. Even in one-step cycle with application of two different hydrides it is possible to hope for a level of adiabatic temperature for cooling of low-temperature hydride about -20 ...-30 °C, Table 2. Maximum HTH temperature - T_{max} and minimum (adiabatic) LTH temperature - T_{min}, maximum P_{max} and minimum P_{min} hydrogen pressure and coefficient of energy performance (COP), as the ratio of energy of generate cold to the energy spent for it are presented in the table. It was supposed that the level of middle temperature T_m does not exceed 40 °C, and design factor (ratio of unit weight to hydride weight reserved in it) is equal to 0.5. Italicized values in Table 2 show that these hydrides are more suitable for car air-conditioning.

HHP performance figures improvement may be achieved applying two-step cycle. In this case the number of hydrides is increased up to three - the third is hydride working at an

TABLE 2. Choice of hydride couples for refrigerating cycle in a vehicle

High-temperature hydride (HTH)	Low-temperature hydride (LTH)	$T_{h\,max}$, °C	$T_{l\,min}$, °C	$P_{max}*10^{-5}$, Pa	$P_{min}*10^{-5}$, Pa	COP
$LaNi_{4.56}Mn_{0.44}$	$Fe_{0.9}Mn_{0.1}Ti$	198.3	-12.2	14.1	0.30	0.348
$TiFe_{0.8}Ni_{0.2}$	$Fe_{0.9}Mn_{0.1}Ti$	157.7	-13.0	14.1	0.27	0.291
$LaNi_{4.6}Al_{0.4}$	FeTi	169.0	-17.1	15.9	0.77	0.307
$ZrMn_{2.7}$	FeTi	160.6	-22.3	15.9	0.59	0.282
ZrMnFe	FeTi	156.1	-25.4	15.9	0.51	0.291
$TiFe_{0.8}Ni_{0.2}$	FeTi	162.4	-36.1	15.9	0.27	0.206
$TiFe_{0.8}Ni_{0.2}$	$TiFe_{0.8}Mn_{0.2}$	110.3	-12.3	3.4	0.27	0.330
ZrCrFe	$TiFe_{0.8}Mn_{0.2}$	139.1	-16.2	3.4	0.21	0.302
ZrMnFe	$TiFe_{0.8}Mn_{0.2}$	100.4	0.0	3.4	0.51	0.465
$TiFe_{0.8}Mn_{0.2}$	$ZrCrFeNi_{0.8}$	101.4	-8.3	22.2	3.37	0.238
$TiFe_{0.8}Mn_{0.2}$	$ZrCr_{0.6}Fe_{1.4}$	103.0	-7.9	2.6	0.27	0.315
$TiFe_{0.8}Mn_{0.2}$	$CeNiCu_2$	103.1	-16.1	23.2	3.37	0.092
$TiFe_{0.8}Mn_{0.2}$	$CeNi_{2.5}Cu_{2.5}$	97.1	-13.2	19.7	3.37	0.144
$TiFe_{0.8}Mn_{0.2}$	$CeNi_{4.5}Al_{0.5}$	95.1	-8.9	18.6	3.37	0.275

intermediate pressure P_{mid}. Realization of such cycle (initial transfer of hydrogen from low-pressure hydride (HTH) to hydride of intermediate pressure, and then transfer of hydrogen to high-pressure hydride (LTH)) can essentially lower LTH temperature T_{min}. The substantial thing is that HTH temperature now can be less than 150°C for achievement of T_{min} at the level of -40 ...-50°C. Two examples of hydrides set are presented in Table 3; they can be used for air-conditioning. It can be seen that transition to two-step cycle has decreased temperature T_{min} in comparison with one-step cycle by 10-15°C. Tables 2 and 3 were specially made for hydrides based mainly on the cheap materials Ti, Fe, Zr, Mn.

TABLE 3. Choice of hydrides tern for air-conditioning cycle in a vehicle

High-temperature hydride	Hydride with intermediate pressure	Low-temperature hydride	$T_{h\,max1}$, °C	$T_{h\,max2}$, °C	$T_{l\,min}$, °C	$P_{max}*10^{-5}$ Па	$P_{mid}*10^{-5}$ Па	$P_{min}*10^{-5}$ Па	COP
$TiFe_{0.8}Mn_{0.2}$	$ZrMn_{2.7}$	FeTi	89.6	98.6	-22.3	15.9	3.4	0.59	0.247
$TiFe_{0.8}Mn_{0.2}$	ZrFeMn	FeTi	89.6	100.4	-25.4	15.9	3.4	0.51	0.236

There is a positive experience of similar HHP models development [7]. Two types of HHP in the specially created and tested refrigerating models of installations were made:

1.One is with use of heat with temperature 180-200°C of products of combustion of liquid fuel (simulating operation of combustion engine) for air-cooling to minus 3°C with periodic mode of operation and one pair of hydride sorbers: $LaNi_{4.6}Al_{0.4}H_x$ as HTH and $MmNi_{4.15}Fe_{0.85}H_y$ as LTH. The work of this installation has confirmed a feasibility of vehicle or ship refrigerator development.

2.The second installation use heated water with temperature 80-90°C as a source of energy. At the cold side of HHP (water was used as coolant) the temperatures 1 - 4°C were achieved, using of antifreeze (i.e. car cooling liquid) allows to obtain temperature up to minus 5°C.

$LaNi_5H_x$ was HTH and $MmNi_{4.15}Fe_{0.85}H_y$ was LTH. The average cold-productivity of installation was 550-660 W. Use of heat of combustion engine cooling liquid as heat carrier and increase of its temperature to 100-110°C allows to expect low temperature at a level of minus 15°C.

There are no principle difficulties with transition to cheaper initial hydrides materials for development of similar HHP (see Table 2 and 3). There is additional requirement to hydride – cyclic stability. Unfortunately, design of developed HHP leaves much to be desired. Because of cycle mode of hydrogen sorption/desorption process in hydrides, it is necessary periodically to heat up and to cool sorption devices. The work of latter occurs at two levels of temperature resulting in a complex system of control of coolant flows and in losses due to preliminary heating and preliminary cooling, i.e. to decreasing of an overall heat engine performance. Continuous operation and simplification of HHP design will promote its use in refrigerating devices including for vehicle.

3.3. STARTING DEVICE FOR COLD SEASON

There is a problem in Russia to start combustion engine in a cold season. This problem may be solved by use of various starting devices based on energy of electric battery. But it is possible to use a device based on hydrides functioning on waste heat of vehicle combustion engine.

Let's assume that two containers filled by hydrides with different chemical composition are connected by the pipeline through the valve. One hydrides is saturated with hydrogen at high pressure, another has no hydrogen and has low pressure. In a cold season it is enough to open the valve, hydrogen comes to a bottle with hydride at low concentration of hydrogen. The temperature of this bottle raises, and released heat through a heat exchanger can be used for combustion engine heating. Heat released by eight kg of hydride (in two bottles) is enough to increase temperature of five litres of motor oil by ~ 60 °C (or 8 litres of antifreeze by ~25 °C). After start and heating of the engine the starting system with use of waste heat can be transferred to an initial condition before the following use. This device, as well as HHP, is also a device of periodic action. But intermittence here absolutely another, than in a case of HHP. At the first stage of the device operation it is necessary to utilize effectively released heat, at the second stage to recharge the device (piping of hydrogen from one bottle to another) and to disconnect it from car systems. The designer's task is to build-in the starting device with the minimum changes into existing vehicle interior.

Such device needs hydrides with high hydrogen capacity, large energy of hydride formation and low cost. But high requirement to cyclic stability of hydrides is not necessary, as the number of device applications during operation (for example, 10 years) presumably will be 300-500 times (for moderate climate) and 1500-2000 times (in conditions of Far North). Such number of cycles does not reduce noticeably sorption capacity of hydride.

4. Conclusions

So, it was shown that the engineering problem of implementation of various hydride devices in vehicle is not hopeless business. But their implementation in motor car construction, as well as in other branches, will depend on to their main shortcomings, which should be overcome for commercial use of hydrides [8]:
1.Low hydrogen capacity (even in comparison with fullerenes, graphite nanotubes and fibres);
2.High sensitivity to gaseous impurities (requiring use of high purity of hydrogen and makes problematic use of hydrogen compressors and other devices which are not using super-pure hydrogen);
3.Difficult initial activation (as an alloy is used at the beginning of technological assembly in the most of hydride devices, instead of hydride, the alloy activation inside such devices is problematic);
4.They are pyrophorus in air (this shortcoming is closely connected with the previous one and complicates hydride technology resulting in necessity to carry out operations in inert gas atmosphere);
5.Disproportionation during intensive cycling (periodicity of hydride devices operation and the increase of their productivity results in fast degradation of hydride initial composition and failure of the device).
All these problems are the problems of material science and chemistry of hydrides. It is necessary to improve existing materials and to develop radically new materials for successful application in various areas including motor car construction.

5. References

1. Mischenko, A.I. (1984) Application of hydrogen for vehicle engines, Naukova dumka, Kiev (in Russian).
2. Materials of a site "Car expert" (2000), *(http://www.expert.ru/auto/) 7* (in Russian).
3. Application for the invention RU98/00190, 07.10.1997 (in Russian).
4. Tarasov, B.P. (1997) Hydride forming metals and intermetallic compounds as catalysts of fullerenes hydrogenation, *Abstracts of the 5-th Int. Conf. "Hydrogen Material Science and Chemistry of Metal Hydrides". (ICHMS'97) (Ukraine, Yalta, 02-08 Sept. 1997)*, 42-43.
5. Dillon, A.C., Heben, M.J. et al (1997) *Nature* 386, 377-379.
6. Shanin, Yu.I. (2000) Hydrides choice for vehicle hydride devices, *Int. Scientific J. for Alternative Energy and Ecology* **1**, 173 (in Russian).
7. Astakhov, B.A., Afanasyev, V.A., Bokalo, S.Yu., Izhvanov, L.A., Solovey, A.I., Frolov, V.P. and Shanin Yu.I. (1999) Development of small-sized refrigerating machines with a metal hydride heat pump, *Abstracts of the Int.Conf. "Hydrogen Material Science and Chemistry of Metal Hydrides", (Ukraine, Yalta, 02-08 Sept. 1999)*, 306-307.
8. Suda, S. and Sandrock, G. (1994) Three decades of intermetallic hydrides - What happened to the applications? , *Zeitschrift fur Physikalische Chemie. Neue Folge* **183**, 149-156.

ANTHROPOLOGICAL IRREVERSIBLE CHANGE OF INTERACTION BETWEEN OUR PLANET AND THE SUN. ROLE OF HYDROGEN AND PRIMARY ENERGY SOURCES

Yu.A. KOCHERZHINSKY
Institute of Hydrogen and Solar Energy,
P.O. Box 195, Kyiv-150, 03150 Ukraine

Abstract

This paper is concerned with three types of natural climate changes: permanent (or continuos) due to the atmosphere cleaning from the carbon dioxide by plants during 3,5 billion years; self-exited periodic (or oscillatory) due to feedbacks between the Sun and the geographic shell of the Earth (2-3 per million years); catastrophic (when some heavenly body collides with the earth surface). The anthropogeneous climate change is now happening due to "backward running" of the permanent climate change caused by the anthropogeneous atmosphere pollution with CO_2 to the concentration, which existed 20 million years ago. Now is the deadline to avoid selfannihilation using immediate replacement of all the present energy sources by the wind-hydrogen power industry.

1. Climate changes, natural and anthropological

In the ecological literature in discussing the threat of the global environmental catastrophe the connection between the state of our planet geographic shell (the upper layer of lithosphere, hydrosphere, the lower layer of atmosphere, biosphere) and the Sun was implied but it was called and discussed nowhere directly. For the first time, the question on the possibility (under certain conditions – inevitability) of developing the global environmental catastrophe in to the space irreversible one was raised in [1].

An average temperature of the geographic shell will remain constant (i.e. the climate will remain invariable) if the amount of the energy radiated by the shell into the space equals the energy gained by the shell from the Sun, inner sources and human activities. Density of the solar energy flow can be considered constant; amount of the energy absorbed by the Earth's surface from the flow depends on albedo, the part of radiation reflected by the surface backwards to space (it has made almost a half, 45%, until now). The Earth's radiation is regulated by atmosphere absorption of the Earth's surface infrared radiation. The Earth's atmosphere is almost completely transparent for the sun beams that mainly consist of visible, more short-wave radiation. The Earth's more long-wave IR radiation is absorbed by 'greenhouse gases' present in the atmosphere, in the first place, by water vapor, carbon dioxide, nitrogen oxides, methane

N. Veziroglu et al. (eds.), Hydrogen Materials Science and Chemistry of Metal Hydrides, 321–326.

[2]. The total heat absorption is determined by the concentration of the above gases in the atmosphere.

In the Earth's history three types of natural climate changes are known: permanent (continual), self-exited periodic (oscillatory) and catastrophic (space-caused) [3]. The nearest neighbors of the Earth (the Mars and the Venus) have carbon dioxide atmosphere. The Earth seems to have such atmosphere as original, too. About 3.5 billions years ago there appeared plants in the Earth. Using the sun energy they began feeding by carbon from the atmosphere and throwing out waste (oxygen) into it. About 2.5 billions years ago life in the Earth was threated because of exhausting the inrenewable feed source – atmospheric carbon dioxide. It is difficult to imagine how population in the Earth felt the threat. (Passing from science to science fiction, one can imagine that the most intellectual Earth inhabitants – I would like they to be not Forest or Seaweeds but Meadows – warned the Earth population. The latter changed their behavior and made that source renewable). Plants gave birth to animals. The latter were deprived of possibility to use the sun energy immediately, as well as carbon from the atmosphere and minerals from soil to feed directly. They were forced to gain energy and organic substances from plants, prolonging the life of organic molecules (which are genuine 'physical persons', the life individuals in such associations as a cell and an organism) and to use oxygen to gain energy for themselves and to renew food (carbon dioxide) for plants.

In that way we have gained a moderate temperature instead of the hot one and the slowly changing atmosphere that allowed us to breathe well.

During 100 years of the industrial revolution we let the carbon out (the plants had been hiding it for many millions years) from under the ground. 'Backward running' of the permanent climate change is occurring at such an unprecedented in the planet history rate that corals which had gone through all natural climatic catastrophes for many millions years did not have enough time to move to cooler places and are died out [4]. We are preparing this for ourselves, too.

Donella H. Meadows, Denis L. Meadows and Jorgen Randers [5, 6] and others [7], are warning us. But, unlike the plants from the first billion years of the Earth life, we do not change our behavior in the proper direction and at the due rate.

Self-exited oscillations of climate with a frequency of 2-3 per million years occur due to numerous feed-backs between the Sun and the geographic shell of the planet and between its individual elements. Some of those feed-backs are positive. Glacier seems to be the most active element. It shows the positive feedback with absorption of the sun energy. Its surface albedo is 8-9 times as high as the earth's or water surfaces without ice or snow. The glacier growth lowers heat absorption, favors the decrease of temperature. As the glacier thaws, the water or land surfaces bare and this enhances heat absorption, favors heating and promotes thawing. The glacier growth and thawing are determined by the relationship between the ice thawing and sublimation, on the one hand, and its building up with snow precipitation, on the other hand. When the atmosphere begins warming (like now), glacier thaws, the Earth albedo decreases that promotes heating and thawing. The air and the land are warmed faster than the ocean does. The ocean is warmed up slower due to high specific heat of water and its huge masses in the ocean. Therefore, at first, the ocean evaporation, cloudiness and precipitation increases insignificantly. The sun beams reach the glacier surface and destroy it, warming up continues. When the ocean is warmed up sufficiently, evaporation is enhanced, the sky becomes overcast totally, the sun beams do not reach the surface. The atmosphere and the land are cooled down. Uninterrupted precipitation builds up glaciers, glaciation involves more and more areas in the northern hemisphere. In the southern hemisphere glacier goes down into the ocean and cools it. The ocean is cooled down, evaporation is diminished, cloudiness and precipitation decreases, the sun beams reach the surface again, heating up (in which the ocean is a backward link again) begins again. Due to the huge masses involved in this process the period of oscillation is very long, several hundred thousand years. So biosphere has enough time to adopt itself to climate changes.

Catastrophic climate changes have space origin and can occur even faster than those induced by human activities. Therefore they have catastrophic consequences on biosphere. When a large meteorite, a small asteroid or a comet nucleus with a diameter of 1 kilometer or more collides with the Earth's surface, it breaks through the Earth's crust and causes the huge magma eruption. Masses of volcanic ashes shut access for the sun beams and give rise to the sudden glaciation. This picture is similar to that obtained by simulation of glaciation caused by atomic warfare (due to clouds of ashes and soot) [8]. If we take into account repeated space catastrophes, Lamarck seems to be right: what we considered a biological evolution is found to be, to a large measure, a survival of marginal species after catastrophes. However, unlike glaciations caused by self-exited climate changes, glaciations induced by catastrophes are relatively short-term. After the last catastrophe that happened near 11500 B.C. when only a few per sent of original Earth's population survived, just after 3 thousand years humanity on our planet began already increasing [9].

What we are doing with our climate now is return for millions years back by the permanent change scale. It results in the uncontrolled interaction (due to a positive feedback) between the Earth's surface temperature and energy absorption. We, people, find ourselves in the state of children capable of making a fire playing with matches but incapable of extinguishing it.

2. Safe energy sources, primary and secondary

In his Welcome to the participants of 7th conference "Hydrogen Material Science and Chemistry of Metal Hydrides" (2001) President of the International Association for Hydrogen Energy, T.N. Veziroglu pointed out the "twin global problem" to which the new energy source "is going to be permanent answer". The problem consists of "(1) the rapid depletion of fossil fuels and (2) the environmental problems caused by their utilization, such as the greenhouse effect, climatic changes, acid rains, ozone layer depletion, pollution and oil spreads" [10, p.9]. One should add the third problem: satisfaction of sharply rising energy requirements caused by depletion of unrenewable sweet water storage that provides the cultivation from 1/4 to 1/3 of all the food in our planet [11]. Therefore it will be necessary to freshen sea water in mass amount that will require unprecedented expenditure of energy.

When we speak about hydrogen energy as an 'energy of the future', we forget that hydrogen is not 'an energy source' (because we have no natural source of hydrogen). It is only a secondary source – an accumulator[1]. We speak about 'renewable sources' in plural. In fact, there is only one practically inexhaustible source of energy – the Sun. There are also transformers to convert its energy: the plants (they cannot be used as fuel because carbon dioxide is evolved on their combustion)[2], running water (practically exhausted and ecologically dangerous: reservoirs flood fertile soil in the plains and give rise to earthquakes in the mountains), and the wind.

The wind power industry is ecologically safe. It does not pollute the air and does not induce heat in it beyond that gained from the Sun. Its energy which, after all transformations in industry, agriculture and as residential, is evolved into atmosphere in form of heat, was previously extracted from the atmosphere (as mechanical one). Wind has an unused potential that exceeds by one order of magnitude the installed power of the all current electric stations and vehicles together [12]. One kilowatt of the installed capacity of a windgenerator is cheaper than that of a device for the direct transformation of the sun radiation into electricity [13]. Unlike a sun battery, a windmill does not stop at night. In addition, each square inch of the surface lighted by the Sun should be given to the green leaves: they produce oxygen and can yield food for us. A sun battery and a windmill have the same defect, i.e. dependence on the weather, inconstancy of (power) output. Until the high-capacitance, cheap and ecologically safe electric accumulators are designed, hydrogen is a promising secondary source. However, there are alternative ones: a compressor (one amateur drives along the town using a standard engine fed with compressed air from a cylinder, he uses gasoline only in intertown journeys), a cryostation for air liquefaction (in Zaporizhzha another amateur feeds the engine with liquid nitrogen - waste of the station that produces oxygen for oxygen blast in metallurgical furnaces). Both of these methods are ecologically cleaner than burning of any fuel including hydrogen because the high temperatures are not developed in the cylinders and, consequently, nitrogen oxides are not formed (they are inevitable on combustion of any fuel in air). These and other variants of secondary source require an impartial technical and economical analysis. Whatever secondary cell we have, when we think about the future energy, we may speak only about windenergy.

3. A dead-line to avoid danger

At present we incur enormous losses because of unprecedented hurricanes, floods, heavy showers and snow-falls due to an increase of the average temperature in our planet by 0.7K (caused by carbon dioxide emission) as compared with the beginning of

---

[1] Being applied for special purposes, hydrogen from hydrocarbons resolves no problem related to mankind survival. In this work 'hydrogen' means 'produced from water'.

[2] One should exclude any technologies that bring into atmosphere the energy beyond that gained from the sun: geothermal energy, nuclear power (it uses splitting materials and nuclear fusion), energy transmitted with mirrors from the space.

XIXth century, Those changes have been occurring for 150 years. Most people supposes that we have enough time to take measures that will allow us to prevent self-annigilation. Authors of the book [6] point out that it is not our case as global changes show exponential behavior: acceleration of the quantity change is proportional to its reached value. Therefore, the slight changes 'suddenly' and 'unexpectedly' become catastrophically impetuous. For example, during 100 years from 1860 to 1960 the content of CO_2 in the atmosphere was increased by 22 millionth volume fractions. During the following 40 years from 1960 to 2000 the content of CO_2 was increased three times as much. It should be known that last 'portion' of 22 millionth has been added only for 7 year. If such a behavior lasts, than what we have to meet soon?

To prevent self-annihilation, it is necessary to realize the world energy revolution, i.e. to replace immediately all present energy sources by windenergy, focusing resources of all nations on this problem what it has been never before.

We simply have no time for an evolution.

4. References

1. Kocherzhinsky, Yu.A. (1999) The Wind-Hydrogen Power Industry as an Optimal Version of the Global Power Strategy at the Beginning of the Global Ecologic Catastrophe, Hydrogen Material Science and Chemistry of Metal Hydrides. *Abstracts Book of VI NATO International Conference (ICHMS'99).* Katsiveli, Yalta, Ukraine, September 02-08, 1999, 376.
2. Sanin, F.P. (1999) Hydrogen, Ozone and Hot-Effect, *ICHMS'99*, 344.
3. Herenchuk, K.I., Bokov, V.A., and Chervanev, I.G. (1984), *General Earth Science*, Moscow, Vysshaya shkola (In Russian).
4. Bright, C. (1999) *Anticipating Environmental 'Surprise'* - State of the World 2000. A Worldwatch Institute Report on Progress Toward a Sustenable Society.- New York, London. W.W. Norton & Company.
5. Meadows, Donella H. et al. (1972) *The Limit to Grows* - New York, Universe books.
6. Meadows, Donella H. et al. (1992) *Beyond the Limits* - White River Junction: Chelsea Green Publishing.
7. Yanch, Erikh (1974) *Prognosis of scientific and technical progress*, Moscow (In Russian).
8. Moiseev, N. (1988) *Ecology of Humanity.* - Moscow, Molodaya Gvardia (In Russian).
9. Gorbovskiy, A. (1971) *Mysteries of the Most Ancient History* - Moscow, Znanie (In Russian).
10. Veziroglu, T.N. (2001) *Hydrogen Material Science and Chemistry of Metal Hydrides*, VII International Conference, Alushta-Crimea-Ukraine, September 16-22, 2001. ADEF, Ukraine, Kiev.
11. Postel, S. (2000) *Redesigning Irrigated Agriculture* - State of the World 2000, 39.

12. Korobko, B.P. (1998) *Role of the New and Renewable Energy Sources in Realization of Conception of Sustanable Development* - Proceedings of the 7th Scientific and Practical Conference of the Ministry of the Power Industry of Ukraine, 31 August - 05 September 1998, Crimea. (In Ukrainian).
13. Dunn, S., Flavin, C. (2000) *Sizing up micropower* - State of the World 2000, 142.

NEXAFS SPECTROSCOPY FOR THE INVESTIGATION OF HYDROGEN ABSORPTION PROPERTIES OF NANOSTRUCTURED CARBON THIN FILMS

C. LENARDI[1,*] , M. Marino[1], E. Barborini[1], P. Piseri[1], V. Briois[2], S. Maffi[3], and P. Milani[1]

[1] *INFM, Dipartimento di Fisica, Università di Milano, Via Celoria 16, I-20133 Milano, Italy*
[2] *LURE, Centre Universitaire Paris-Sud, Batiment 209D, I-91405 Orsay Cedex, France*
[3] *CESPEL-CNR, Dipartimento di Chimica Fisica Applicata, Politecnico di Milano, Via Mancinelli 7, I-20131 Milano, Italy*

Carbon based materials are interesting systems for hydrogen storage, which is a technological challenge of primary interest for mobile power units in many different applications. We used NEXAFS spectroscopy to study the capability of hydrogen absorption by porous nanostructured carbon deposited in the form of thin films by supersonic cluster beam deposition (SCBD). A method has been developed for the quantitative analysis of the spectra. In this way the effectiveness of various hydrogen loading methods and the performance of films grown using different cluster precursors have been investigated.

* Corresponding author. Tel:+39 02 58357672; Fax +39 02 58357487; e-mail: cristina.lenardi@mi.infn.it

N. Veziroglu et al. (eds.), Hydrogen Materials Science and Chemistry of Metal Hydrides, 327–339.

1. Introduction

Hydrogen has been recognized as one of the most promising candidates for substituting the current fossil fuel in particular for transportation vehicles. The main hindrance to the use of hydrogen is the difficulty in finding a suitable storage medium. Solid storage approaches could represent more efficient, safe and cost-effective systems with respect to the liquefaction or compression of H_2. Metal hydrides and gas adsorption from porous materials have prompted considerable interest thanks to their potential capability of storing high volumetric density of hydrogen. In this perspective high specific surface area and lightweight carbon-based materials seem to be very attractive for hydrogen uptake applications. Many studies have been worked out on nanotubes[1–3], alkali-doped nanotubes[4], graphite nanofibers[5, 6]. These experimental researches have stimulated a number of theoretical works[7, 8] in an attempt to explain the fundamental mechanism of the hydrogen adsorption process. A recent review on experimental results can be found in Dillon *et al.*[9] and on theoretical progress in the field in Meregalli *et al.* [10]. A few groups have reported values of adsorption larger than 6.5 wt.%, i.e. more than the target value set by the U.S. Department of Energy (DOE). These results are rather controversial and up to now not reproducible. Further experimental and theoretical investigations are then necessary.

A fundamental parameter in the storage capability of solid materials is the size of the pores. In fact the adsorption of hydrogen is strictly related to their dimension since they should be small enough for favoring a strong interaction between the hydrogen molecules and the walls of the internal cavities. In a previous paper of ours [11], we have shown that our films have a typical specific surface area of 665 m^2/g with a pore size distribution prevalently in the micropore region (diameter less than 20 Å) with a peak at 34 Å in the mesopore region (diameter between 20 Å and 500 Å). The porosity of nanostuctured carbon ns-C films indicates that this kind of material could be considered a good candidate as storage medium for fuel cell systems.

We have also probed the adsorption of H_2 by means Temperature Programmed Desorption (TPD) analysis[11]. A population of physisorbed sites has been observed at about 230°C and of chemisorbed sites at about 500°C.

These results have suggested to performe more systematic investigation on the adsorbing properties of nanostructured carbon films. In this work we present x-ray absorption studies on the capability of the ns-C films to adsorb and release hydrogen. For this purpose we have developed a method for a quantitative determination of hydrogen uptake from the carbon *K*-edge NEXAFS spectra.

2. Experimental

The nanostructured carbon thin films have been grown by depositing cluster beams produced in a supersonic expansion. With respect to effusive beams used in MBE, this technique provide higher deposition rate, control on the cluster mass distribution, and high directionality of the beam. In Refs.[12, 13] detailed descriptions of the deposition apparatus are presented. It schematically consists of three differentially evacuated chambers separated by skimmers of 2 mm of diameter. The apparatus operates in high vacuum regime (base pressure 1×10^{-7} Torr). During the deposition in the first chamber, which hosts the cluster source, the pressure is maintained in the range between 2×10^{-5} and 2×10^{-4} Torr. The second chamber is equipped with a manipulator on which is mounted a sample holder that can intersect the beam along the axis of the apparatus. A linear Time of Flight Mass Spectrometer (TOF/MS) is hosted in the third chamber and placed collinear to the beam axis.

The source is a Pulsed Microplasma Cluster Source (PMCS)[14] in which a confined plasma discharge ablates a rod of graphite. The plasma is obtained by the ionization of helium by means of pulsed discharge between graphite electrodes. The ion sputtering of the cathode produces vaporized carbon atoms which are quenched by a pulse of helium and condense in clusters, which are carried out of the source in a seeded supersonic beam through a nozzle of 2 mm of diameter. A mass selection of the outgoing particles can be obtained by using a focusing nozzle. It consists of a disk with eight holes named focuser and a cylindrical nozzle. On the disk side facing the cylindrical nozzle the plane is machined so that a narrow gap is produced between the disk and the cylindrical nozzle. This opening is called focuser gap L_{gap}. According to the focusing nozzle configuration only a certain range of cluster mass can follow the carrier gas streamlines and concentrate on the cylindrical nozzle axis. Thus the focusing device can be used as a bandpass filter which allows only clusters of suitable mass to leave the source. The beam results to be highly collimated (angular semi-aperture $\theta/2 = 10$ mrad) leading to a relevant increase of the deposition rate (5 nm/s at 300 mm distance from the nozzle). A complete description of the focuser is reported in Ref.[15]. Films deposited with the focuser show a different surface morphology with respect to deposits obtained without the focuser. This effect is related to the different spatial and mass distributions of the precursors. We have investigated if this structural and morphological properties can affect the hydrogen uptake capability. We have verified that the hydrogen absorption is lower in films deposited without the focuser and, since the high deposition rate is a fundamental demand for industrial applications, we have preferred to carried out a systematic examination of nanostructured carbon films grown with the focuser by using different focuser gap, namely 0.25, 0.35 and 0.40 mm.

A few samples have been exposed to humid ambient air (100% RH) at 90°C for 45 h. Hydrogen loading of two samples, grown on thin iron substrate, has been performed electrochemically: one by means of permeation of hydrogen through the rear of the sample placed in contact with a solution of tiosulphide, which forms H_2S (pH 2.7); the other one by means of potentiostatic cathodic polarization of the electrode immersed in a 0.01 N solution of boric acid.

The NEXAFS experiments have been carried out at LURE Super-ACO storage ring on the SACEMOR experiment equipped with a high energy 10 m TGM monochromator (200 - 800 eV) of the beamline SA72. The energy resolution of the NEXAFS measurements was 0.2 eV at carbon K-edge. NEXAFS spectra have been acquired in Total Electron Yield (TEY) mode. In order to reduce the overall counts and to increase the sensitivity to the surface, a few spectra have been acquired in Partial Electron Yield mode (PEY) using a retarding grid at the voltage of 50 V. No relevant differences have been observed with respect to corresponding spectra acquired in TEY mode. The NEXAFS signal has been obtained from the ratio between the sample signal collected by a channeltron and that of an 85% transmission gold coated grid giving the intensity of the incident beam. The photon incidence angle on the samples was of 55° (the so-called "magic angle") in order to obtain spectra independent on the synchrotron light polarization.

3. Data handling

A first description of the calibration and normalization procedure of the acquired data has been presented in Ref.[16]. In the following we give a brief description of the method that we have developed for a quantitative evaluation of the hydrogen uptake from the analysis of carbon *K*-edge spectra. A complete presentation of the method will be given in a following paper.

3.1. PEAK FITTING AND QUANTITATIVE ANALYSIS

A typical carbon *K*-edge NEXAFS spectrum is shown in Fig. 1a). The background is identified with a straight line that fits the behavior of the low-energy end of the spectrum. Then, after background subtraction, the spectra are rescaled in such a way that the high-energy tail is fitted by a decreasing exponential curve normalized to 1 at the ionization potential (IP) edge (dashed line in Fig. 1a).

The main practical difficulty in the application of this procedure is given by the fact that for many spectra the almost flat low-energy region, which may be suitable for the identification of the background, has a very limited extension. This fact could involve a significant uncertainty in the slope of the straight line extrapolated from that region. For this reason, the procedure has been first applied for a particular spectrum of our collection that presented a sufficiently wide low-energy plateau, and that for this reason has been chosen as a reference. Then, taking advantage of the fact that the great majority of the spectra have basically the same overall behavior, apart from the near edge features, we have prepared a numerical routine that, using as an input the background-subtracted and rescaled reference spectrum, calculates the background and the rescaling factor of all the other spectra so as to make them best overlap to the reference one in the high-energy region. In this way both the low and the high-energy regions are simultaneously used in the calculation of the background, thus removing the difficulty mentioned above and providing a reliable criterion for a consistent normalization of all the spectra of the collection. The spectra are then fitted in the near-edge region as the superposition of an erf function, representing the step at the ionization threshold, a set of three pseudo-Voigt functions representing the peaks in the sub-threshold region, and an asymmetric gaussian function describing the main peak above the threshold (see Fig. 1b). All the spectra can be conveniently fitted by assuming a ionization potential close to 289.4 eV. In the sub-threshold region it is possible to recognize the peak related to the $\pi^*_{C=C}$ transition, located at 285.34 $\pm$0.16 eV (the number

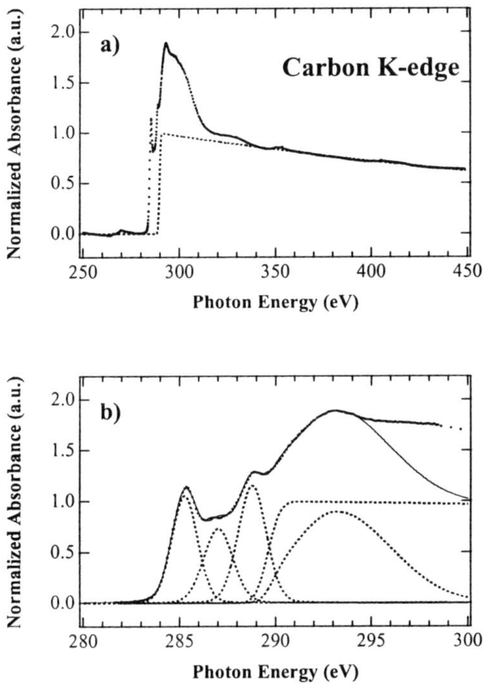

Figure 1. NEXAFS spectrum of a nanostructured carbon film: a) complete spectrum (solid circles) and fit of the continuum normalized to 1 at the ionization potential edge (dashed line); b) example of fit of the near-edge region.

after the sign ± indicates the standard deviation of the distribution of the peak maxima for our entire set of spectra), a peak at about 287.10 eV and a peak associated with the σ^*_{C-H} transition at 288.70 ±0.07 eV. It is reasonable to assume that the area of this last peak can provide some quantitative indication about the amount of H chemically bonded to C in the sample. This assumption is supported by the analysis of EELS spectra of hydrocarbons reported in the literature. We have considered in fact the curves for the oscillator strength reported by Hitchcock and Mancini[17] and, after performing an analogous fit, we have reported the areas of the peaks $\pi^*_{C=C}$ and σ^*_{C-H} (normalized with respect to the area of a fixed larger portion of the spectra) as a function respectively of the number of $\pi_{C=C}$ and σ_{C-H} bonds per C atom present in the molecule of each hydrocarbon. For both types of bonds a linear relationship was found to hold with satisfactory approximation. For our NEXAFS data, the direct comparison of spectra acquired on the same samples before and after heat treatments and hydrogen exposures clearly exhibits a definite trend in the variation of the height of the σ^*_{C-H} and $\pi^*_{C=C}$ peaks. We have assumed the average amount $n_{C=C}$ of $\pi_{C=C}$ bonds per C atom in each sample to be proportional to the ratio between the area $A_{C=C}$ of the $\pi^*_{C=C}$ peak and the area

A_0 of a fixed larger portion of the spectra:

$$n_{\mathrm{C=C}} = \alpha_{\mathrm{C=C}} \frac{A_{\mathrm{C=C}}}{A_0} \ , \tag{1}$$

where $\alpha_{\mathrm{C=C}}$ is a proportionality factor independent of the particular sample considered. A spectrum acquired on a specimen of Highly Oriented Pyrolytic Graphite (HOPG) was assumed to be representative for a 100% graphitic sample (i.e. $n_{\mathrm{C=C}} = 0.5$). Therefore it was possible to determine the $\alpha_{\mathrm{C=C}}$ as

$$\alpha_{\mathrm{C=C}} = 0.5 \frac{A_0^{\mathrm{HOPG}}}{A_{\mathrm{C=C}}^{\mathrm{HOPG}}} \ , \tag{2}$$

and to estimate by comparison the *sp*^{2} percentage in all the other samples, according to the criterion already proposed by Fallon *et al.*[18]:

$$f_{sp^2} = \frac{A_{\mathrm{C=C}}}{A_0} \frac{A_0^{\mathrm{HOPG}}}{A_{\mathrm{C=C}}^{\mathrm{HOPG}}} \ . \tag{3}$$

We have further assumed the number $n_{\mathrm{C-H}}$ of $\sigma_{\mathrm{C-H}}$ bonds to be linearly related to the area $A_{\mathrm{C-H}}$ of the $\sigma^*_{\mathrm{C-H}}$ peak:

$$n_{\mathrm{C-H}} = \alpha_{\mathrm{C-H}} \frac{A_{\mathrm{C-H}}}{A_0} - \beta \ , \tag{4}$$

where the term β represents some additional contribution not related to the presence of H, which has been assumed to be the same for all the samples. The parameter $\alpha_{\mathrm{C-H}}$ in this relation has been determined by assuming its ratio to $\alpha_{\mathrm{C=C}}$ to be the same as for the corresponding parameters previously derived from the EELS spectra of Hitchcock and Mancini. This gave:

$$\alpha_{\mathrm{C-H}} = 7.4\, \alpha_{\mathrm{C=C}} \ . \tag{5}$$

In this way it was possible to estimate the difference in H content between any two NEXAFS spectra of our collection. In Table 4 we have summarized the results of the quantitative analysis of the spectra, i.e. the *sp*^{2} fraction and the hydrogen content. Note that we have arbitrarily set to zero the hydrogen content of the sample V35b, since it presented the smallest C–H peak and it is reasonable to assume that the thermal treatment may remove all the adsorbed hydrogen from this sample. In practice this assumption fixes the value of the constant β in eq. (4).

4. Results and Discussion

As described in the previous section, the carbon *K*-edge spectra show three main features in the pre-edge part of the spectrum: the $\pi^*_{\mathrm{C=C}}$ resonance at 285.3 eV, a peak at 287.1 eV principally related to oxygen contamination, and the $\sigma^*_{\mathrm{C-H}}$ resonance at 288.7 eV. The relative intensities of these features show a remarkable

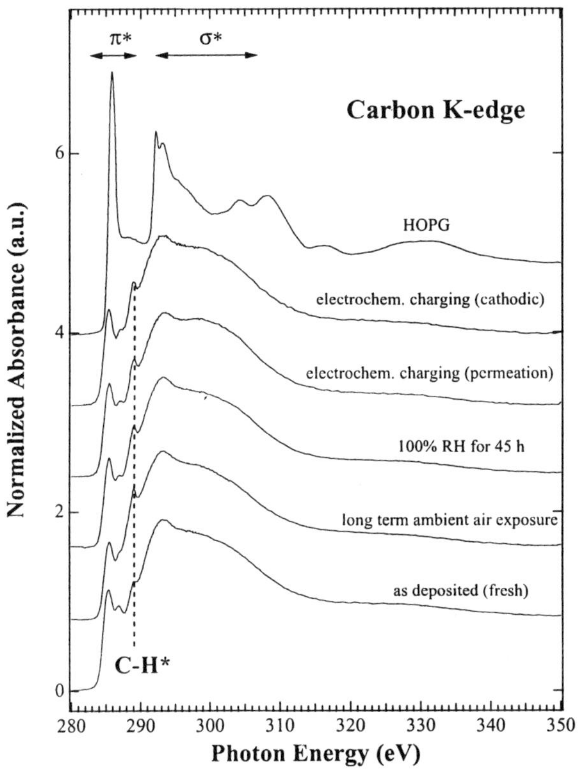

Figure 2. Carbon *K*-edge NEXAFS spectra of carbon films grown with a 0.35 mm focuser gap: as deposited, long term air exposure, 100% RH, electrochemical loading. For comparison also the reference spectrum of HOPG is reported.

correlation with the extent of their exposure to hydrogen rich environments. In particular the last peak, due to transitions $1s \rightarrow \sigma^*_{C-H}$, is a mark of the presence of hydrogen bonded to carbon. The chemisorption of hydrogen is due to the dissociation of hydrogen containing physisorbed species (water, H_2, hydrocarbons) induced by the imping x-ray photons or by a catalytic action of the films themselves. Thus from the carbon *K*-edge spectra it is not possible to determine the source of the adsorbed hydrogen. Nevertheless, after a heat treatment of the sample in order to degas the adsorbed species, and a controlled exposure to a very pure hydrogen flux, the procedure above presented provides a reliable method for the quantification of the hydrogen uptake. The results reported in Table 4 should be then considered in this perspective. Fig. 2 shows a series of carbon *K*-edge NEXAFS spectra of ns-C films grown using a focuser with a gap of 0.35 mm. At a first sight, the σ^*_{C-H} resonance of the freshly deposited film (sample F35) is the smallest one, whereas it increases for the other samples. The higher peak appears to be that one of the sample exposed to ambient air for a few weeks (sample F35_old).

TABLE I. Calculated sp^2 fraction and hydrogen content of different series of samples for various values of the focuser gap.

Sample		Treatment	sp^2 fraction	H at.%	H_2 wt.%
HOPG		none	1.00	17.77	1.77
focuser gap 0.25 mm					
F25		freshly deposited	0.56	28.39	3.20
V25	a	100% RH, 45 h, at 90 °C	0.50	36.53	4.58
	b	heated at 400°C	0.88	12.76	1.20
	c	H_2 exposure 1.2 bar, 3h, RT	0.70	17.82	1.78
	d	humid air exposure	0.58	17.13	1.69
focuser gap 0.35 mm					
F35_old		exposure to ambient air (few weeks)	0.43	41.69	5.62
F35		freshly deposited	0.57	26.43	2.91
V35	a	100% RH, 45 h, at 90 °C	0.48	36.22	4.52
	b	heated at 400°C	0.81	0.00(*)	0.00(*)
	c	H_2 exposure 1.2 bar, 3h, RT	0.64	15.67	1.53
	d	humid air exposure	0.60	19.64	2.00
EL_Cath		H_2 electrochem. charge (cathodic polariz.)	0.53	36.34	4.54
EL_Perm		H_2 electrochem. charge (permeation)	0.53	34.49	4.20
focuser gap 0.40 mm					
F40		freshly deposited	0.57	32.93	3.93
V40	a	100% RH, 45 h, at 90 °C	0.41	46.71	6.81
	b	heated at 400°C	0.84	11.97	1.12
	c	H_2 exposure 1.2 bar, 3h, RT	0.72	19.33	1.96
	d	humid air exposure	0.59	26.51	2.92

(*) It has been arbitrarily set to zero (see text).

Electrochemical charging (samples EL_Cath and EL_Perm) and exposure to an environment with a 100% RH for 45 h at 90 °C (sample V35a) seem to bring to a comparable amount of C–H bonds. The calculated quantity of adsorbed hydrogen, as reported in Table 4, confirms the observed trend: the H_2 wt.% for the "old"

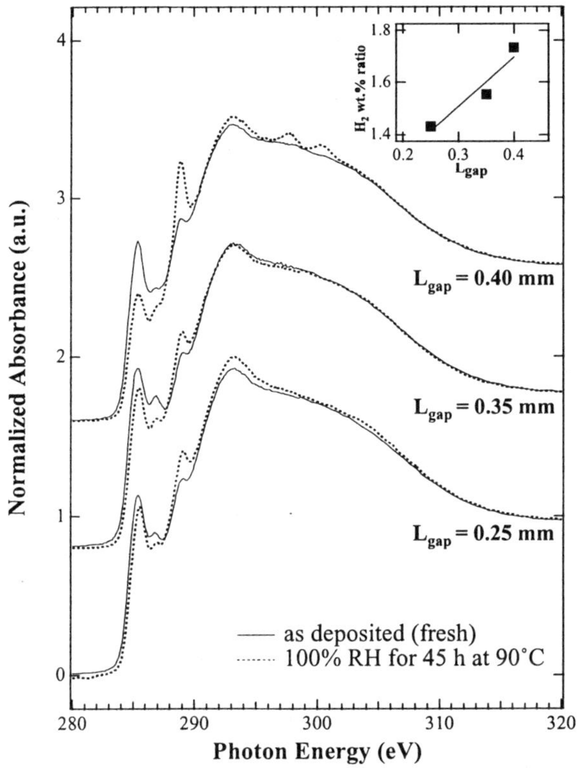

Figure 3. Comparison between freshly deposited samples (solid line) and samples exposed to 100% RH atmosphere at 90 °C for 45 h (dashed lines) for different focuser gaps. In the inset the ratio between the H_2 weight percentages of the exposed samples and of the freshly deposited ones is shown.

sample is 5.6, whereas for the "fresh" one is 2.9, and about 4.5 for that exposed to 100% RH and those electrochemically loaded. A first observation can be done: a key parameter for the sorption of hydrogen containing species is the exposure time since the loading kinetics appears to be rather slow.

Fig. 3 shows a direct comparison between freshly deposited samples (solid line, samples F25, F35 and F40) and those heated at 90 °C in 100% RH atmosphere (dashed lines, samples V25a, V35a and V40a) for different focuser gaps. For all the samples exposed to the humid environment the $\sigma^*_{C=C}$ peak appears to be more prominent and the $\pi^*_{C=C}$ resonance smaller and narrower. These results indicate that hydrogen chemisorption sites are most favorably present in distorted graphitic structure. In fact the reduction of the amount and of the energy dispersion of π^* sites suggests that the hydrogen is preferentially bonded to sp^3 hybridized carbon, at the expense of the most distorted double bondings between carbon atoms. These observations are confirmed by the calculated values of sp^2 fractions

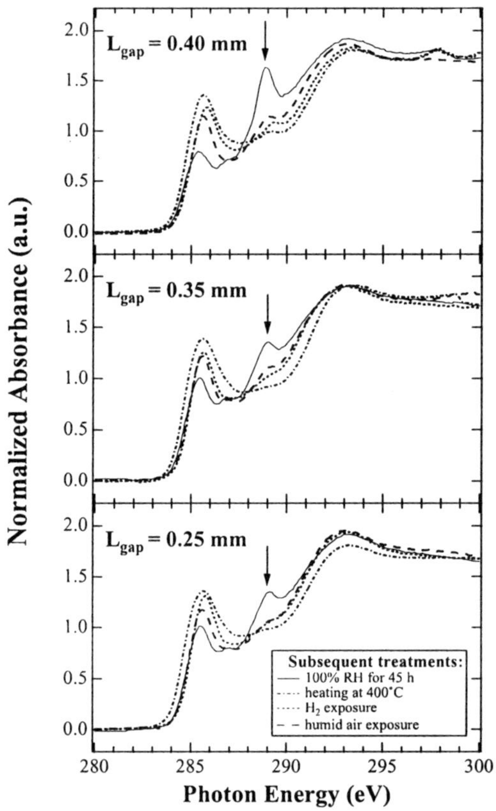

Figure 4. NEXAFS carbon *K*-edge spectra of various samples exposed to 100% RH (solid lines) and subsequently subjected to heat treatment (dot-dashed lines), exposure to pure hydrogen (short-dashed lines) and exposure to a humid ambient air (long-dashed lines). The arrow indicates the position of the $\sigma^*_{\mathrm{C-H}}$ peak.

reported in Table 4. The films grown with a focuser gap of 0.40 mm take up a greater quantity of hydrogen with respect to the other samples. In general the ratio between the hydrogen content in weight of the exposed sample and that of the freshly deposited one grows as function of the focuser gap, as shown in the inset of the Fig. 3. This is a first indication as the adsorption properties of the ns-C films depend also on the properties of the cluster beam.

The NEXAFS spectra of various samples subjected to heat treatments and subsequent exposure to pure hydrogen are shown in Fig. 4. The samples exposed to 100% RH (solid lines) have been heated up to 400°C for about 1 h in UHV. The treatment removes almost the whole $\sigma^*_{\mathrm{C-H}}$ resonance peak from the spectra (dot-dashed lines, samples V25b, V35b and V40b). In fact the values of the hydrogen contents are the lowest (for this reason, as explained in the previous section, the value for the sample V35b has been arbitrarily set to zero). A well known effect due to the heating is the graphitization of the films leading to a more intense and wide π^* peak, as the increment of sp^2 fraction testifies (about 0.8 for all the heated samples). Subsequent exposure to pure hydrogen at 1.2 bar for 3 hours at room temperature has the effect of partially restoring the $\sigma^*_{\mathrm{C-H}}$ peak (short-dashed lines, samples V25c, V35c, V40c). The narrowing effect on π^* resonance is attested by the reduction of the sp^2 fraction, which decreases to about 0.7. The increments of the hydrogen content have been found to be about 0.6, 1.5 and 0.8 wt.% for the sample V25c, V35c and V40c respectively. It should be pointed out that, although the loading conditions were not particularly severe, the quantity of absorbed hydrogen is sizable.

Finally the samples have been extracted from the UHV chamber and briefly exposed to a humid ambient air (long-dashed lines, samples V25d, V35d, V40d). All the samples show an increase of $\sigma^*_{\mathrm{C-H}}$ peak and an associated reduction of the $\pi^*_{\mathrm{C=C}}$ resonance, corresponding to a sp^2 fraction of about 0.6.

5. Conclusions

NEXAFS measurements have been carried out to study the hydrogen sorption of nanostructured carbon films and the induced modifications in their hybridization state. A method based on the fitting of the near-edge structures and on the evaluation of the peak area has been developed in order to obtain quantitative results. In particular the evolution of the $\pi^*_{\mathrm{C=C}}$ and $\sigma^*_{\mathrm{C-H}}$ resonances has been examined as a function of the adopted focuser and post-deposition treatments (heating, hydrogen and humidity exposures). The results provide the evidence that these films are able to adsorb hydrogen in a reversible way. In the case of exposure to pure hydrogen the adsorbed quantity is found to be about 1 wt.%. In general the absorption is accompanied by a decrease of the height and FWHM of the $\pi^*_{\mathrm{C=C}}$ peak, that means loss of sp^2 sites. This behaviour indicates that the hydrogen chemisorption takes place preferentially at the expense of the more distorted π bonds.

The obtained results show the capacity of nanostructured carbon films to adsorb hydrogen and encourage to consider them as possible candidates for hydrogen storage devices. Nevertheless further investigations are necessary. In particular more effective hydrogen charging method, e.g. high pressure exposure, will be tested, as well as the inclusion of metals, which seems to favor the uptake capability of carbon based materials.

6. Acknowledgments

We acknowledge partial financial support from INFM under Advanced Research Project CLASS and from MURST under project COFIN99. The authors wish also to thank C. Laffon and Ph. Parent for their technical support at LURE facility.

References

1. Liu, C., Fan, Y.Y., Liu, M., Cong, H.T., Cheng, H.M., and Dresselhaus, M.S. (1999) Hydrogen storage in single–walled carbon nanotubes at room temperature, *Science* **286**, 1127–1129.
2. Dillon, A.C., Gennett, T., Alleman, J.L., Jones, K.M., Parilla P.A., and Heben, M.J. (1999) Carbon nantubes materials for hydrogen storage, *Proceeding of the 1999 U.S DOE Hydrogen Program Review 1999*, NREL/CP–570–26938.
3. Ding, R.G., Lu, G.Q., Yan, Z.F., and Wilson M.A. (2001) Recent advances in the preparation and utilization of carbon nanotubes for hydrogen storage, *J. Nanosci. Nanotech.* **1**, 7–29.
4. Chen, P., Wu, X., Lin, J., and Tan, K.L. (1999) High H_2 uptake by alkali–doped carbon nanotubes under ambient pressure and moderate temperatures, *Science* **285**, 91–93.
5. Park, C., Anderson, P.E., Chambers, A., Tan, C.D., Hidalgo, R., and Rodriguez, N.M. (1999) Further studies of the Interaction of hydrogen with graphite Nanofibers, *J. Phys. Chem. B* **103**, 10572–10581.
6. Chambers, A., Park, C., Baker, R.T.K., and Rodriguez, N.M. (1998) Hydrogen storage in graphite nanofibers, *J. Phys. Chem. B* **102**, 4253–4256.
7. Wang, Q., and Johnson, J.K. (1999) Molecular simulation of hydrogen adsorption in single–walled carbon nanotubes and idealized carbon slit pores, *J. Chem. Phys.* **110**, 577–586.
8. Lee, S.M., Park, K.S., Choi, Y.C., Park, Y.S., Bok, J.M., Bae, D.J., Nahm, K.S., Choi, Y.G., Yu, S.C., Kim, N., Frauenheim, T., and Lee, Y.H. (2000) Hydrogen adsorption and storage in carbon nanotubes, *Synth. Met.* **113**, 209–216.
9. Dillon, A.C., and Heben, M.J. (2001) Hydrogen storage using carbon adsorbents: past, present and future, *Appl. Phys. A*, **72**, 133–142.
10. Meregalli, V. and Parrinello, M. (2001) Review of theoretical calculations of hydrogen storage in carbon– based materials, *Appl. Phys. A* **72**, 143–146.
11. Lenardi, C., Barborini, E., Briois, V., Lucarelli, L., Piseri, P., and Milani, P. (2001) NEXAFS characterization of nanostructured carbon thin films exposed to hydrogen, *Diamond Relat. Mater.* **10**, 1195–1200.
12. Barborini, E., Piseri, P., Li Bassi, A., Ferrari, A.C., Bottani, C.E., and Milani, P. (1999) Synthesis of carbon films with controlled nanostructure by separation of neutral clusters in supersonic beams, *Chem. Phys. Lett.* **300**, 633–638.
13. Milani, P., Piseri, P., Barborini, E., Lenardi, C., Castelnovo, C., and Podestà, A. (2001) Cluster assembling of nanostructured carbon for advanced applications, Eds. G. Benedek, P. Milani, V. G. Rachenko 2001; NATO-ASI series Kluwer 53–70.
14. Barborini, E., Piseri, P., and Milani, P. (1999) A pulsed microplasma source of high intensity supersonic carbon cluster beams, *J. Phys. D: Appl. Phys.* **32**, L105–L109.
15. Piseri, P., Podestà, A., Barborini, E., and Milani, P. (2001) Production and characterisation of highly intense and collimated cluster beams by inertial focusing in supersonic expansions, *Rev. Sci. Instrum.* **72**, 2261–2267.

16. Lenardi, C., Piseri, P., Briois, V., Bottani, C., Li Bassi, A., and Milani, P. (1999) NEXAFS and Raman Characterization of Amorphous and Nanostructured Carbon Films, *J. Appl. Phys.* **85**, 7159–7167.
17. Hitchcock, A.P. and Mancini, D.C. (1994) Bibliography and database of inner shell excitation spectra of gas phase atoms and molecules, *J. Electron Spectrosc. and Rel. Phenom.* **67**, 1–132.
18. Fallon, P.J., Veersamy, V.S., Davis, C.A., Robertson, J., Amaratunga, G.A.J., Milne, W.J., and Koskinen, J. (1993) Properties of filtered-ion-beam-deposited diamondlike carbon as a function of ion energy, *Phys. Rev. B* **48**, 4777–4782.

INTERNATIONAL ASSOCIATION FOR HYDROGEN ENERGY

ORGANIZING COMMITTEE OF ICHMS'2001

RESOLUTION

"HYDROGEN ENERGY AND HYDROGEN MATERIALS SCIENCE"

ICHMS'2001

We, the participants of ICHMS'2001,

Being aware of our share of the responsibility for the fate of the planet Earth, address the Parliaments and Governments of all countries, and,

Taking into account that fossil fuels still remain the main source of energy on Earth and their stocks are being quickly depleted, and that many objective factors require the creation of a new energy system such as the solar-hydrogen-based one;

Realizing that because of toxic spills, pollution, acid rains, global warming, ozone layer depletion and other calamities, the combustion products of fossil fuels cause increasing damage to both the biosphere (the medium where life as it exists today can exist) and the very life on Earth;

Assuming that it would be wise to conserve the diminishing stocks of fossil fuels for non-energy applications where they may appear irreplaceable (such as lubricants, fibers, plastics, fertilizers);

Knowing that the metallurgy consumes a lot of energy and requires a re-assessing of many technological processes;

Taking into account that there is a concept of smooth transition to a new energy system utilizing all currently known secondary sources of energy;

Realizing that the new energy system, in contrast to the two currently existing ways of energy consumption (i.e., electricity and heat), is based on hydrogen utilization, thereby demanding creation of pumping stations, corresponding monitoring and measuring devices, hardware, pipelines, etc.;

Foreseeing the transition of transportation means to electrical vehicles fueled by various energy sources, resulting in rapid development of attendant devices and technologies (fuel cells, converters, etc.), as well as infrastructure of fueling stations, etc.;

Taking into account that during the transition to the hydrogen fuel, we will have to create electrolysis devices, as well as high-power fuel cells;

Noting that the International Association for Hydrogen Energy's unifying scientists from many countries accumulate and promote the ideas of transition to the solar-hydrogen-based energy system;

We propose the founding principles of the development concept of materials science in the areas of energy and hydrogen for the coming decade, which will allow the focusing of the efforts of material scientists on the problems related to the development of new materials in order to:

> Create new, more efficient, less energy-consuming, and ecologically friendly industrial technologies;

Replace the existing materials with new ones, produced by less energy-consuming methods and exhibiting more advanced properties;

During the next 50 years, make a smooth and natural transition to the solar-hydrogen-based energy system, by creating materials for efficient conversion and utilization of solar energy , as well as for safe storage and transportation.

The development of such materials with tailored properties and utilization of the newest technologies will lead to the beginning of the next, qualitatively higher level cycle of mankind's evolution.

We believe that this will be the first development cycle in the history of mankind promoted not by military objectives, but by science-based politics of human activity on Earth, and by deep understanding of the necessity to conserve the natural habitat of the humanity.

Presented at the 7th International Conference
"Hydrogen Materials Science and Chemistry of Metal Hydrides",
approved and endorsed by the attendees.

Alushta, Ukraine, September 16-22, 2001

Subject Index